Sensors and Advanced Sensing Techniques for Computer Vision Applications

Sensors and Advanced Sensing Techniques for Computer Vision Applications

Guest Editors

Christos Nikolaos E. Anagnostopoulos
Stelios Krinidis

MDPI

Basel • Beijing • Wuhan • Barcelona • Belgrade • Novi Sad • Cluj • Manchester

Guest Editors

Christos Nikolaos E. Anagnostopoulos
Cultural Technology and Communication
University of the Aegean
Mytilene
Greece

Stelios Krinidis
Management Science and Technology
Democritus University of Thrace
Kavala
Greece

Editorial Office
MDPI AG
Grosspeteranlage 5
4052 Basel, Switzerland

This is a reprint of the Special Issue, published open access by the journal *Sensors* (ISSN 1424-8220), freely accessible at: https://www.mdpi.com/journal/sensors/special_issues/1ITCAVQVZT.

For citation purposes, cite each article independently as indicated on the article page online and as indicated below:

Lastname, A.A.; Lastname, B.B. Article Title. *Journal Name* **Year**, *Volume Number*, Page Range.

ISBN 978-3-7258-3261-3 (Hbk)
ISBN 978-3-7258-3262-0 (PDF)
https://doi.org/10.3390/books978-3-7258-3262-0

Contents

About the Editors

Christos Nikolaos E. Anagnostopoulos

Christos Nikolaos E. Anagnostopoulos obtained a Mechanical Engineering Diploma and PhD in Electrical and Computer Engineering from the National Technical University of Athens. He is a Professor of Informatics and is the Director of the Intelligent Systems lab and Intelligent Computer Systems MSc Program in the Cultural Technology and Communication Department at the University of the Aegean. His research interests include computer vision, computer graphics, 3D reality modeling/digitization, mixed reality, and artificial intelligence for the development of applications in cultural informatics and digital culture. He has published more than 250 papers in scientific journals and conferences relating to the above subjects and other fields in informatics (Google Scholar h-index: 27; citations > 5000). From 2004, he has served several academic departments as an adjunct Lecturer, a Lecturer, and/or an Assistant/Associate/Professor (University of the Aegean, University of Piraeus, National Technical University of Athens, and Hellenic Open University). Under his supervision, five students have successfully completed their PhD thesis in the fields of computer vision, artificial intelligence, and 3D visual representation using deep learning. In addition, he has served as an external reviewer and evaluator in National and International Research and Development Calls and Projects. As of 2020, he is among the top 2% of the most influential researchers in the field of artificial intelligence and image processing (Stanford University, Elsevier), based on the peer-reviews his work has received.

Stelios Krinidis

Stelios Krinidis is Assistant Professor at the department of Management Science and Technology (MST) at the Democritus University of Thrace (DUTH). He received his Diploma degree and PhD degree in Computer Science from the Computer Science Department of the Aristotle University of Thessaloniki (AUTH), Thessaloniki, Greece, in 1999 and 2004, respectively. His main research interests include computational intelligence, signal processing and analysis, pattern recognition, data analysis, data analytics, energy flexibility, non-intrusive load monitoring, building performance, building optimization, decision making techniques, visual analytics, etc. He has authored 64 papers in international scientific peer review journals and 118 papers in international and national peer review conferences. His work has been cited more than 2.800 times according to Google Scholar, while his h-index and g-index are equal to 20 and 50, respectively. He has also been involved in 30 research projects funded by the EC and the Greek secretariat of Research and Technology. In nine of them, he acted as the Scientific Responsible; in eight, he acted as the Deputy Scientific Responsible; in nine, he acted as Senior Researcher; and in four of them, he served as an assistant researcher.

sensors

MDPI

Editorial

Sensors and Advanced Sensing Techniques for Computer Vision Applications

Christos-Nikolaos Anagnostopoulos [1,*] and Stelios Krinidis [2,*]

[1] Department of Cultural Technology and Communication, University of the Aegean (UAEGEAN), 81100 Mytilene, Greece
[2] Management Science and Technology Department, Democritus University of Thrace (DUTh), 65404 Kavala, Greece
* Correspondence: canag@aegean.gr (C.-N.A.); krinidis@mst.duth.gr (S.K.)

1. Introduction and Current Trends in the Field

Computer vision is a multidisciplinary field that enables machines to interpret and understand visual information from the world, simulating human vision. It encompasses a variety of tasks, including object detection, image segmentation, and image understanding, all of which have seen significant advancements due to the integration of Artificial Intelligence techniques. The evolution of computer vision can be traced back to its early days in the 1950s, focusing primarily on two-dimensional image analysis. However, the advent of deep learning, particularly convolutional neural networks (CNNs), has revolutionized the field, allowing for more complex and accurate interpretations of visual data [1–3].

Recent studies highlight the transformative impact of deep learning on various applications within computer vision. For instance, object detection has significantly improved using CNNs, which facilitate the identification and localization of objects within images [1,4]. Moreover, image segmentation, a critical aspect of computer vision, has evolved with deep learning methods, enabling precise delineation of objects in images, which is essential for applications ranging from autonomous driving to medical imaging [5]. The integration of CNNs has not only enhanced performance but has also expanded the scope of applications, including real-time systems for precision agriculture and intelligent manufacturing [6,7].

Recent advancements in sensors and advanced sensing techniques have also significantly influenced the field of computer vision, enabling more efficient and effective visual perception systems. One of the most notable trends is the development of in-sensor computing techniques, which allow for data processing directly within the sensor hardware. By processing visual information at the sensor level, systems can significantly reduce the amount of data that needs to be transmitted, thus, enhancing the speed and efficiency of computer vision applications. Recent studies have demonstrated the potential of ferroelectric photosensors for in-sensor artificial neural networks, which can perform computations directly on the sensed data [8,9]. This paradigm shift is particularly beneficial for time-critical applications such as autonomous driving and similar critical decision-making tasks, where rapid decision making is essential.

In addition, the integration of AI into sensor technology is another significant trend. Recent advancements in AI sensors have led to the development of systems that can learn from their environment and adapt to changing conditions [10]. These sensors utilize machine learning algorithms to enhance their performance in tasks such as object detection and recognition. For instance, Complementary Metal Oxide Semiconductor (CMOS) image sensors are enhanced with AI capabilities, allowing them to perform complex computations and improve their accuracy in computer vision tasks [11]. This integration not only

Received: 13 December 2024
Accepted: 20 December 2024
Published: 24 December 2024

Citation: Anagnostopoulos, C.-N.; Krinidis, S. Sensors and Advanced Sensing Techniques for Computer Vision Applications. *Sensors* **2025**, *25*, 35. https://doi.org/10.3390/s25010035

improves the functionality of sensors but also expands their applicability across various domains, including healthcare, agriculture, and smart cities.

2. Scope of Special Issue and Contributions

The Special Issue on "Sensors and Advanced Sensing Techniques for Computer Vision Applications" aimed to address all topics related to the challenging problems of computer vision and pattern recognition in conjunction with the emerging field of deep learning. As a result of the open call for papers, papers related to deep learning, neural networks, and soft computing have been accepted after a rigorous peer review process and assessed for their technical merit and relevance. The accepted articles cover applied issues in the following fields:

- Deep learning for 2D/3D object recognition and classification;
- Autonomous navigation and robotic agents;
- Data augmentation in computer vision;
- Image fusion, segmentation, and classification from different sensors;
- Parallel Machine Learning;
- Photogrammetry and 3D point clouds;
- Multidisciplinary applications of deep learning, pattern recognition, and computer vision for driving assisting systems and aircraft industry;

More specifically:

In contribution 1, Kumari et al. introduced a method that combines advanced imaging technology (Multi-modal Raman Light Sheet Microscopy) with AI to improve the visualization of complex 3D biological structures and, more specifically, cell cultures and spheroids. Using a special microscopy system, detailed images without needing additional markers are captured, while a deep learning model (Zero-Shot Deconvolution Networks or ZS-DeconvNet) enhances the resolution and sharpness without adding artifacts. This approach provides significant potential for advancing high-resolution imaging in biomedical research and other related fields.

In contribution 2, Trung et al. presented a non-contact method for measuring human height in various postures using computer vision and Deep Learning (MediaPipe library and the YOLOv8 model). By analyzing images from a smartphone camera, the proposed method identifies body joint points with advanced algorithms and calculates height using a regression model. Tested on 166 individuals in different postures, the method achieves high accuracy with minimal error. Future improvements aim to expand its capabilities to more positions and scenarios, increasing its usefulness across healthcare, sports, and other fields.

In contribution 3, Giakoumidis et al. proposed an innovative method (ARM4CH) for automating the 3D modeling of cultural heritage monuments using robotic agents (quadrupedals and UAVs) equipped with advanced sensors. These robotic agents may perform the scanning process systematically and accurately, reducing the need for human expertise and intervention. The approach is designed to improve efficiency and to act a key enabler to applications like digital twins for monitoring and managing cultural sites and spaces. ARM4CH aligns with Industry 4.0 principles and sets the groundwork for future real-world testing.

In contribution 4, Xiang et al. presented a new method for merging infrared and visual images into a single, detailed image by combining their unique features. A specially designed filter (Local Extrema-Driven Image Filter) extracts and processes bright and dark features from both images, which are then fused using advanced techniques. The final image integrates these features along with structural and intensity-based elements, producing superior results. Tests on a standard dataset demonstrate that this method

outperforms or at least has equal results compared to eleven state-of-the-art image-fusion methods in terms of quality and accuracy.

In contribution 5, Borstelmann et al. introduced a cutting-edge method to add color to near-infrared (NIR) images, addressing the challenges posed by differences in light properties between NIR and visible light. Traditional methods struggle due to the lack of paired training data, so the researchers use diffusion models, a powerful alternative to Generative Adversarial Networks (GANs). The framework translates NIR intensities into visible light, achieving impressive results. Experiments demonstrate that even simple implementations rival GANs, while more advanced versions outperform them. This work bridges the fields of diffusion models, NIR colorization, and visible-NIR fusion, advancing techniques for biodiversity monitoring, capturing wildlife activities day and night.

In contribution 6, Petracchi et al. focused on accelerating the processing of hyperspectral imaging (HSI), which is widely used in fields like medicine for diagnostics and surgery guidance. To address the challenge of processing large HSI datasets quickly, the researchers parallelized three popular machine-learning algorithms—SVM, Random Forest, and XGBoost—using GPU-based CUDA technology. Results show significant speed improvements, especially for SVM and XGBoost, making them more effective for classifying hyperspectral skin cancer images. The authors illustrate the parallelization techniques adopted for each approach, highlighting the suitability of Graphical Processing Units (GPUs) to enhance HSI applications, when the issue of rapid disease detection is critical.

In contribution 7, Jarahizadeh et al. compared three popular software tools, namely AgiSoft Metashape, PIX4DMapper, and DJI Terra, for processing UAV data to create 3D models of forested areas. Using datasets collected at different flight altitudes and angles, the researchers evaluated the tools in terms of point cloud density, reconstruction quality, computational time, and tree detection accuracy. The results report that AgiSoft and Pix4D produced denser point clouds, but DJI Terra excelled in generating more complete models with fewer gaps, particularly for trees, power lines, and poles. DJI Terra also presented faster processing times and provided more accurate height contours. The overall findings highlight that DJI Terra is a reliable choice for 3D modeling and tree detection in forestry and urban planning applications.

In contribution 8, Matei et al. introduced a method for designing and efficiently implementing 2D Far Infrared Range (FIR) circular filter banks. The filters are created using a frequency transformation of a 1D prototype following a Gaussian shape, designed to meet specific frequency specifications (peak frequency and bandwidth). The resulting filters are accurate and computationally efficient as a result of a factored transfer function and a polyphase structure combined with block filtering. Two types of filter banks—uniform and non-uniform—are developed, with an example demonstrating precise image reconstruction using the uniform filter bank. The proposed example is reported to achieve low computational complexity, making it practical for system-level applications.

In contribution 9, Cardone et al. presented a fast and efficient fuzzy-based framework for segmenting remote sensing images, implemented on a GIS platform. The method uses the Fast Generalized Fuzzy C-means algorithm to detect spatial relationships between pixels and a validity index to determine the optimal number of clusters. The process generates segmented images and a thematic map where pixel classifications are based on their highest membership degree, with reliability estimates provided for each class. Tested on imagery from Naples, Italy, the method produced results consistent with expert analyses while maintaining high computational speed, making it suitable for large-scale, high-resolution datasets.

In contribution 10, Bhatti et al. introduced a cost-effective, automated method for measuring facial skin temperature using a combination of a low-cost thermopile sensor

matrix and a 2D image sensor. By fusing temperature and image data through an affine transformation, the system can assign temperature readings to specific facial regions identified via face recognition. Throughout the paper, the advantages of the proposed method are described. A participant study shows that the method achieves accuracy comparable to commercial infrared forehead thermometers, offering a non-contact and precise alternative without requiring manual alignment.

In contribution 11, Ruiz-Beltrán et al. introduced an eye image detection system implemented on an MPSoC (multiprocessor system-on-chip), which includes a block in the programmable logic (PL) to assess the focus quality of the images. The system can discard images that are out of focus during processing. The solution, designed using Vitis High-Level Synthesis (VHLS), works with a 16 MP sensor and processes over 57 fps. Experiments using the CASIA-Iris-distance V4 database show that the system can successfully discard unfocused images, improving efficiency by eliminating up to 97% of blurry images, which reduces the computational load on subsequent processing steps like segmentation and iris pattern extraction. The overall goal of the study is to make iris recognition systems less intrusive and more user friendly.

In contribution 12, Katunin et al. presented a new approach for the automatic quantification of hidden corrosion using image processing of D-Sight images during periodic inspections. The performance of the algorithm was demonstrated through the inspection of a Mi family military helicopter. The nondimensional quantitative measurement introduced in this study was aligned with qualitative analysis by inspectors that performed qualitative analysis, confirming its effectiveness. The proposed method enables the automation of the inspection process and aids inspectors in assessing the extent and progression of hidden corrosion. The results of the study are of great importance to the aircraft industry (and many more), since hidden corrosion remains a major challenge in aircraft maintenance services.

Ultimately, contribution 13 is a review article that studies the latest trends in object detection, recognition, and tracking algorithms for Advanced Driver Assistance Systems (ADASs). ADASs use a range of sensors, including cameras, radars, and lidars, to perceive the environment and detect and track objects on the road, such as vehicles, pedestrians, cyclists, obstacles, and traffic signs. Specifically, Malligere Shivanna et al. survey the latest object detection, recognition, and tracking algorithms used in ADASs, discussing analytically their functionalities and the datasets employed. The review paper also highlights the need for further research in challenging environments, such as those with low visibility or high traffic density and concludes by exploring the future directions for these algorithms in ADASs.

3. Conclusions

As Guest Editors, we feel very delighted and satisfied with the final outcome of this Special Issue (SI) and we anticipate that fellow researchers and members of the scientific community will enjoy studying the articles included in it. Moreover, we would like to express our special thanks to the managing team of the Sensors journal for the continuous efforts and support during all the editing stages in this SI, including the initial preparation and planning, as well as the submission and review process of all the candidate manuscripts. Finally, we feel honored to receive outstanding research papers from the contributing authors, and at the same time we are also grateful to the reviewers for their help, their timely feedback, and their valuable comments.

Conflicts of Interest: The authors declare no conflict of interest.

List of Contributions

1. Kumari, P.; Keck, S.; Sohn, E.; Kern, J.; Raedle, M. Advanced Imaging Integration: Multi-Modal Raman Light Sheet Microscopy Combined with Zero-Shot Learning for Denoising and Super-Resolution. *Sensors* **2024**, *24*, 7083. https://doi.org/10.3390/s24217083.
2. Nguyen Trung, P.; Nguyen, N.; Nguyen Phan, K.; Pham Van, H.; Hoang Van, T.; Nguyen, T.; Gandjbakhche, A. A Non-Contacted Height Measurement Method in Two-Dimensional Space. *Sensors* **2024**, *24*, 6796. https://doi.org/10.3390/s24216796.
3. Giakoumidis, N.; Anagnostopoulos, C. ARM4CH: A Methodology for Autonomous Reality Modelling for Cultural Heritage. *Sensors* **2024**, *24*, 4950. https://doi.org/10.3390/s24154950.
4. Xiang, W.; Shen, J.; Zhang, L.; Zhang, Y. Infrared and Visual Image Fusion Based on a Local-Extrema-Driven Image Filter. *Sensors* **2024**, *24*, 2271. https://doi.org/10.3390/s24072271.
5. Borstelmann, A.; Haucke, T.; Steinhage, V. The Potential of Diffusion-Based Near-Infrared Image Colorization. *Sensors* **2024**, *24*, 1565. https://doi.org/10.3390/s24051565.
6. Petracchi, B.; Torti, E.; Marenzi, E.; Leporati, F. Acceleration of Hyperspectral Skin Cancer Image Classification through Parallel Machine-Learning Methods. *Sensors* **2024**, *24*, 1399. https://doi.org/10.3390/s24051399.
7. Jarahizadeh, S.; Salehi, B. A Comparative Analysis of UAV Photogrammetric Software Performance for Forest 3D Modeling: A Case Study Using AgiSoft Photoscan, PIX4DMapper, and DJI Terra. *Sensors* **2024**, *24*, 286. https://doi.org/10.3390/s24010286.
8. Matei, R.; Chiper, D. Analytic Design Technique for 2D FIR Circular Filter Banks and Their Efficient Implementation Using Polyphase Approach. *Sensors* **2023**, *23*, 9851. https://doi.org/10.3390/s23249851.
9. Cardone, B.; Di Martino, F.; Miraglia, V. A Novel Fuzzy-Based Remote Sensing Image Segmentation Method. *Sensors* **2023**, *23*, 9641. https://doi.org/10.3390/s23249641.
10. Bhatti, F.; Engel, G.; Hampel, J.; Khalil, C.; Reber, A.; Kray, S.; Greiner, T. Non-Contact Face Temperature Measurement by Thermopile-Based Data Fusion. *Sensors* **2023**, *23*, 7680. https://doi.org/10.3390/s23187680.
11. Ruiz-Beltrán, C.; Romero-Garcés, A.; González-García, M.; Marfil, R.; Bandera, A. Real-Time Embedded Eye Image Defocus Estimation for Iris Biometrics. *Sensors* **2023**, *23*, 7491. https://doi.org/10.3390/s23177491.
12. Katunin, A.; Synaszko, P.; Dragan, K. Automated Identification of Hidden Corrosion Based on the D-Sight Technique: A Case Study on a Military Helicopter. *Sensors* **2023**, *23*, 7131. https://doi.org/10.3390/s23167131.
13. Malligere Shivanna, V.; Guo, J. Object Detection, Recognition, and Tracking Algorithms for ADASs—A Study on Recent Trends. *Sensors* **2024**, *24*, 249. https://doi.org/10.3390/s24010249.

References

1. Zhao, Z.; Zheng, P.; Xu, S.; Wu, X. Object detection with deep learning: A review. *IEEE Trans. Neural Netw. Learn. Syst.* **2019**, *30*, 3212–3232. [CrossRef] [PubMed]
2. Cao, S. Analysis of object recognition trends based on deep learning. *Appl. Comput. Eng.* **2023**, *5*, 292–299. [CrossRef]
3. Denghui, Z.; Song, L.; Feng, Y.; Yang, Q. Research status of damage identification algorithm based on deep learning. *E3s Web Conf.* **2021**, *233*, 04039. [CrossRef]
4. Nakkach, C.; Zrelli, A.; Ezzeddine, T. Deep learning algorithms enabling event detection: A review. In Proceedings of the 2nd International Conference on Industry 4.0 and Artificial Intelligence (ICIAI 2021), Hammamet, Tunisia, 28–30 November 2021. [CrossRef]
5. Wang, Y. Overview of image segmentation methods based on deep learning. In Proceedings of the Volume 13184, Third International Conference on Electronic Information Engineering and Data Processing (EIEDP 2024), Kuala Lumpur, Malaysia, 15–17 March 2024. 131842V. [CrossRef]
6. Li, Y. Application of computer vision in intelligent manufacturing under the background of 5g wireless communication and industry 4.0. *Math. Probl. Eng.* **2022**, *2022*, 1–9. [CrossRef]
7. Milioto, A.; Lottes, P.; Stachniss, C. Real-Time Semantic Segmentation of Crop and Weed for Precision Agriculture Robots Leveraging Background Knowledge in CNNs. In Proceedings of the 2018 IEEE International Conference on Robotics and Automation (ICRA), Brisbane, QLD, Australia, 21–25 May 2018; pp. 2229–2235. [CrossRef]

8. Wan, T.; Shao, B.; Ma, S.; Zhou, Y.; Li, Q.; Chai, Y. In-sensor computing: Materials, devices, and integration technologies. *Adv. Mater.* **2023**, *35*, 2203830. [CrossRef] [PubMed]
9. Wang, Y.; Cai, Y.; Wang, F.; Yang, J.; Yan, T.; Li, S.; Wang, Z. A three-dimensional neuromorphic photosensor array for nonvolatile in-sensor computing. *Nano Lett.* **2023**, *23*, 4524–4532. [CrossRef] [PubMed]
10. Zhang, Z.; Wang, L.; Lee, C. Recent advances in artificial intelligence sensors. *Adv. Sens. Res.* **2023**, *2*, 2200072. [CrossRef]
11. Lee, S.; Peng, R.; Wu, C.; Li, M. Programmable black phosphorus image sensor for broadband optoelectronic edge computing. *Nat. Commun.* **2022**, *13*, 1485. [CrossRef]

sensors

Article

Self-Supervised and Zero-Shot Learning in Multi-Modal Raman Light Sheet Microscopy

Pooja Kumari [1,*], Johann Kern [2] and Matthias Raedle [1]

[1] CeMOS Research and Transfer Center, Mannheim University of Applied Sciences, 68163 Mannheim, Germany; m.raedle@hs-mannheim.de

[2] Universitätsklinikum Mannheim, Universität Heidelberg, 68167 Mannheim, Germany; johann.kern@medma.uni-heidelberg.de

* Correspondence: p.kumari@hs-mannheim.de

Abstract: Advancements in Raman light sheet microscopy have provided a powerful, non-invasive, marker-free method for imaging complex 3D biological structures, such as cell cultures and spheroids. By combining 3D tomograms made by Rayleigh scattering, Raman scattering, and fluorescence detection, this modality captures complementary spatial and molecular data, critical for biomedical research, histology, and drug discovery. Despite its capabilities, Raman light sheet microscopy faces inherent limitations, including low signal intensity, high noise levels, and restricted spatial resolution, which impede the visualization of fine subcellular structures. Traditional enhancement techniques like Fourier transform filtering and spectral unmixing require extensive preprocessing and often introduce artifacts. More recently, deep learning techniques, which have shown great promise in enhancing image quality, face their own limitations. Specifically, conventional deep learning models require large quantities of high-quality, manually labeled training data for effective denoising and super-resolution tasks, which is challenging to obtain in multi-modal microscopy. In this study, we address these limitations by exploring advanced zero-shot and self-supervised learning approaches, such as ZS-DeconvNet, Noise2Noise, Noise2Void, Deep Image Prior (DIP), and Self2Self, which enhance image quality without the need for labeled and large datasets. This study offers a comparative evaluation of zero-shot and self-supervised learning methods, evaluating their effectiveness in denoising, resolution enhancement, and preserving biological structures in multi-modal Raman light sheet microscopic images. Our results demonstrate significant improvements in image clarity, offering a reliable solution for visualizing complex biological systems. These methods establish the way for future advancements in high-resolution imaging, with broad potential for enhancing biomedical research and discovery.

Keywords: deep learning; unsupervised learning; zero-shot learning; self-supervised learning; super-resolution; denoising; light sheet microscopy; Raman scattering; Rayleigh scattering; fluorescence; spheroid; multi-mode

Citation: Kumari, P.; Kern, J.; Raedle, M. Self-Supervised and Zero-Shot Learning in Multi-Modal Raman Light Sheet Microscopy. *Sensors* **2024**, *24*, 8143. https://doi.org/10.3390/s24248143

Academic Editors: Christos Nikolaos E. Anagnostopoulos and Stelios Krinidis

Received: 8 November 2024
Revised: 12 December 2024
Accepted: 18 December 2024
Published: 20 December 2024

1. Introduction

Biomedical imaging plays a crucial role in advancing our understanding of complex biological systems, particularly three-dimensional (3D) structures such as cell cultures, spheroids, and organoids. These 3D structures have become fundamental models in drug discovery, cancer research, and histology, offering insights into tissue organization and cellular interactions. Traditional imaging techniques, such as fluorescence microscopy and electron microscopy, have long been used to visualize these structures. However, many of these methods require invasive sample preparation or the use of external labeling agents, which can introduce artifacts and affect the biological systems under observation.

The major advantages of 3D cell cultures over 2D cell cultures are that cell cultures in free three-dimensional space grow much more similarly to real organs and react to

pharmaceuticals than flat cell cultures. The use of 3D cell cultures thus shows a strict way to avoid animal experiments to an ever greater extent. As a result, there is a growing need for imaging techniques that allow for non-invasive, label-free visualization of biological samples in their native states [1–3].

Raman light sheet microscopy has emerged as a powerful, non-invasive, and label-free imaging modality that enables the study of 3D biological structures without altering the biological samples. This technique combines Rayleigh scattering, Raman scattering, and fluorescence detection, offering complementary spatial and molecular information about cellular architecture and interactions. The multi-modal approach of Raman light sheet microscopy provides a unique advantage by capturing rich data from biological systems, including high-resolution images of both spatial organization and molecular composition. The individual image modes contrast different information of the cell structure without using dye markers: image-based elastic scattering shows the arrangement of cell walls and nuclei; fluorescence imaging indicates, e.g., the spatial distribution of collagen or diffused pharmaceuticals; and Raman imaging shows, e.g., the distribution of hydrocarbons or amino acids, or even water. The label-free nature of this technique minimizes the risk of sample perturbation, making it ideal for studying delicate biological structures like spheroids and organoids in their natural state [4,5]. Even growing processes in living cells could be observed.

Despite its advantages, Raman light sheet microscopy faces several technical challenges, particularly related to image noise and resolution limitations. Raman scattering, although valuable for providing molecular information, generates extremely weak signals that are prone to noise. This, coupled with low signal intensity, often limits the achievable resolution, making it difficult to capture fine subcellular details in 3D samples [6,7]. Traditional methods, such as deconvolution and spectral unmixing, have been employed to enhance image quality, but these techniques often require complex preprocessing steps or rely on large amounts of labeled data, which can be impractical in real-time imaging [8,9].

Recent advances in machine learning, particularly deep learning techniques, offer promising solutions to overcome these limitations. While deep learning has been successfully applied to enhance resolution and reduce noise in medical imaging, these methods typically require extensive amounts of high-quality, labeled data for training. However, for applications like Raman light sheet microscopy, obtaining such labeled datasets can be impractical. This has led to the exploration of zero-shot and self-supervised learning approaches, which can enhance image quality without the need for labeled data [10,11].

Techniques such as Noise2Void, Deep Image Prior (DIP), and ZS-DeconvNet represent state-of-the-art approaches in zero-shot and self-supervised learning. These methods operate by leveraging the inherent structure and statistical properties of the data itself, enabling them to perform denoising and super-resolution without labeled training datasets. Their ability to enhance image clarity and resolution in an unsupervised manner makes them particularly well suited for multi-modal Raman light sheet microscopy [1,12,13].

This paper presents a comprehensive comparative evaluation of zero-shot and self-supervised learning algorithms for denoising and super-resolution in multi-modal Raman light sheet microscopy. By systematically evaluating these algorithms across different imaging modalities, we aim to identify the most effective techniques for improving image quality while preserving biological fidelity. This work aims to advance the field of high-resolution biomedical imaging and facilitate more accurate visualization of complex biological systems [1,10,14].

2. Materials and Methods

2.1. Biological Samples

Biological samples, including 3D spheroids and cell cultures, were prepared using established protocols to ensure optimal imaging conditions while maintaining cellular integrity. Spheroids were embedded in a low-scattering hydrogel matrix, providing optical transparency, and preserving physiological conditions during imaging. This method mini-

mized light scattering, ensuring high-quality imaging while maintaining an environment conducive to cellular function.

For this study, spheroids were generated from two HPV-negative head and neck squamous cell carcinoma (HNSCC) cell lines: UMSCC-11B, derived from laryngeal carcinoma, and UMSCC-14C, from oral cavity carcinoma. Both cell lines were cultured in Eagle's minimum essential medium (EMEM) supplemented with 10% fetal bovine serum (FBS) and 1% Penicillin/Streptomycin. Cultures were maintained at 37 °C in a humidified 5% CO_2 atmosphere. Cells were detached using Trypsin/EDTA, counted using a Neubauer hemocytometer, and seeded into ultra-low attachment (ULA) 96-well plates at densities of 2.5×10^4 or 5×10^4 cells per well to generate spheroids. These spheroids were cultured for up to eight days, with media changes on days 3, 5, and 8, reaching a diameter of 300–400 µm.

To investigate drug treatment effects, spheroids were treated with cisplatin (50 µM) on day 4 of culture, while control spheroids were treated with DMSO. After 72 h of incubation, both treated and untreated spheroids were fixed in 4% formalin to preserve their structural integrity for subsequent imaging. Samples were mounted in a 3D-printed hydrogel carrier designed for optimal alignment with the light sheet and detection objective, allowing multi-view imaging at 37 °C with 5% CO_2 and ensuring sample viability during extended imaging sessions.

2.2. Raman Light Sheet Microscope

The Raman light sheet microscope developed in this study integrates Rayleigh scattering, fluorescence, and Raman scattering modalities into a single, high-precision platform. The system utilizes dual-laser architecture, featuring a 660 nm and a 785 nm continuous wave laser, coaxially aligned through a series of broadband coated mirrors (M1-M5) and a beam splitter. These beams are then passed through achromatic doublet lenses to generate a well-defined, static light sheet that illuminates the sample chamber. (Figure 1a).

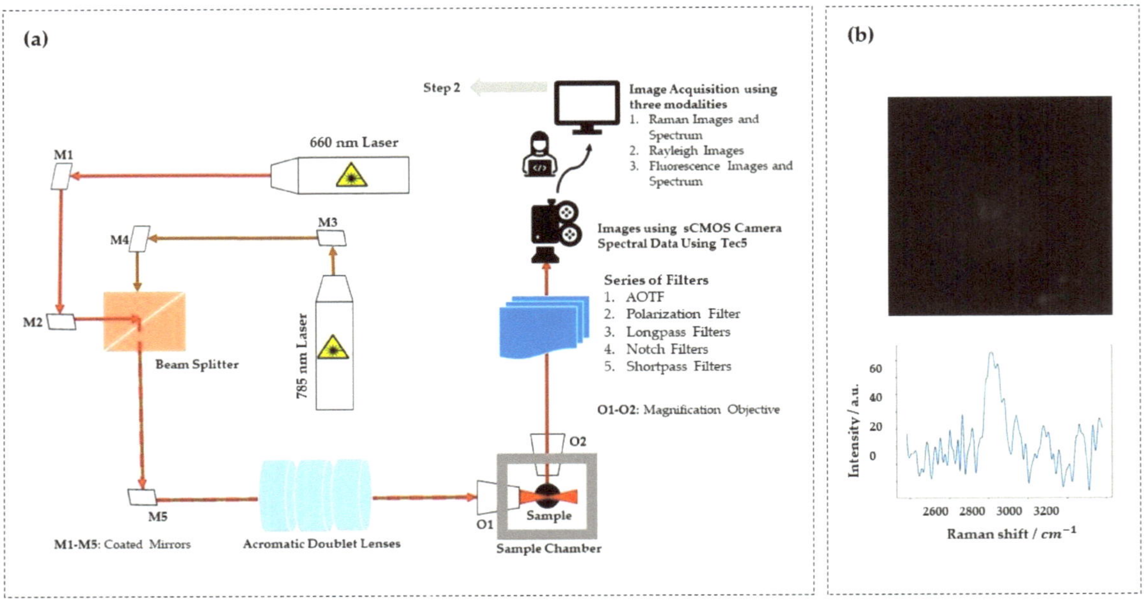

Figure 1. (**a**) Schematic of the multi-modal Raman light sheet microscope and (**b**) Raman image and spectral data acquired using a multi-modal Raman light sheet microscope with a 660 nm excitation laser and an acousto-optic tunable filter (AOTF) set at 817 nm.

The illumination optics generate a vertically oriented light sheet in the sample chamber, while the orthogonally positioned detection optics collect scattered or emitted photons. The axial resolution of the light sheet, defined by the beam waist, was determined to be approximately 8 μm for both the 785 nm and 660 nm lasers. This was measured using a BP209-VIS/M Scanning-Slit Optical Beam Profiler (Thorlabs Inc., Newton, MA, USA), ensuring uniform illumination and minimal scattering artifacts during imaging. This configuration supports precise optical slicing with a step size of 10 μm along the optical axis, crucial for capturing high-resolution subcellular structures in sequential imaging. The system's effective imaging field of view, calculated as 635 μm × 635 μm, accommodates spheroid samples with diameters up to this size, enabling complete optical sectioning for various experimental conditions. These parameters ensure high fidelity in multi-modal Raman imaging, particularly when paired with robust sample positioning for accurate alignment during sequential acquisitions. At the detection interface, a sCMOS camera records the emitted or scattered light after it passes through a configurable filter assembly. The filter set includes an acousto-optic tunable filter (AOTF), polarization filters, and a combination of longpass, notch, and shortpass filters. The AOTF enables precise spectral selection, allowing for fine-tuning of the transmitted wavelengths for each modality. This flexibility is critical for switching between the Rayleigh, Raman, and fluorescence modes without physically altering the optical setup. The polarization filters further improve contrast by rejecting unwanted light, enhancing the efficiency of inelastic scattering detection.

The detection system comprises a sCMOS camera paired with a modular filter assembly that includes an acousto-optic tunable filter (AOTF), polarization filters, and various longpass, notch, and shortpass filters. This modular system ensures that specific wavelengths are selected for each modality without physical realignment, enabling seamless transition between imaging modes. A multi-axis stage ensures precise sample positioning, with submicron resolution along the X, Y, and Z axes, and rotational control. This fine control is essential for maintaining stable and consistent imaging, particularly when working with 3D biological samples such as spheroids.

Figure 1b presents the Raman image and spectral data acquired using a 660 nm laser for excitation and an acousto-optic tunable filter (AOTF) for detection, centered at a wavelength of 817 nm.

2.3. Image Acquisition and Data Management

The multi-modal imaging system allows for comprehensive data acquisition using Rayleigh, fluorescence, and Raman modalities. For Rayleigh scattering, imaging was performed with a 785 nm laser at 1 mW power and 100 ms exposure, with an AOTF wavelength of 775 nm. Additionally, Rayleigh imaging at 660 nm was conducted under the same conditions, but with an AOTF wavelength of 650 nm. This enabled high-contrast Rayleigh data collection with minimal interference from other signals (Table 1).

Table 1. Imaging parameters for multi-modal acquisition.

Modality	Laser	Power	Exposure Time	AOTF Wavelength
Rayleigh (Modality 1)	785 nm	1 mW	100 ms	775 nm
	660 nm	1 mW	100 ms	650 nm
Fluorescence (Modality 2)	660 nm	130 mW	5000 ms	694 nm
Raman (Modality 3)	660 nm	130 mW	5000 ms	817 nm

For fluorescence imaging, a 660 nm laser was employed with a power output of 130 mW and an exposure time of 5000 ms, with the AOTF adjusted to 694 nm. This longer exposure time was critical to accommodate the inherently weaker fluorescence signals. Raman spectroscopy was conducted using the same 660 nm laser at 130 mW, with an exposure time of 5000 ms and an AOTF wavelength set to 817 nm, ensuring an optimal signal-to-noise ratio in capturing Raman spectra.

Data acquisition was synchronized across all modalities, and each dataset was meticulously archived with detailed metadata, including laser power, wavelength, exposure time, and filter settings. This approach ensured reproducibility and traceability, allowing for comparative evaluation across different modalities. The system's capability to rapidly switch between imaging modes via the tunable filter set facilitated efficient data collection, significantly reducing downtime between modality transitions.

By integrating these three imaging modalities into a single experimental setup, the Raman light sheet microscope provided a robust platform for high-resolution, multi-modal data acquisition. This comprehensive data management strategy further ensured that collected datasets could be efficiently processed and analyzed for detailed insights into the samples.

2.4. *Image Processing and Enhancement Using Deep Learning*

2.4.1. Preprocessing

Before applying zero-shot and self-supervised learning algorithms, multi-modal original images (Figure 2a) undergo background subtraction to eliminate unwanted signals and sand-noise reduction using filters like Gaussian or median filtering. These steps ensure clean, noise-reduced images, crucial for improving the performance and accuracy of the subsequent learning models.

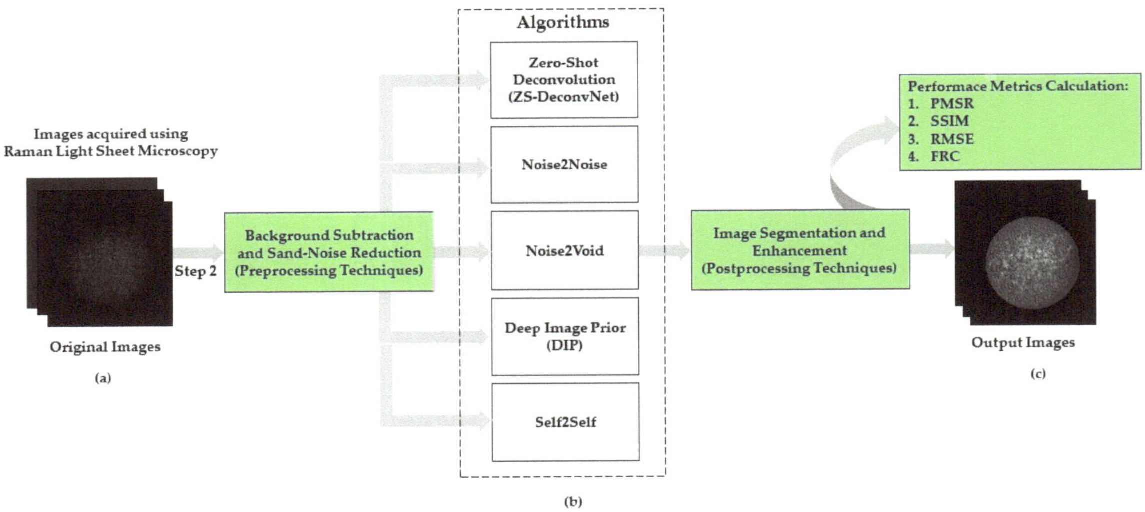

Figure 2. (**a**) Original images acquired using Multi-Modal Raman Light Sheet Microscopy (**b**) Implementation of zero-shot and self-supervised learning algorithms (ZS-DeconvNet, Noise2Noise, Noise2Void, DIP and Self2Self) on Original Images after Preprocessing Techniques (**c**) Denoised Output Images evaluated using metrics (PSNR, SSIM, RMSE, FRC).

2.4.2. Zero-Shot and Self-Supervised Learning Algorithms

To address the inherent blurring and noise challenges in multi-modal Raman light sheet microscopy, we applied the zero-shot deconvolution network (ZS-DeconvNet), an unsupervised deep learning model designed to perform deconvolution directly on noisy

images without the need for labeled training datasets. This approach allowed us to significantly enhance the resolution of microscopy images and recover finer subcellular structures that are otherwise obscured by system-induced blurring. In this study, we compared a series of advanced self-supervised and zero-shot learning methods aimed at denoising and enhancing resolution in multi-modal Raman light sheet microscopy. Each method was selected based on its ability to improve image clarity without requiring large, labeled datasets—an important consideration given the difficulty of acquiring clean, high-resolution ground truth images in microscopy. The following sections describe the methodologies employed: Zero-Shot Deconvolution Network (ZS-DeconvNet), Noise2Noise, Noise2Void, Deep Image Prior (DIP), and Self2Self (Figure 2b).

Zero-Shot Deconvolution Network (ZS-DeconvNet)

The zero-shot deconvolution network (ZS-DeconvNet) is a deep learning model designed to perform image deconvolution directly from noisy and corrupted images without the need for clean reference images during training. Operating in a zero-shot manner, ZS-DeconvNet adapts to each specific image stack during the deconvolution process, making it highly suitable for enhancing microscopy images, where acquiring high-quality, labeled training data is challenging or impractical. ZS-DeconvNet is particularly effective in applications such as multi-modal Raman light sheet microscopy, where images are frequently degraded by system-induced blur and noise. The algorithm learns to reverse the blurring process directly from the noisy input, recovering sharp details while preserving the biological structures. Specifically, it addresses the challenge of image blurring caused by the system's point-spread function (PSF) in multi-modal microscopy. As a fully convolutional neural network (CNN), ZS-DeconvNet operates within a zero-shot learning framework, allowing the model to learn and reverse the convolutional effects of the PSF from the observed image data, without relying on any external clean datasets [15–17].

Mathematical Formulation:

The observed microscopy image Y can be modeled as the convolution of the latent sharp image X with the point-spread function (PSF) h, combined with noise n:

$$Y = h * X + n \tag{1}$$

ZS-DeconvNet optimizes the network parameters θ by minimizing the loss between the convolved network output $f_\theta(h * Y)$ and the noisy observation Y:

$$\hat{\theta} = argmin_{\theta} ||Y - f_\theta(h * Y)||^2 \tag{2}$$

By minimizing this loss, the network learns to perform deconvolution and recover the underlying image structure.

Network Architecture:

The architecture consists of an encoder–decoder structure with skip connections to retain spatial details. The encoder compresses the input image, while the decoder restores the image resolution:

1. Encoder: Three layers of Conv2D, BatchNormalization, ReLU activation, and MaxPooling2D progressively downsample the image.
2. Decoder: The decoder mirrors the encoder with UpSampling2D layers followed by concatenation with the corresponding encoder layers, allowing detailed feature recovery.

Training Process:

ZS-DeconvNet was trained using pairs of corrupted images generated by applying random Gaussian noise to the input. The model was optimized using Adam for 100 epochs with mean squared error (MSE) as the loss function. The results demonstrated significant improvements in image sharpness and noise reduction, particularly in resolving subcellular structures [1,17,18].

Noise2Noise

Noise2Noise is a self-supervised learning algorithm designed for denoising tasks where clean reference images are unavailable. Instead of learning from noisy–clean pairs, Noise2Noise uses pairs of noisy images to learn to suppress noise while preserving the underlying image content [19].

Mathematical Formulation:

Noise2Noise operates under the assumption that noise present in the image is independent across acquisitions, but the underlying clean image remains constant. Given two noisy observations Y_1 and Y_2 represent two noisy observations of the same underlying clean image X, the model learns a mapping f_θ to predict one noisy observation from another:

$$Y_1 = X + n_1 \tag{3}$$

$$Y_2 = X + n_2 \tag{4}$$

where n_1 and n_2 are independent noise realizations and f_θ is the convolutional neural network (CNN) with learnable weights θ. The network f_θ is trained to predict Y_2 from Y_1 with the objective of minimizing the mean squared error (MSE) between the predicted and observed noisy images:

$$L(\theta) = \frac{1}{N}\sum_{i=1}^{N} \left|\left| f_\theta(Y_1) - Y_2 \right|\right|^2 \tag{5}$$

By optimizing this loss, the network learns to denoise the input image by recovering the shared structure between the noisy pairs while ignoring the noise.

Network Architecture:

The model used a U-Net-like encoder–decoder structure, similar to ZS-DeconvNet, but with a focus on learning from noisy image pairs rather than reconstructing from blurred images. The same Conv2D, BatchNormalization, and ReLU activation layers were used in both the encoder and decoder sections, with skip connections to ensure detail preservation.

Training Process:

Training was conducted on pairs of noisy images, with random Gaussian noise added to simulate real-world noise in microscopy. The model was trained for 100 epochs using the Adam optimizer and MSE loss, achieving high-quality denoising without requiring clean training data [20].

Noise2Void

Noise2Void is a self-supervised learning approach designed specifically for denoising tasks in the absence of paired noisy–clean image data. Unlike traditional supervised methods, Noise2Void learns to restore clean images from a single noisy image by exploiting the local structure of the image itself. This is achieved by predicting pixel values based on the context provided by neighboring pixels, masking the central pixel during training to prevent the model from directly learning the noise pattern [21].

Mathematical Formulation:

Let Y represent the observed noisy image and X the underlying clean image, with the relationship modeled as:

$$Y = X + n \tag{6}$$

where n is the noise. Noise2Void operates by applying a blind-spot strategy, where the central pixel in the receptive field is masked, and the network is trained to predict the value of the masked pixel using the surrounding pixels as context. The loss function for training is defined as:

$$L(\theta) = \frac{1}{N}\sum_{i=1}^{N} \left|\left| f_\theta\left(Y_{\setminus i}\right) - Y_i \right|\right|^2 \tag{7}$$

Here, $Y_{\setminus i}$ represents the image with pixel i masked, and $f_\theta\left(Y_{\setminus i}\right)$ is the CNN's prediction for the masked pixel value. By minimizing the difference between the predicted and actual pixel values, the model learns to denoise the image while preserving structural details.

Network Architecture:

Noise2Void used a U-Net-like architecture, similar to the one used for Noise2Noise. The encoder–decoder architecture was designed to capture both local and global context from the input image, allowing the network to predict the missing pixel values.

Training Process:

Blind-spot masking was applied to the input images during training, ensuring that the model never learns from the pixel it is supposed to predict. This forces the network to rely on surrounding context, making it effective for noise suppression in microscopy images. The model was trained for 1000 epochs using the Adam optimizer with early stopping to prevent overfitting [22,23].

Deep Image Prior (DIP)

The deep image prior (DIP) is a self-supervised learning approach that leverages the structure of convolutional neural networks (CNNs) as an implicit regularizer for image restoration tasks, such as denoising and super-resolution, without the need for pretrained models or large labeled datasets. Unlike traditional deep learning models, DIP directly trains on a single noisy image, using the architecture of the CNN itself to impose regularization on the output. This makes DIP particularly useful for microscopy applications, where obtaining clean, labeled ground-truth images is difficult [24].

Mathematical Formulation:

Given a noisy observation Y, DIP aims to recover the underlying clean image X by optimizing the CNN $f_\theta(z)$, where z is a fixed random input. The optimization objective is expressed as:

$$\hat{\theta} = arg\min_{\theta} \| f_\theta(z) - Y \|^2 \quad (8)$$

Here, $f_\theta(z)$ is the CNN with learnable weights θ, and z is typically a fixed random noise input. The key innovation of DIP is that the network is not pretrained on external datasets; rather, it learns to denoise the image directly during the optimization process. The network's architecture naturally regularizes the output by capturing image priors such as smoothness and continuity, which are inherent in most natural images.

Network Architecture:

The architecture for DIP is an encoder–decoder CNN similar to other methods described but uses random noise as input. Skip connections are used to retain fine details, and the decoder reconstructs the image from its compressed representation. The final output is generated by applying a sigmoid activation to constrain the pixel values between 0 and 1.

Training Process:

The model was trained directly on the noisy microscopy image, using early stopping to prevent overfitting. Training typically converged after 1000 epochs, yielding results that improved both image clarity and structural preservation [25].

DIP achieves denoising by regularizing the optimization process. As training progresses, the network gradually captures the underlying image structure, and denoising occurs as a natural outcome of the training process. The network converges to a solution where the output $f_\theta(z)$ represents a denoised version of the input image.

Self2Self

Self2Self is a self-supervised denoising technique that employs dropout as a form of regularization, allowing it to learn directly from noisy images without requiring clean or paired data. Unlike other denoising algorithms that rely on multiple noisy images or clean references, Self2Self is designed to operate on a single noisy image, using dropout to mask out pixels and learning to predict missing values based on the remaining context. This makes Self2Self highly effective in applications where acquiring multiple noisy samples or clean labels is not feasible, such as multi-modal Raman light sheet microscopy [26,27].

Mathematical Formulation:

Given a noisy observation Y, which is modeled as:

$$Y = X + n \tag{9}$$

where X is the latent clean image and n is the noise, Self2Self applies random dropout to the input image, masking a portion of pixels during training. The objective is to train the network to reconstruct the clean image by predicting the dropped pixels using the remaining visible ones. The dropout introduces randomness, which acts as a form of regularization, preventing the network from overfitting to the noise.

The training loss function is defined as:

$$L(\theta) = \frac{1}{N}\sum_{i=1}^{N}\left|\left| f_\theta\left(Y_{dropout}\right) - Y\right|\right|^2 \tag{10}$$

Here, $f_\theta\left(Y_{dropout}\right)$ is the network's prediction for the masked pixels and Y is the noisy input image. The dropout mask ensures that the network is forced to learn useful features of the underlying clean image rather than overfitting to the noise.

Network Architecture:

The Self2Self network architecture is similar to DIP but incorporates dropout layers to mask a portion of the input pixels during training. This forces the network to learn useful features from the unmasked pixels while avoiding overfitting to noise.

Training Process:

Self2Self was trained with random dropout applied to the noisy image during each training step. The model was optimized using Adam for 1000 epochs, and early stopping was applied to halt training when the loss stopped improving. Self2Self demonstrated effective denoising while preserving key subcellular structures [28].

2.4.3. Model Training and Implementation

All models were implemented using the TensorFlow deep learning framework. Training and inference were conducted on a high-performance computing system equipped with NVIDIA A100 GPUs, enabling fast, parallelized processing of the large multi-modal microscopy datasets.

Training Details

Each algorithm was trained on the same set of input images, consisting of noisy or corrupted microscopy data. Hyperparameters, including the learning rate and batch size, were fine-tuned to optimize each model's performance. The following key hyperparameters and parameters were used across all models, as shown in Table 1.

Each model was trained using these parameters and hyperparameters until convergence, defined as no improvement in the loss function for 10 consecutive epochs. For models like ZS-DeconvNet and Noise2Noise, 100 epochs were sufficient due to their efficiency in learning image structures from noisy pairs. Due to the large size of the 3D microscopy image stacks, a batch size of 1 was used across all models. This allowed efficient memory usage on the GPUs while maintaining high-performance processing, particularly for models that require the handling of large, high-dimensional data. These parameters were selected same for fair performance comparison.

2.4.4. Evaluation Metrics

The performance of the denoising and image enhancement algorithms was evaluated using a series of quantitative metrics to assess the quality of the output images, particularly focusing on the preservation of biological structures and overall noise reduction.

Peak Signal-to-Noise Ratio (PSNR)

PSNR was calculated to measure the quality of the denoised images relative to the original noisy inputs. A higher PSNR value indicates better noise suppression, with less distortion introduced during the image restoration process. PSNR is defined as [28,29]:

$$PSNR = 10 \times \log_{10}\left(\frac{MAX_I^2}{MSE}\right) \tag{11}$$

where MAX_I is the maximum possible pixel value of the image and MSE is the mean squared error between the original and processed image. Higher PSNR values suggest better noise suppression, meaning the deblurred image is closer to the clean image, which is ideal for microscopy, where noise can obscure fine details of subcellular structures. PSNR values above 30 dB generally indicate good image quality with significant noise reduction. For microscopy images, values between 30 and 50 dB are common, with values closer to 50 dB indicating near-perfect noise reduction and high-quality image restoration [27,30].

Structural Similarity Index (SSIM)

The structural similarity index (SSIM) is a metric designed to evaluate the perceived quality of an image by comparing luminance, contrast, and structural information between two images. SSIM ensures that the structural integrity of the deblurred images is preserved, especially critical for maintaining the accuracy of biological structures such as cells and organelles [31].

SSIM is calculated as:

$$SSIM(x,y) = \frac{(2\mu_x\mu_y + C_1)(2\sigma_{xy} + C_2)}{\left(\mu_x^2 + \mu_y^2 + C_1\right)\left(\sigma_x^2 + \sigma_y^2 + C_2\right)} \tag{12}$$

where:

μ_x and μ_y are the mean intensities of images x and y.

σ_x^2 and σ_y^2 are the variances of the images x and y, respectively.

σ_{xy} is the covariance between the images.

C_1 and C_2 are constants to stabilize the division when the denominator is close to zero.

SSIM values above 0.85 indicate that the structural content of the image is well preserved, and there is minimal distortion. Ideal SSIM values for high-quality biological microscopy images range between 0.90 and 0.99, indicating that the structural similarity between the noisy and denoised images is high [19,26].

Root Mean Squared Error (RMSE)

Root mean squared error (RMSE) measures the pixel-wise error between the noisy input and the denoised output. Lower RMSE values indicate better performance, with fewer deviations between the noisy and denoised images. RMSE is particularly useful for quantifying how accurately the algorithm has reconstructed the image from noisy input [21].

$$RMSE = \sqrt{\frac{1}{N}\sum_{i=1}^{N}\left(I_{noisy,i} - I_{denoised,i}\right)^2} \tag{13}$$

where:

N is the total number of pixels in the image.

$I_{original}$ and $I_{deblurred}$ are the pixel intensities in the original and deblurred images, respectively.

Lower RMSE values indicate better performance. In microscopy, RMSE values below 0.10 are preferred, with values closer to 0.01–0.05 representing excellent noise reduction and minimal pixel-wise error [25,32].

Fourier Ring Correlation (FRC) Analysis

Fourier ring correlation (FRC) is used to assess how well the spatial frequency components of the denoised image match those of the original noisy image. FRC is essential for evaluating the retention of high-frequency information, which corresponds to the sharpness and fine details in the image [24,25].

$$FRC = \frac{\sum_{i\epsilon R(f)} FFT_1(i) \cdot \overline{FFT_2(i)}}{\sqrt{\sum_{i\epsilon R(f)} |FFT_1(i)|^2 \cdot \sum_{i\epsilon R(f)} |FFT_1(i)|^2}} \quad (14)$$

where:

FFT_1 and FFT_2 are the Fourier transforms of the original and denoised images.

$R(f)$ represents the pixels corresponding to the frequency f.

The numerator measures the cross-power spectrum between the two images, and the denominator normalizes it by accounting for the energy of each image.

Higher FRC values indicate that the denoising algorithm has preserved high-frequency information, which is indicative of maintaining sharpness and fine details in the images. FRC values closer to 1 indicate better retention of structural details at high spatial frequencies. Preferred values typically range between 0.7 and 1.0, with higher values indicating excellent resolution and detail preservation in the denoised images [23,33].

2.4.5. Postprocessing

After applying zero-shot and self-supervised learning algorithms, postprocessing techniques such as image segmentation and enhancement are employed to further refine the output. Image segmentation isolates key regions of interest within the image, while enhancement improves visual clarity and contrast, highlighting critical features for evaluation. These steps ensure that the final images are optimized for interpretation and subsequent quantitative evaluation using PSNR, SSIM, RMSE, and FRC.

3. Results

We evaluated five zero-shot and self-supervised learning models—ZS-DeconvNet, Noise2Noise, Noise2Void, Deep Image Prior (DIP), and Self2Self—across four distinct multi-modal Raman light sheet microscopy modalities. The results were analyzed based on quantitative metrics, including PSNR, SSIM, RMSE, and FRC, combined with detailed visual comparisons.

3.1. Modality 1: Laser: 785 nm; Rayleigh Scattering; AOTF: 775 nm; Sample Type: 14C (Section 2.1)

In the first modality, Rayleigh scattering was imaged using a 785 nm laser, with scattered light filtered through an acousto-optic tunable filter (AOTF) centered at 775 nm to enhance spectral selectivity and minimize background noise. This setup introduces strong noise in regions with low signal intensity. This setup is used for imaging 14C-untreated samples (refer to Section 2.1), where the goal is to enhance image clarity without losing important structural details.

3.1.1. Image Comparison

Figure 3 presents a visual comparison of the denoised images produced by ZS-DeconvNet (Figure 3b), Noise2Noise (Figure 3c), Noise2Void (Figure 3d), DIP (Figure 3e), and Self2Self (Figure 3f), with the original noisy image (Figure 3a) included for reference. Each model exhibits varying levels of noise suppression and structural recovery.

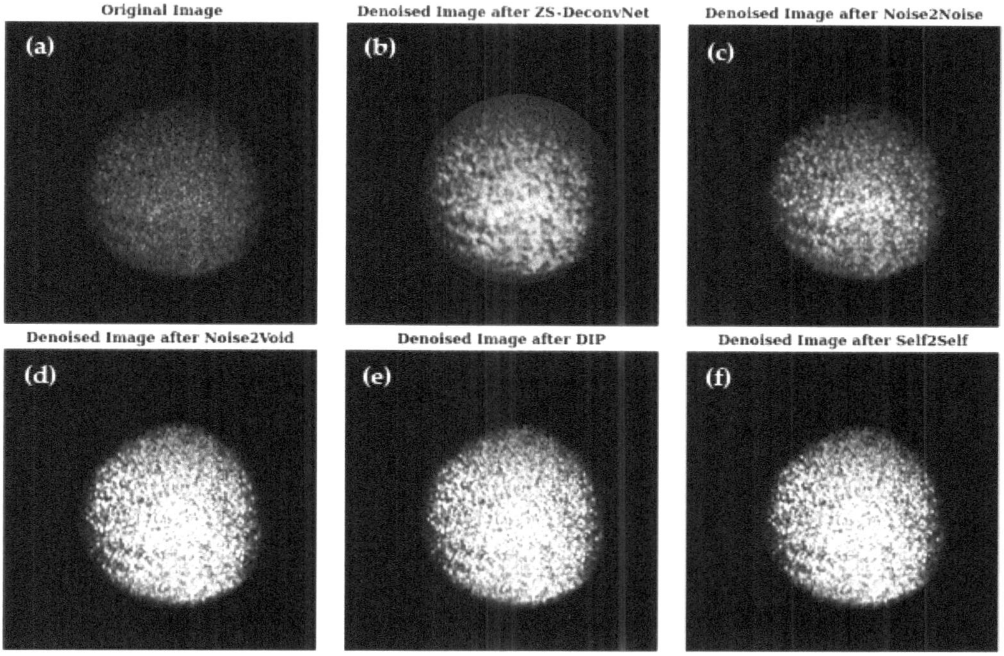

Figure 3. Visual comparison of original and denoised images for all zero-shot and self-supervised learning models for the 785 nm laser and Rayleigh scattering for the untreated 14C samples (refer to Section 2.1).

3.1.2. Quantitative Evaluation

To quantitatively assess the model's performance, we computed PSNR, SSIM, RMSE, and FRC metrics, as shown in Figure 2c. These metrics provide a comprehensive understanding of how well each model balances noise suppression, structural preservation, and high-frequency detail recovery. The PSNR (Figure 4a), SSIM (Figure 4b), RMSE (Figure 4c), and FRC (Figure 4d) analyses for the Rayleigh modality using the 785 nm laser on 14C-untreated samples demonstrate the relative strengths and weaknesses of the tested models. Noise2Void, DIP, and Self2Self emerged as the top performers, achieving the highest PSNR (>40 db), SSIM (close to 1.0), and very low RMSE, maintaining high FRC values, indicating excellent noise suppression and structural preservation. ZS-DeconvNet was moderately effective in noise reduction but struggled to retain structural integrity. Noise2Noise performed better than ZS-DeconvNet with higher PSNR, higher SSIM, lower RMSE, and better information retention, as shown by the FRC curve.

3.2. *Modality 2: Laser: 785 nm; Fluorescence Scattering; AOTF: 694 nm; Sample Type: 14C (Section 2.1)*

In the first modality, fluorescence scattering at 694 nm using a 660 nm laser introduces strong noise in regions with low signal intensity. This setup was used for imaging 14C-untreated samples (refer to Section 2.1), where the goal is to enhance image clarity without losing important structural details.

3.2.1. Image Comparison

Figure 5 presents a visual comparison of the denoised images produced by ZS-DeconvNet (Figure 5b), Noise2Noise (Figure 5c), Noise2Void (Figure 5d), DIP (Figure 5e), and Self2Self (Figure 5f), with the original noisy image (Figure 5a) included for reference. Each model exhibits varying levels of noise suppression and structural recovery.

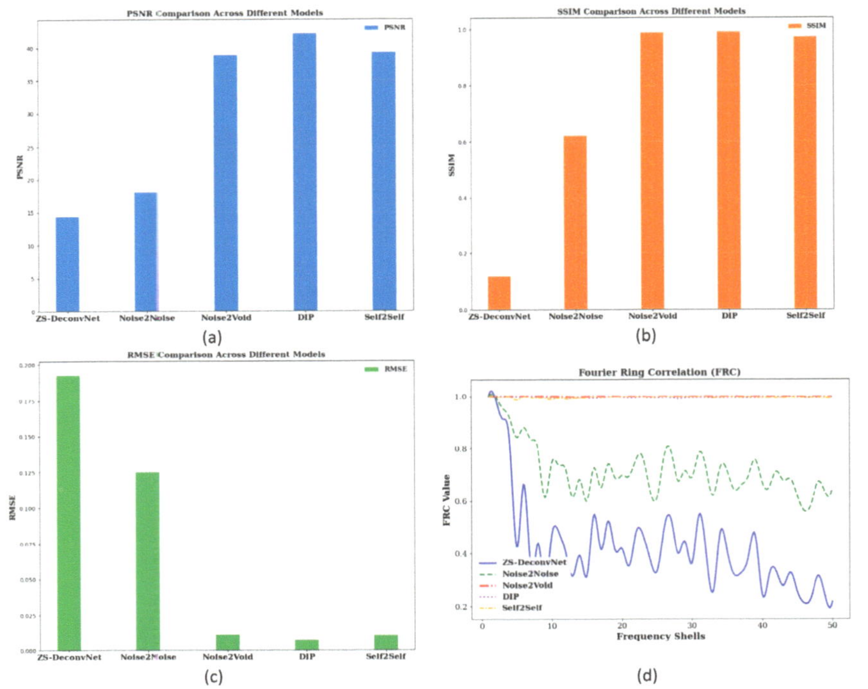

Figure 4. PSNR (**a**), SSIM (**b**), and RMSE (**c**) histograms and FRC curves (**d**) for the 14C-untreated samples and Rayleigh scattering using the 785 nm laser.

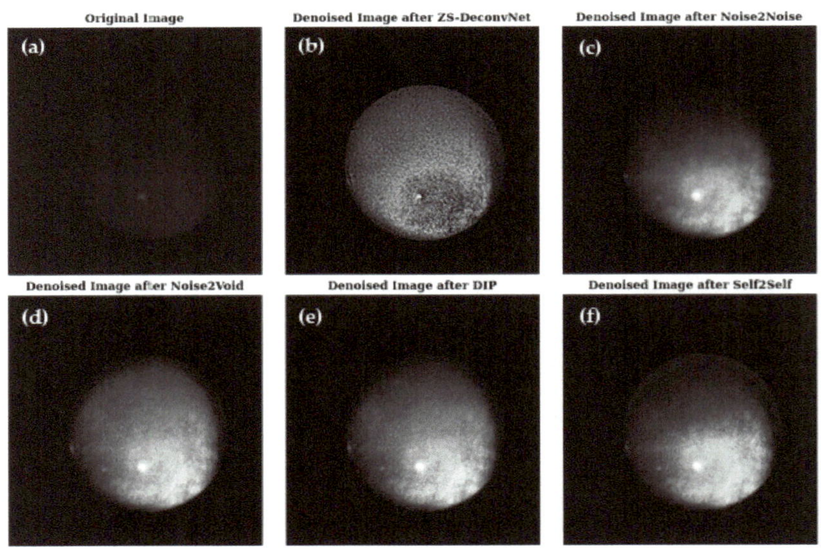

Figure 5. Visual comparison of original and denoised images for all zero-shot and self-supervised learning models for the 660 nm laser and fluorescence scattering for the untreated 14C samples (refer to Section 2.1).

3.2.2. Quantitative Evaluation

To quantitatively assess the model's performance, we computed PSNR, SSIM, RMSE, and FRC metrics, as shown in Figure 2c. These metrics provide a comprehensive understanding of how well each model balances noise suppression, structural preservation, and high-frequency detail recovery. The PSNR (Figure 6a), SSIM (Figure 6b), RMSE (Figure 6c), and FRC (Figure 6d) analyses for the fluorescence modality using the 785 nm laser on 14C-untreated samples demonstrate the relative strengths and weaknesses of the tested models. Noise2Void, DIP, and Self2Self again emerged as the top performers, achieving the highest PSNR (~40 db), SSIM (close to 1.0), and very low RMSE (0.004–0.01), maintaining high FRC values, indicating excellent noise suppression and structural preservation. ZS-DeconvNet performed better in noise reduction but struggled to retain structural integrity. Noise2Noise performed well in noise reduction but failed to retain structural integrity.

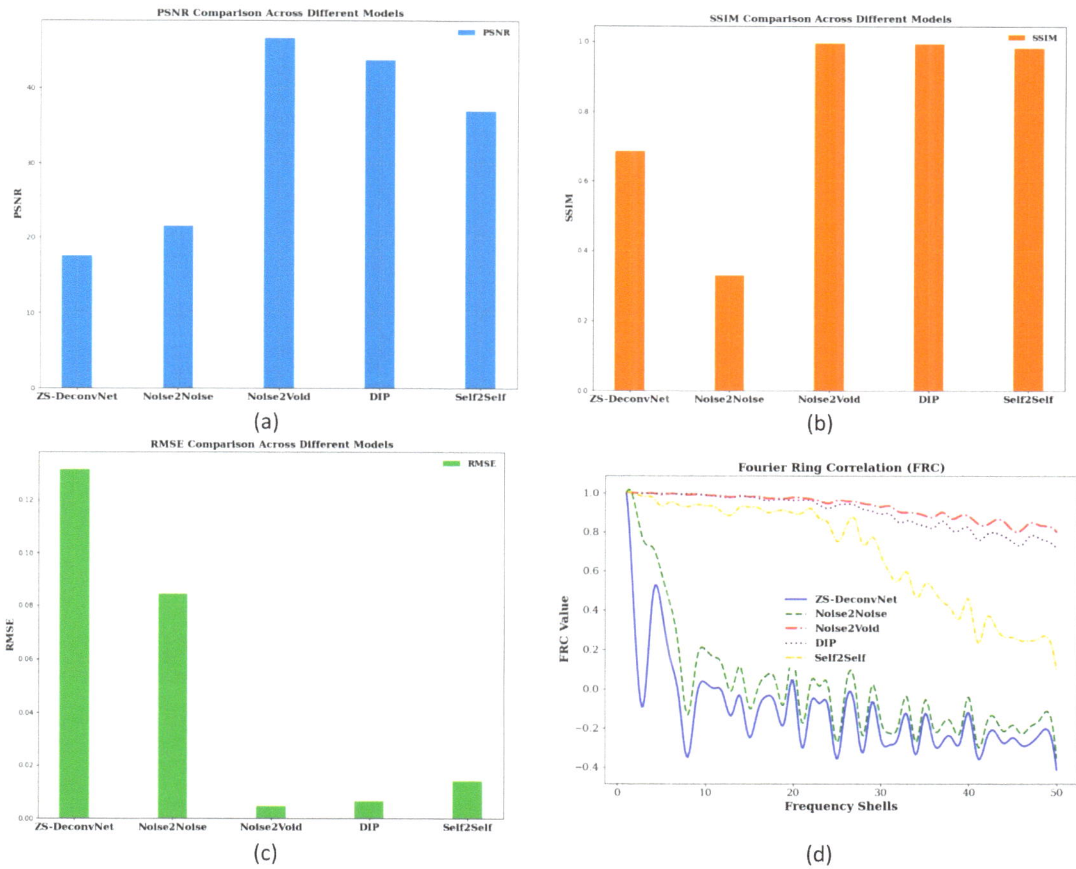

Figure 6. PSNR (**a**), SSIM (**b**), and RMSE (**c**) histograms and FRC curves (**d**) for the 14C-untreated samples and fluorescence scattering using the 785 nm laser.

3.3. *Modality 3: Laser: 660 nm; Raman Scattering; AOTF: 817 nm; Sample Type: Treated 11B (Section 2.1)*

In this modality, we evaluate Raman signals from 11B-treated samples, which have high noise levels and complex subcellular structures. The challenge lies in suppressing noise while retaining subtle structural details.

3.3.1. Image Comparison

Figure 7 presents a visual comparison of the denoised images produced by ZS-DeconvNet (Figure 7b), Noise2Noise (Figure 7c), Noise2Void (Figure 7d), DIP (Figure 7e), and Self2Self (Figure 7f) with the original noisy image (Figure 7a) included for reference. Each model exhibits varying levels of noise suppression and structural recovery.

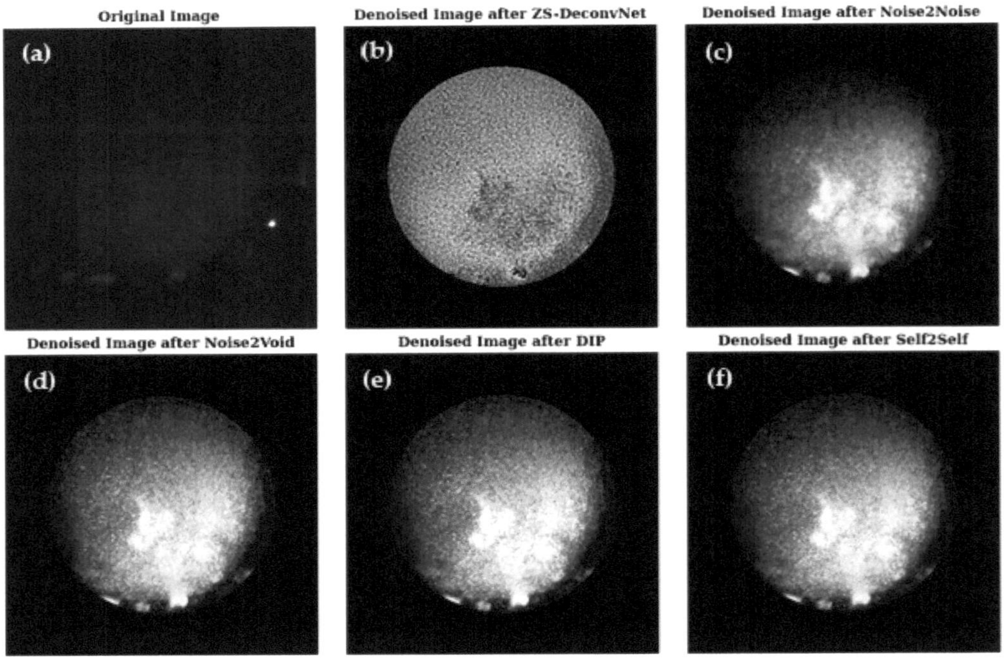

Figure 7. Visual comparison of original and denoised images for all zero-shot and self-supervised learning models for the 660 nm laser and Raman scattering for the treated 11B samples (refer to Section 2.1).

3.3.2. Quantitative Evaluation

The PSNR (Figure 8a), SSIM (Figure 8b), RMSE (Figure 8c), and FRC (Figure 8d) analyses for the Raman modality using the 660 nm laser on 11B-treated samples demonstrate the relative strengths and weaknesses of the tested models. ZS-DeconvNet emerged as winner for this modality providing the clearest results—PSNR (~30 db), SSIM (~0.9), and very low RMSE (0.004–0.01), maintaining acceptable FRC values, indicating excellent noise suppression and structural preservation. Noise2Noise, Noise2Void, DIP, and Self2Self also showed improvement in image quality.

3.4. Modality 4: Laser: 660 nm; Raman Scattering; AOTF: 817 nm; Sample Type: Untreated 11B (Section 2.1)

In this modality, we evaluate Raman signals from 11B-untreated samples, which have high noise levels and complex subcellular structures. The challenge lies in suppressing noise while retaining subtle structural details.

3.4.1. Image Comparison

Figure 9 presents a visual comparison of the denoised images produced by ZS-DeconvNet (Figure 9b), Noise2Noise (Figure 9c), Noise2Void (Figure 9d), DIP (Figure 9e)

and Self2Self (Figure 9f) with the original noisy image (Figure 9a) included for reference. Each model exhibits varying levels of noise suppression and structural recovery.

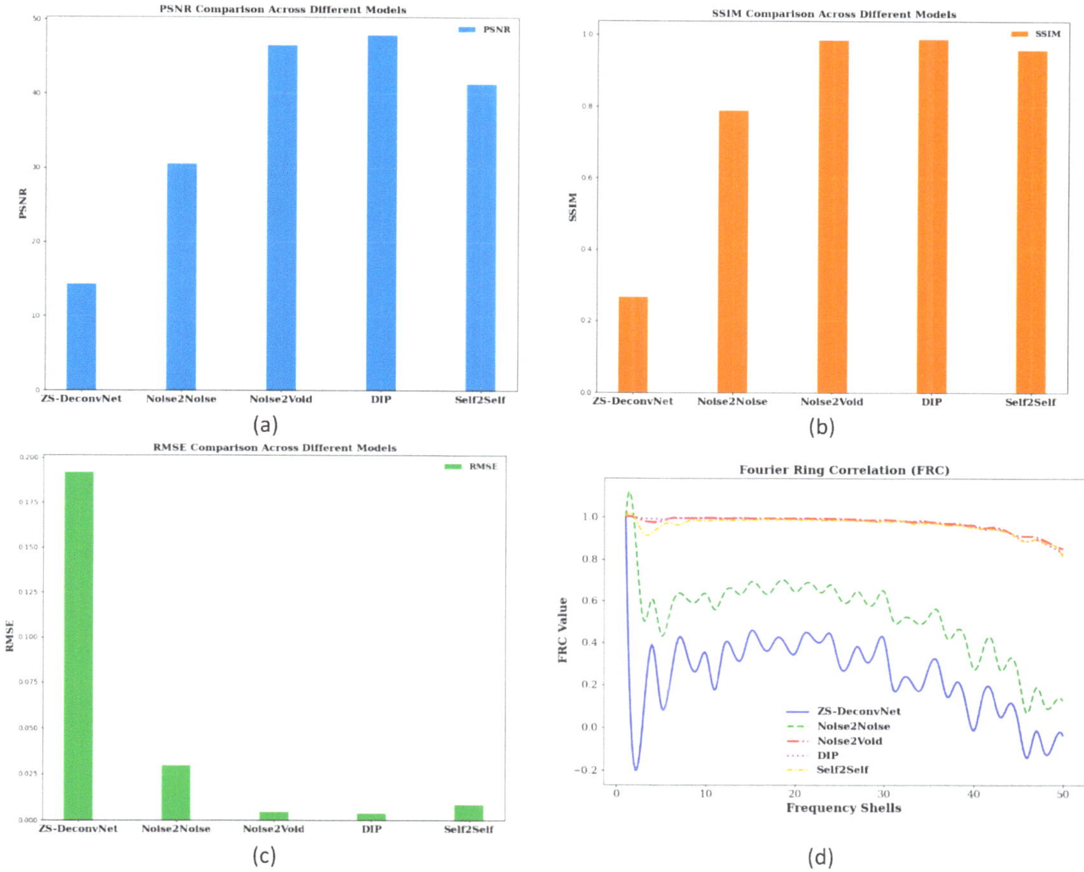

Figure 8. PSNR (**a**), SSIM (**b**), and RMSE (**c**) histograms and FRC curves (**d**) for the 11B-treated samples and Raman scattering using the 660 nm laser.

3.4.2. Quantitative Evaluation

Noise2Void, DIP and Self2Self again emerged as the top performers, achieving the highest PSNR (>40 db, Figure 10a), SSIM (close to 1.0, Figure 10b), and very low RMSE (0.004–0.01, Figure 10c), maintaining high FRC values (Figure 10d), indicating excellent noise suppression and structural preservation. ZS-DeconvNet performed better in noise reduction but struggled to retain structural integrity. Noise2Noise performed well in noise reduction but failed to retain structural integrity.

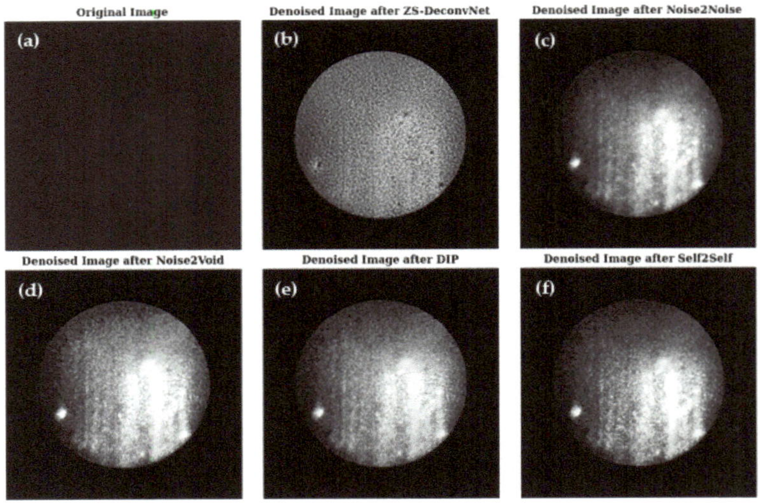

Figure 9. Visual comparison of original and denoised images for all zero-shot and self-supervised learning models for the 660 nm laser and Raman scattering for the untreated 11B samples (refer to Section 2.1).

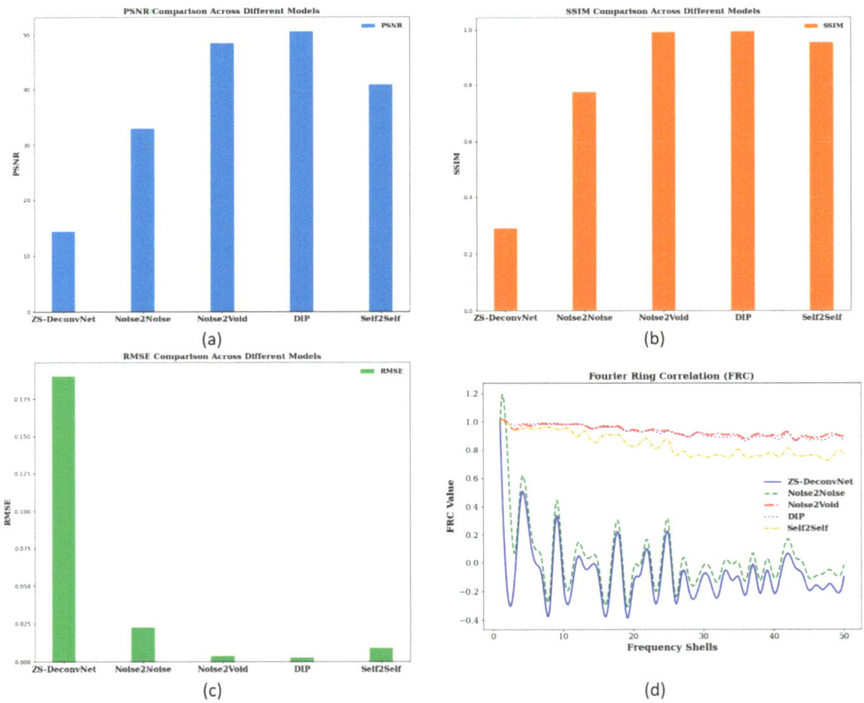

Figure 10. PSNR (**a**), SSIM (**b**), and RMSE (**c**) histograms and FRC curves (**d**) for the 11B-untreated samples and Raman scattering using the 660 nm laser.

3.5. Training Convergence: Loss vs. Epoch Curves

Figure 11 illustrates the loss vs. epoch curves for the ZS-DeconvNet, Noise2Noise, Noise2Void, DIP, and Self2Self models. The graph shows the first 25 epochs for clarity, although the models were trained for different total epochs: ZS-DeconvNet and Noise2Noise were trained for 100 epochs, while Noise2Void, DIP, and Self2Self were trained for 1000 epochs each (refer to Table 2). All models demonstrate a consistent reduction in loss as the training progresses, reflecting effective optimization across the different architectures. ZS-DeconvNet and Noise2Void exhibit similar convergence profiles, with a steady decline in loss throughout the epochs. Noise2Noise shows a faster reduction in loss during the initial epochs, while DIP and Self2Self exhibit a more gradual but continuous loss minimization over the training period.

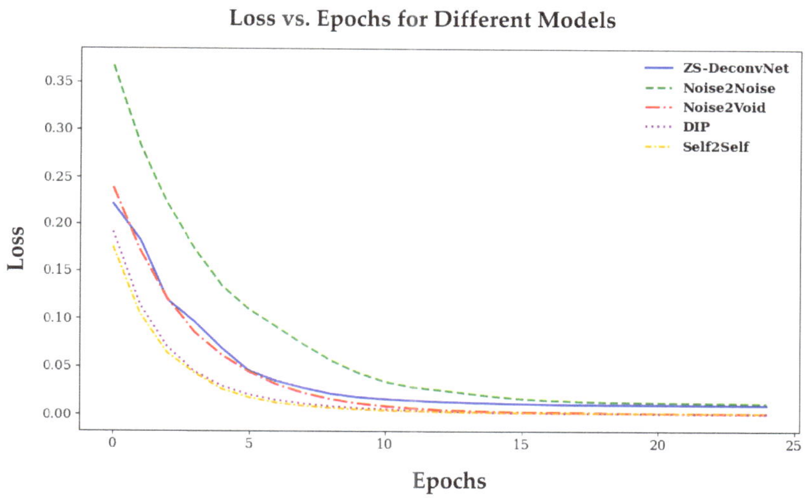

Figure 11. Loss vs. epoch curves for all zero-shot and self-supervised learning algorithms, reflecting overall training performance across all modalities described in Section 3.1.

Table 2. Selected parameters and hyperparameters of zero-shot and self-supervised algorithms for training.

Models	Learning Rate	Epochs	Optimizer	Loss Function	Batch Size
ZS-DeconvNet	0.001	100	Adam	MSE	1
Noise2Noise	0.001	100	Adam	MSE	1
Noise2Void	0.001	1000	Adam	MSE	1
DIP	0.001	1000	Adam	MSE	1
Self2Self	0.001	1000	Adam	MSE	1

These results indicate that all models successfully converge, with steady improvements in loss as the number of epochs increases, signifying effective learning and optimization of the respective models.

4. Discussion

In this study, we performed an extensive comparative evaluation of five state-of-the-art zero-shot and self-supervised learning methods for denoising and super-resolution in multi-modal Raman light sheet microscopy. The models—ZS-DeconvNet, Noise2Noise, Noise2Void, Deep Image Prior (DIP), and Self2Self—were evaluated across key quantitative metrics, including PSNR, SSIM, RMSE, and FRC. Across all modalities tested, we observed consistent trends in performance with DIP, Noise2Void, and Self2Self demonstrating superior capabilities in preserving both image fidelity and structural detail, while ZS-DeconvNet and Noise2Noise showed limitations under certain conditions.

4.1. Model Performance Across Modalities and Samples

A key finding of this study is the consistent performance of DIP, Noise2Void, and Self2Self across different imaging modalities—Rayleigh scattering, fluorescence, and Raman imaging—and varying sample conditions (both treated and untreated). These models consistently achieved high PSNR and SSIM values, indicating their robustness in noise suppression while retaining essential structural information.

In modality 1 (Table 1) (Laser: 785 nm, Rayleigh scattering, AOTF: 775 nm, Sample: Untreated 14C), DIP achieved the highest PSNR (41.83 dB) and FRC (0.994) values, indicating excellent noise reduction and preservation of high-frequency details critical for molecular imaging. Noise2Void performed similarly, with a PSNR of 38.96 dB and FRC of 0.997, demonstrating its ability to effectively handle different noise profiles without requiring paired data or external labels. These results are consistent with previous research highlighting the ability of these models to generalize well across various noisy datasets, especially in biomedical imaging [15,23].

Self2Self (Table 3), which uses dropout-based regularization, performed robustly with a PSNR of 39.15 dB and FRC of 0.994. Its adaptability to both treated and untreated samples made it an excellent choice for high-noise environments, where it maintained high-frequency information, crucial for subcellular structure preservation [34].

Table 3. Quantitative comparison of zero-shot and self-supervised learning models using PSNR, SSIM, RMSE, and FRC metrics for modality 1 (Laser: 785 nm, Rayleigh scattering, AOTF: 775 nm, Sample: Untreated 14C).

Model	PSNR (dB)	SSIM	RMSE	FRC
ZS-DeconvNet	14.06	0.07	0.19	0.430
Noise2Noise	18.40	0.33	0.12	0.720
Noise2Void	38.96	0.96	0.01	0.997
DIP	41.83	0.95	0.008	0.994
Self2Self	39.15	0.91	0.01	0.994

Similarly, for modality 2 (Table 4) (Laser: 660 nm, Fluorescence scattering, AOTF: 694 nm, Sample: Untreated 14C), the DIP, Noise2Void, and Self2Self models performed better than ZS-DeconvNet and Noise2Noise.

Table 4. Quantitative comparison of zero-shot and self-supervised learning models using PSNR, SSIM, RMSE, and FRC metrics for modality 2 (Laser: 660 nm, Fluorescence scattering, AOTF: 694 nm, Sample: Untreated 14C).

Model	PSNR (dB)	SSIM	RMSE	FRC
ZS-DeconvNet	13.34	0.076	0.215	0.050
Noise2Noise	31.30	0.150	0.027	0.171
Noise2Void	44.60	0.828	0.006	0.924
DIP	44.74	0.810	0.006	0.936
Self2Self	41.52	0.735	0.008	0.889

The performance of these three models (DIP, Noise2Void, and Self2Self) across both the treated 11B and untreated 11B samples showed remarkable consistency (Tables 5 and 6) for Raman scattering using the 660 nm laser. In the treated 11B samples, where biomolecular markers and fluorescence signals were enhanced, the models excelled in noise suppression while preserving structural integrity, as reflected in SSIM values ranging from 0.90 to 0.99. In the untreated samples, which presented higher noise levels, Self2Self, DIP, and Noise2Void still maintained FRC values near 1.0, indicating their strong performance in retaining high-resolution details despite the more challenging noise conditions.

Table 5. Quantitative comparison of zero-shot and self-supervised learning models using PSNR, SSIM, RMSE, and FRC metrics for modality 3 (Laser: 660 nm, Raman scattering, AOTF: 817 nm, Sample: Treated 11B).

Model	PSNR (dB)	SSIM	RMSE	FRC
ZS-DeconvNet	14.34	0.267	0.192	0.319
Noise2Noise	30.51	0.790	0.0298	0.592
Noise2Void	46.49	0.986	0.005	0.993
DIP	47.87	0.988	0.004	0.992
Self2Self	41.35	0.959	0.0085	0.980

Table 6. Quantitative comparison of zero-shot and self-supervised learning models using PSNR, SSIM, RMSE, and FRC metrics for modality 3 (Laser: 660 nm, Raman scattering, AOTF: 817 nm, Sample: Untreated 11B).

Model	PSNR (dB)	SSIM	RMSE	FRC
ZS-DeconvNet	14.43	0.290	0.189	−0.066
Noise2Noise	32.96	0.773	0.022	0.041
Noise2Void	48.43	0.991	0.004	0.934
DIP	50.62	0.993	0.003	0.928
Self2Self	40.81	0.954	0.009	0.836

In contrast, ZS-DeconvNet and Noise2Noise showed significantly lower performance in the treated 11B and untreated 11B samples, where the absence of enhanced signals made noise suppression more difficult. ZS-DeconvNet, with a PSNR of 14.34 dB for the treated 11B samples and 14.43 dB for the untreated 11B samples and an FRC of 0.319 and −0.066, demonstrated strong smoothing but failed to retain fine structural details, particularly in the 11B-untreated Raman imaging. Noise2Noise performed little better than ZS-DeconvNet.

4.2. Generalization Across Modalities and Noise Profiles

An important observation from this study is the generalizability of DIP, Noise2Void, and Self2Self across all tested modalities and varying noise levels. These models demonstrated the ability to retain high-resolution structural features, as indicated by their consistently high FRC values (0.994–0.997), even in challenging conditions like Rayleigh scattering and untreated samples with weaker Raman signals. This is critical for multi-modal microscopy where noise characteristics vary significantly depending on both the imaging technique and sample type.

The self-supervised nature of these models—particularly Noise2Void and Self2Self—allowed them to adapt to diverse noise profiles without requiring paired training data, making them especially suitable for applications where labeled datasets are unavailable or difficult to generate [15,34]. In contrast, models like Noise2Noise, which rely on paired noisy data, struggled in real-world scenarios where such data are scarce.

The high PSNR values achieved by DIP and Noise2Void across all imaging conditions indicate that these models are well suited for tasks requiring high-fidelity restoration of subcellular structures. Their performance in untreated samples further highlights their

potential for applications where noise levels are unpredictable, such as live-cell imaging or dynamic microscopy [18].

In summary, DIP, Noise2Void, and Self2Self proved to be the most robust across all modalities and noise conditions, making them excellent candidates for real-time imaging applications where maintaining high-resolution detail is paramount. Noise2Noise and ZS-DeconvNet, while effective in certain modalities, struggled with more complex noise profiles, limiting their broader applicability in multi-modal imaging tasks.

5. Conclusions

This study provides an in-depth comparative evaluation of five advanced zero-shot and self-supervised learning models—ZS-DeconvNet, Noise2Noise, Noise2Void, Deep Image Prior (DIP), and Self2Self—for denoising and super-resolution in multi-modal Raman light sheet microscopy applied to the visualization of 3D cell cultures. Across diverse modalities, including Rayleigh scattering, fluorescence, and Raman imaging, and across both treated and untreated samples, DIP, Noise2Void, and Self2Self consistently delivered the best results, achieving high PSNR, SSIM, and FRC values. DIP achieved the highest PSNR of >40 dB across different modalities and samples, demonstrating its unique ability to balance noise suppression with the preservation of fine structural details. Leveraging the inherent structure of convolutional neural networks as a prior, DIP operates in completely unsupervised manner, without requiring explicit training data, enabling adaptive regularization that excels in scenarios where labeled datasets are unavailable. Noise2Void and Self2Self provided similarly strong performance, with FRC values near 1.0, indicating their robustness in maintaining high-frequency structural information.

In contrast, ZS-DeconvNet and Noise2Noise showed limited effectiveness, particularly in high-noise environments, with ZS-DeconvNet using corrupted noise pairs within a zero-shot framework, struggled with the complex noise patterns characteristic of spheroid imaging, often resulting in oversmoothing and the loss of high-frequency details; similarly, Noise2Noise's reliance on paired noisy datasets restricted its adaptability, highlighting the need for further optimization in these models. These findings reinforce the potential of self-supervised learning techniques, particularly in contexts where acquiring large labeled datasets is impractical.

The broader applicability of these methods to other imaging modalities, such as MRI, CT, and super-resolution microscopy, presents a promising direction for future research. Expanding these models to handle 3D volumetric data and exploring hybrid architectures could further enhance their utility in real-time biological imaging. Overall, this study demonstrates the versatility and effectiveness of self-supervised and zero-shot learning models for improving image quality in biomedical microscopy.

Author Contributions: Conceptualization, P.K.; methodology, P.K.; software, P.K.; validation, P.K.; formal analysis, P.K.; investigation, P.K.; resources, J.K.; data curation, P.K.; writing—original draft preparation, P.K.; writing—review and editing, M.R. and P.K.; visualization, P.K.; supervision, J.K. and M.R.; project administration, P.K.; funding acquisition, M.R. CeMOS—Center for Science and Transfers designed the entire set of experiments including the setup. CeMOS—Center for Science and Transfer built the setup and conducted the experiments. Also, it analyzed the obtained images and spectra and drafted this manuscript. All coauthors contributed to discussion, interpretation, and final writing. All authors have read and agreed to the published version of the manuscript.

Funding: We would like to acknowledge funding support from the graduate program Perpharmance (BW6_07) provided by the Ministry of Science, Research, and the Arts (MWK) of Baden–Württemberg.

Institutional Review Board Statement: Not Applicable.

Informed Consent Statement: Not Applicable.

Data Availability Statement: The data used to support the results of this study are included within the article. In addition, some of the data in this research are supported by the references mentioned in the manuscript. If you have any queries regarding the data, the data of this research is available from the corresponding author upon request.

Acknowledgments: In this article, the authors draw on contributions from many members of the CeMOS Research and Transfer Center, especially Johann Strischakov and Shaun Keck. We also thank Emma Sohn from Universitätsklinikum Mannheim. All images and plots without source were created at the CeMOS—Center for Science and Transfer, University of Applied Science Mannheim, 68163 Mannheim, Germany.

Conflicts of Interest: The authors declare no conflicts of interest.

References

1. Kumari, P.; Keck, S.; Sohn, E.; Kern, J.; Raedle, M. Advanced Imaging Integration: Multi-modal Raman Light Sheet Microscopy Combined with Zero-Shot Learning for Denoising and Super-Resolution. *Sensors* **2024**, *24*, 7083. [CrossRef] [PubMed]
2. Manser, S.; Keck, S.; Vitacolonna, M.; Wühler, F.; Rudolf, R.; Raedle, M. Innovative Imaging Techniques: A Conceptual Exploration of Multi-modal Raman Light Sheet Microscopy. *Micromachines* **2023**, *14*, 1739. [CrossRef]
3. Mueller, T.; Eberle, F.; Michel, J.; Alshafee, M.; Humeau, A. Light Sheet Raman Micro-Spectroscopy for Label-Free Biomolecular Analysis. *Nat. Commun.* **2016**, *7*, 10948.
4. Oshima, Y.; Minamikawa, T.; Okuda, K.; Yoshikawa, Y. Combining Light Sheet Microscopy with Raman Spectroscopy for Cell Imaging. *J. Biomed. Opt.* **2012**, *17*, 046006.
5. Geraldes, C.F.G.C.; Sousa, L.; Ferreira, D.; Parra, M. Introduction to Infrared and Raman-Based Biomedical Molecular Imaging and Comparison with Other Modalities. *Molecules* **2020**, *25*, 5547. [CrossRef]
6. Chen, M.; Chefd'hotel, C.; Bogunovic, H.; Collins, D. Automated Detection and Analysis in Histopathology Using Deep Learning. *Biophys. Rev.* **2014**, *6*, 95–109.
7. Jiao, L.; Xu, Y.; Jiang, S.; Zhang, Y. Deep Learning for Segmentation in Microscopic Images. *Biophys. Rev.* **2019**, *11*, 169–184.
8. Kraus, O.; Vorobyov, I.; Sapp, M.; Pereira, P. DeepLoc: Deep Learning-Based Localization of Protein Subcellular Compartments. *EMBO J.* **2017**, *36*, 3210–3225.
9. Maitra, A.; Desai, S.; Mathur, A. Raman Spectroscopy in the Diagnosis of Esophageal Adenocarcinoma. *Photonix* **2020**, *7*, 121.
10. Li, X.; Su, J.; Zhang, H.; He, X. U-Net-Based CNN for Parasite Detection in Microscopic Images. *Biophys. Rev.* **2019**, *11*, 85–96.
11. Kobayashi, T.; Saito, M.; Uchida, H.; Okamoto, Y. Deep Learning in Drug Response Monitoring via Morphological Changes in *Cells. Biophys. Rev.* **2017**, *9*, 235–248.
12. Lu, F.K.; Basu, S.; Igras, V.; Lee, H.S. Label-Free, Bond-Selective Imaging of Biological Tissues with Coherent Raman Scattering Microscopy. *Nat. Photonics* **2020**, *14*, 148–153.
13. Su, W.; Wang, H.; Dong, X.; Zhang, J. Advanced Super-Resolution Techniques in Optical Microscopy. *Nat. Methods* **2021**, *18*, 1251–1260.
14. Schaub, F.; Dai, Z.; Zhang, Q. Deep Learning-Based Super-Resolution in Medical Imaging: From Theory to Practice. *IEEE Trans. Med. Imaging* **2019**, *38*, 1650–1662.
15. Krull, A.; Buchholz, T.-O.; Jug, F. Noise2Void—Learning Denoising from Single Noisy Images. In Proceedings of the IEEE Conference on Computer Vision and Pattern Recognition, Long Beach, CA, USA, 15–20 June 2019; pp. 2129–2137.
16. Lefkimmiatis, S. Universal Denoising Networks: A Novel CNN Architecture for Image Denoising. In Proceedings of the IEEE Conference on Computer Vision and Pattern Recognition, Salt Lake City, UT, USA, 18–22 June 2018; pp. 3204–3213.
17. Li, T. Zero-shot learning enables instant denoising and super-resolution in real-time across multiple imaging modalities. *Nat. Commun.* **2024**, *15*, 48575. [CrossRef]
18. Lehtinen, J.; Munkberg, J.; Hasselgren, J.; Laine, S.; Karras, T.; Aittala, M.; Aila, T. Noise2Noise: Learning Image Restoration without Clean Data. In Proceedings of the 35th International Conference on Machine Learning (ICML), Stockholm, Sweden, 10–15 July 2018; pp. 2965–2974.
19. Ronneberger, O.; Fischer, P.; Brox, T. U-Net: Convolutional Networks for Biomedical Image Segmentation. In Proceedings of the International Conference on Medical Image Computing and Computer-Assisted Intervention, Munich, Germany, 5–9 October 2015; pp. 234–241.
20. Batson, J.; Royer, L. Noise2Self: Blind Denoising by Self-Supervision. In Proceedings of the 36th International Conference on Machine Learning (ICML), Long Beach, CA, USA, 9–15 June 2019; pp. 524–533.
21. Pelt, D.M.; Sethian, J.A. A mixed-scale dense convolutional neural network for image analysis. *Proc. Natl. Acad. Sci. USA* **2018**, *115*, 254–259. [CrossRef]
22. Zhang, K.; Zuo, W.; Gu, S.; Zhang, L. Learning Deep CNN Denoiser Prior for Image Restoration. In Proceedings of the IEEE Conference on Computer Vision and Pattern Recognition, Honolulu, HI, USA, 21–26 July 2017; pp. 3929–3938.
23. Tripathi, S.; Sharma, N. Denoising of magnetic resonance images using discriminative learning-based deep convolutional neural network. *Technol. Health Care* **2022**, *30*, 145–160. [CrossRef]
24. Ulyanov, D.; Vedaldi, A.; Lempitsky, V. Deep Image Prior. In Proceedings of the IEEE Conference on Computer Vision and Pattern Recognition, Salt Lake City, UT, USA, 18–23 June 2018; pp. 9446–9454.
25. Kremer, J.R.; Mastronarde, D.N.; McIntosh, J.R. Computer visualization of three-dimensional image data using IMOD. *J. Struct. Biol.* **1996**, *116*, 71–76. [CrossRef]

26. Chheda, R.R.; Priyadarshi, K.; Muragodmath, S.M.; Dehalvi, F.; Kulkarni, U.; Chikkamath, S. EnhanceNet: A Deep Neural Network for Low-Light Image Enhancement with Image Restoration. In *Proceedings of 4th International Conference on Recent Trends in Machine Learning, IoT, Smart Cities and Applications. ICMISC 2023*; Gunjan, V.K., Zurada, J.M., Eds.; Lecture Notes in Networks and Systems; Springer: Singapore, 2024; Volume 873. [CrossRef]
27. Wang, Z.; Bovik, A.C.; Sheikh, H.R.; Simoncelli, E.P. Image quality assessment: From error visibility to structural similarity. *IEEE Trans. Image Process.* **2004**, *13*, 600–612. [CrossRef]
28. Blanchet, S.; Moisan, L. An explicit sharpness index related to visual image quality. *Signal Process. Image Commun.* **2006**, *21*, 487–509.
29. You, C.; Li, G.; Zhang, Y.; Zhang, X.; Shan, H.; Li, M. CT Super-resolution GAN constrained by the identical, residual, and cycle learning ensemble (GAN-CIRCLE). *IEEE Trans. Med. Imaging* **2020**, *39*, 188–203. [CrossRef] [PubMed]
30. Zhang, K.; Zuo, W.; Zhang, L. FFDNet: Toward a fast and flexible solution for CNN-based image denoising. *IEEE Trans. Image Process.* **2018**, *27*, 4608–4622. [CrossRef] [PubMed]
31. Pan, X.; Gao, J.; He, K. Deepl0ck: A deep denoising model based on self-similarity learning. *Int. J. Adv. Res. Artif. Intell.* **2019**, *8*, 48–54.
32. Scheres, S.H.; Chen, S. Prevention of overfitting in cryo-EM structure determination. *Nat. Methods* **2012**, *9*, 853–854. [CrossRef]
33. Cheng, Y.; Grigorieff, N. FRC-based criterion for the resolution of cryo-EM maps. *J. Struct. Biol.* **2015**, *186*, 199–203.
34. Quan, T.M.; Nguyen-Duc, T.; Jeong, W.-K. Self2Self with Dropout: Learning Self-Supervised Denoising from Single Image. In Proceedings of the IEEE/CVF Conference on Computer Vision and Pattern Recognition, Seattle, WA, USA, 13–19 June 2020; pp. 1890–1898.

sensors

MDPI

Article

Advanced Imaging Integration: Multi-Modal Raman Light Sheet Microscopy Combined with Zero-Shot Learning for Denoising and Super-Resolution

Pooja Kumari [1,*], Shaun Keck [1], Emma Sohn [2], Johann Kern [2] and Matthias Raedle [1]

1 CeMOS Research and Transfer Center, University of Applied Science, 68163 Mannheim, Germany; s.keck@hs-mannheim.de (S.K.); m.raedle@hs-mannheim.de (M.R.)
2 Universitätsklinikum Mannheim, Universität Heidelberg, 68167 Mannheim, Germany; emma.sohn@uni-heidelberg.de (E.S.); johann.kern@medma.uni-heidelberg.de (J.K.)
* Correspondence: p.kumari@hs-mannheim.de

Abstract: This study presents an advanced integration of Multi-modal Raman Light Sheet Microscopy with zero-shot learning-based computational methods to significantly enhance the resolution and analysis of complex three-dimensional biological structures, such as 3D cell cultures and spheroids. The Multi-modal Raman Light Sheet Microscopy system incorporates Rayleigh scattering, Raman scattering, and fluorescence detection, enabling comprehensive, marker-free imaging of cellular architecture. These diverse modalities offer detailed spatial and molecular insights into cellular organization and interactions, critical for applications in biomedical research, drug discovery, and histological studies. To improve image quality without altering or introducing new biological information, we apply Zero-Shot Deconvolution Networks (ZS-DeconvNet), a deep-learning-based method that enhances resolution in an unsupervised manner. ZS-DeconvNet significantly refines image clarity and sharpness across multiple microscopy modalities without requiring large, labeled datasets, or introducing artifacts. By combining the strengths of multi-modal light sheet microscopy and ZS-DeconvNet, we achieve improved visualization of subcellular structures, offering clearer and more detailed representations of existing data. This approach holds significant potential for advancing high-resolution imaging in biomedical research and other related fields.

Keywords: raman scattering; rayleigh scattering; zero-shot deconvolution networks; denoising; fluorescence; light sheet; microscopy; spheroid; multimode; hyperspectral; deep learning; super-resolution

Citation: Kumari, P.; Keck, S.; Sohn, E.; Kern, J.; Raedle, M. Advanced Imaging Integration: Multi-Modal Raman Light Sheet Microscopy Combined with Zero-Shot Learning for Denoising and Super-Resolution. *Sensors* **2024**, *24*, 7083. https://doi.org/10.3390/s24217083

Academic Editor: Christos Nikolaos E. Anagnostopoulos

Received: 6 October 2024
Revised: 29 October 2024
Accepted: 30 October 2024
Published: 3 November 2024

1. Introduction

Advances in imaging technologies have transformed the study of complex biological systems, particularly in the analysis of three-dimensional (3D) cellular structures. In biomedical research and histology, the ability to accurately visualize and analyze 3D cell cultures, such as spheroids, is critical for understanding cellular behavior, interaction, and function [1]. High-resolution imaging tools and techniques are vital for gaining insights into cellular organization and molecular dynamics, which are essential for fields like drug development and disease modeling. Multi-modal Raman Light Sheet Microscopy has emerged as a highly effective tool for these purposes, combining elastic and inelastic light scattering, including Rayleigh and Stokes Raman scattering, along with fluorescence detection, to provide high-resolution, marker-free imaging of biological samples. The principle of Multi-modal Raman Light Sheet Microscopy is shown in Figure 1. This technique facilitates the reconstruction of comprehensive 3D images, capturing both spatial and molecular information crucial for studies in tissue engineering, cancer biology, and drug development [2,3]. The Multi-modal Raman Light Sheet Microscope is specifically designed to overcome some of the key challenges in biological imaging, such as maintaining the native state of live tissues and cell cultures during imaging [4]. By utilizing the intrinsic

molecular properties of Rayleigh and Raman scattering, this technique eliminates the need for external fluorescent markers, thereby reducing potential sample perturbations and preserving physiological conditions [2,5]. This is especially valuable in dynamic, live-cell imaging, where maintaining cellular viability is critical. Moreover, by combining multiple imaging modalities, this system offers a detailed, multi-layered view of both structural and biochemical aspects of the sample, making it a versatile tool in a variety of biomedical applications [6].

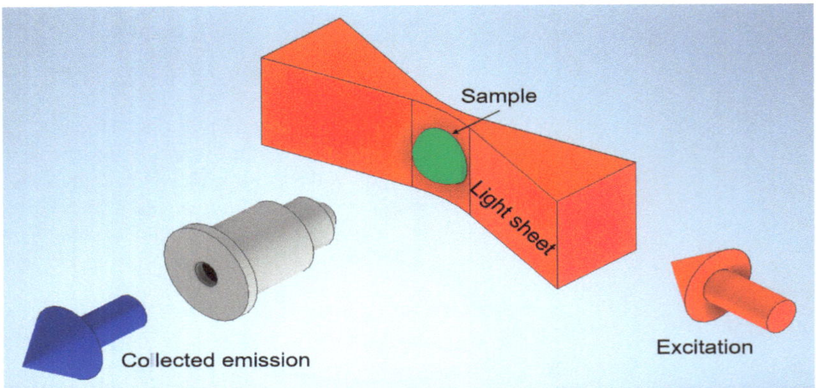

Figure 1. Principle of light sheet microscopy. Excitation and collection axes are orthogonally oriented with the sample placed at their intersection. A laser beam is shaped into a sheet and illuminates a thin section of the sample in the focal plane of the detection objective. The objective images the plane onto a camera chip [2].

However, while Multi-modal Raman Light Sheet Microscopy offers significant advancements in imaging 3D structures, its resolution is still constrained by the diffraction limit of light, and conventional imaging methods are often hampered by noise and signal degradation, particularly in low-light or long-term imaging conditions [2]. To address these challenges, computational super-resolution techniques [7,8] have been developed, with recent breakthroughs in machine learning offering new avenues for enhancing imaging performance [7,9]. Among these, Zero-Shot Deconvolution Networks (ZS-DeconvNet) have shown considerable promise in improving image resolution in real-time and in an unsupervised manner, without the need for large training datasets [10]. ZS-DeconvNet utilizes a CNN-based encoder–decoder structure to achieve computational super-resolution, enhancing spatial resolution by denoising and recovering high-frequency details beyond the optical limits of traditional microscopy. This approach allows visualization of sub-diffraction structures without additional hardware, thereby significantly increasing imaging detail and accuracy. ZS-DeconvNet enhances the resolution of microscope images without requiring ground truth data or additional data acquisition steps, making it particularly suitable for imaging dynamic biological processes [10].

In this study, we incorporate ZS-DeconvNet into multi-modal Raman light sheet microscopy to create a highly advanced imaging platform capable of delivering high-resolution images. The novelty of this approach lies in ZS-DeconvNet's zero-shot learning capability, allowing adaptive image enhancement across multiple microscopy modalities without pre-training or modality-specific tuning. This multimodal adaptability provides a unified solution for image enhancement, efficiently overcoming modality-specific challenges in fluorescence, Raman, and other microscopy techniques. By integrating cutting-edge computational techniques with Multi-modal Light Sheet Microscopy, we aim to significantly improve both the spatial and molecular resolution of biological imaging [11–13]. This combined approach provides new opportunities for real-time visualization of complex cellular structures and dynamic processes, with far-reaching implications for biomedical

research, cellular biology, and therapeutic development. Spheroids derived from head and neck squamous cell carcinoma (HNSCC) are particularly valuable in cancer research due to their resemblance to in vivo tumor architecture and behavior, including response to chemotherapeutics. In this study, we use UMSCC-11B cells, which are derived from HPV-negative HNSCC cell lines, to demonstrate the capabilities of Multi-modal Raman Light Sheet Microscopy enhanced by ZS-DeconvNet. Additionally, we explore the effect of the chemotherapeutic agent cisplatin on these spheroids, providing insights into both imaging and drug-response dynamics.

2. Materials and Methods

2.1. The Enhanced Multi-Modal Raman Light Sheet Microscopy with Zero-Shot Denoising Integration

This study utilizes an advanced Multi-modal Raman Light Sheet Microscope combining Rayleigh scattering, Raman scattering, and fluorescence emission for high-resolution, two-dimensional imaging of biological samples such as 3D spheroids and cell cultures [2,14–16]. The system integrates Zero-Shot Deconvolution Networks (ZS-DeconvNet), an advanced unsupervised deep learning technique to enhance image quality by significantly reducing noise without the need for additional training datasets or reference images [10,17].

2.2. Multi-Model Light Sheet Microscope Design

The Multi-modal Raman Light Sheet Microscope is based on the OpenSPIM platform [18], with significant enhancements to incorporate multi-modal imaging capabilities. The optical system consists of three primary components: the beam-shaping and illumination optics, the spectral selection and imaging optics, and the precision sample positioning system (Figure 2) (Table 1).

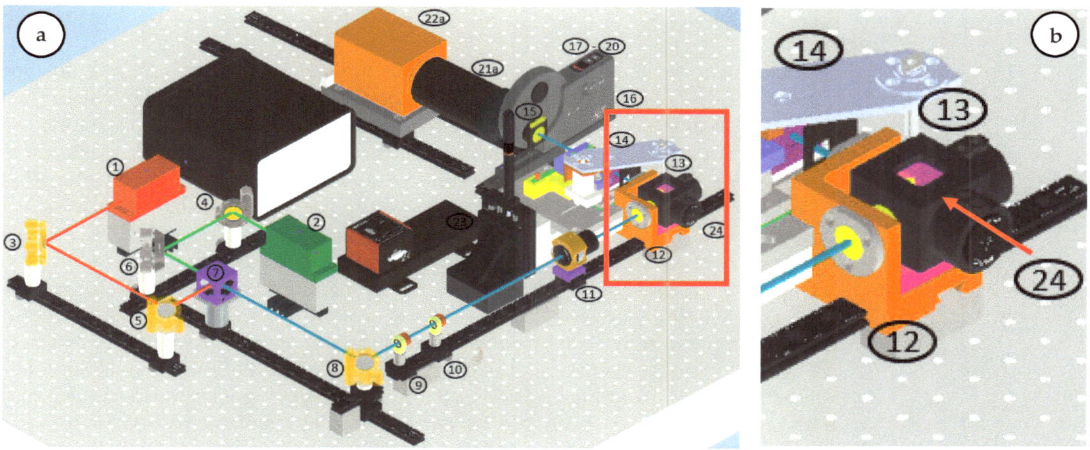

Figure 2. (**a**) Isometric view of the Raman light sheet microscope CAD model with connected sCMOS camera. The colored lines indicate the optical path of the illuminating lasers. Red: 660 nm beam propagation. Green: 785 nm beam propagation. Blue coaxial superimposed 660 nm and 785 nm beam propagation [2]. (**b**) Sample chamber (24), where sample is placed.

Table 1. Optical and mechanical components used in the Raman light sheet microscope.

No.	Component Specification	Manufaturer
1	LuxX Laser 785 nm, adjustable laser power 0.5–200 mW	Omicron GmbH (Dudenhofen Germany)
2	LuxX Laser 660 nm, adjustable laser power 0.5–130 mW	Omicron GmbH
3,5,8	Broadband mirror, Ø25.4 mm, EO2 coated, mounted in Polaris K1 Kinematic Mirror Mount	Thorlabs GmbH (Lübeck, Germany)
4,6	Broadband mirror, Ø25.4 mm, EO3 coated, mounted in Polaris K1 Kinematic Mirror Mount	Thorlabs GmbH
7	BrightLine laser dichroic beamsplitter, 25 mm × 36 mm, reflection band 350–671 nm, transmission band 702–1200 nm	Semrock (New York, NY, USA)
9	Mounted achromatic doublet lens, Ø12.7 mm, focal length 25 mm, anti-reflex coating 400–1100 nm	Thorlabs GmbH
10	Mounted achromatic doublet lens, Ø12.7 mm, focal length 50 mm, anti-reflex coating 400–1100 nm	Thorlabs GmbH
11	Mounted cylindrical achromatic doublet lens, Ø25.4 mm, focal length 50 mm, anti-reflex coating 650–1050 nm	Thorlabs GmbH
12	UMPLFLN10XW water dipping objective, magnification 10×, numerical aperture 0.3, working distance 3.5 mm	Evident (Hamburg, Germany)
13	UMPLFLN20XW water dipping objective, magnification 20×, numerical aperture 0.5, working distance 3.5 mm	Evident
14	Acousto-Optic Tunable Filter (AOTF), spectral range 550–1000 nm	Brimrose
15	Polarization filter	Thorlabs GmbH
16	6-position motorized filter wheel	Thorlabs GmbH
17	Longpass filter, 660 nm	Semrock
18	Notch filter, 660 nm	Semrock
19	Shortpass filter, 660 nm	Semrock
20	Longpass filter, 785 nm	Semrock
21a	Tube lens U-TLU and C-mount (U-TV0.5XC-3)	Evident
21b	Aspheric condenser lens, Ø25 mm, focal length 20 mm, anti-reflex coating 650–1050 nm	Thorlabs GmbH
22a	sCMOS camera ORCA Flash 4.0 LT+	Hamamatsu (Herrsching, Germany)
22b	CXY1 two-axis translating lens mount, Ø550 µm optic fiber	Thorlabs GmbH
23	USB-4D stage (X, Y, Z, R)	Picard-Industries (Albion, NY, USA)
24	Sample chamber, aluminum mounting frame, acrylic water chamber	CeMOS Research and Transfer Center (Mannheim, Germany)
25	MultiSpec® Raman spectrometer	tec5 GmbH (Steinbach, Germany)

- Excitation Lasers: Two continuous-wave lasers, with emission wavelengths of 660 nm and 785 nm, are used to excite Rayleigh and Raman scattering, respectively. The 660 nm laser is optimized for fluorescence imaging while minimizing autofluorescence, and the 785 nm laser enhances Raman scattering signals. Adjustable power outputs (0.5–130 mW for 660 nm and 0.5–200 mW for 785 nm) allow fine control of illumination intensity. Both laser beams are collimated and aligned coaxially using a series of broadband mirrors and dichroic splitters.
- Beam Shaping: The laser beams are expanded using a Keplerian telescope system formed by achromatic doublets, which increase the illuminated field of view without compromising beam focus. A cylindrical lens focuses the expanded beam into a static light sheet, projected into the sample chamber through a 10× water immersion objective.
- Imaging Optics: Photons scattered and emitted by the sample are collected by a 20× water immersion detection objective, positioned orthogonally to the light sheet for optimal detection. An Acousto-Optic Tunable Filter (AOTF) enables precise spectral selection with a 2 nm bandwidth, allowing for fine control over the wavelengths collected. Additional long-pass and short-pass filters further refine the detected signal. A high-sensitivity sCMOS camera (Hamamatsu ORCA Flash 4.0 LT+) is used for image acquisition, operating in a 1024 × 1024 pixel mode optimized for low-light conditions.

2.3. *Sample Preparation and Positioning*

Biological samples, including 3D spheroids and cell cultures, were prepared following standard protocols. Samples were embedded in a low-scattering hydrogel matrix, ensuring both optical transparency and the preservation of physiological conditions during imaging. This approach minimized scattering while maintaining an environment conducive to cellular function.

2.3.1. Cell Culture and Spheroid Formation

For this study, spheroids were generated using HPV-negative head and neck squamous cell carcinoma (HNSCC) cell lines, UMSCC-11B, provided by Dr. Thomas Carey from the University of Michigan. UMSCC-11B was derived from a laryngeal carcinoma.

Monoculture Spheroids

The UMSCC-11B cell lines were cultured in Eagle's Minimum Essential Medium (EMEM, Lonza, United States (Walkersville, Maryland)), supplemented with 10% fetal bovine serum (FBS) and 1% Penicillin/Streptomycin (Pen/Strep). Cells were incubated at 37 °C in a humidified atmosphere with 5% CO_2. Once cells reached confluency, they were washed with Dulbecco's phosphate-buffered saline (DPBS) and detached using Trypsin/EDTA. The total number of cells was determined using a Neubauer hemocytometer.

For spheroid formation, UMSCC-11B cells were seeded into 96-well ultra-low attachment (ULA) round-bottom plates (ThermoFisher Scientific, Mannheim, Germany) at a density of 2.5×10^4 or 5×10^4 cells per well. Spheroids were cultured for up to eight days, with media changes on days 3, 5, and 8. After spheroids reached the desired size, typically around 300–400 µm in diameter, they were prepared for subsequent imaging experiments.

Drug Treatment of Spheroids

To explore the effects of drug treatment, spheroids were treated with the chemotherapy drug cisplatin (Selleck Chemicals) during the course of their formation. On day 4 of spheroid culture, cisplatin was added to designated wells at concentrations of 50 µM or 100 µM, while control spheroids were treated with an equivalent volume of dimethyl sulfoxide (DMSO). Following drug treatment, the spheroids were incubated for an additional 48 or 72 h to assess the impact on morphology and viability.

Spheroid Fixation

At the end of the treatment period, both treated and untreated spheroids were fixed in 4% formalin for subsequent imaging using multi-modal Raman light sheet microscopy. Fixation ensured the structural integrity of the spheroids during the imaging and analysis processes.

2.3.2. Sample Mounting

For imaging, the spheroids were mounted in a custom-designed 3D-printed hydrogel carrier (Figure 3a,b). This carrier was optimized to precisely align the spheroids within the light sheet and the focal plane of the detection objective. Additionally, the carrier's design allowed for rotational adjustments, facilitating multi-view imaging from different angles.

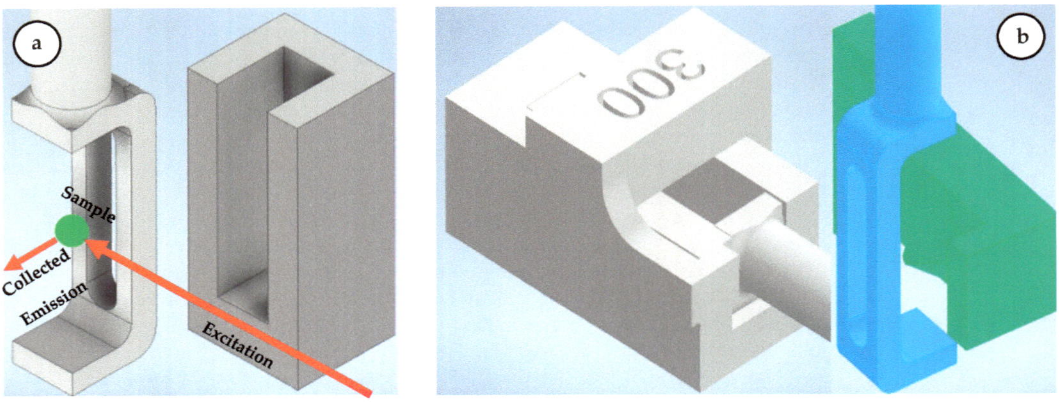

Figure 3. (**a**) CAD model of a multi-view sample carrier and corresponding frame for embedding spheroid samples in hydrogels. This sample is located in the sample chamber shown in Figure 2b. (**b**) Sample holder system consisting of gel chamber with cylindrical extension, casting frame and negative mold for precise, reproducible embedding of spheroids in hydrogel.

2.3.3. Environmental Control

During imaging, the sample chamber was maintained at 37 °C, with a regulated atmosphere containing 5% CO_2 to preserve sample viability over the course of extended imaging sessions. A 4D positioning stage enabled precise movement along the X, Y, and Z axes, with rotational adjustments for consistent sample alignment and accurate positioning within the microscope's field of view.

2.4. Zero-Shot Deconvolution Network (ZS-DeconvNet)

ZS-DeconvNet is an advanced machine-learning-based denoising algorithm designed to enhance image quality by reducing noise and preserving fine structural details. Unlike traditional supervised deep learning methods, ZS-DeconvNet operates in an unsupervised manner, requiring no ground-truth data or pre-trained models [10,19].

Mathematical Model of ZS-DeconvNet

The network minimizes noise in the acquired images using a self-supervised learning framework. The objective function for ZS-DeconvNet is formulated as:

$$L(\theta) = ||I_{raw} - I_{denoised}(\theta)||^2 + \lambda \Big| |\nabla I_{denoised}(\theta)| \Big|$$

where:

I_{raw} represents the noisy input image.

$I_{denoised}(\theta)$ represents the output of the neural network after applying the network's parameters (θ).

λ is a regularization term controlling the smoothness of the denoised image.

$\nabla I_{denoised}(\theta)$ represents the image gradients, ensuring that edges are preserved during denoising.

The Zero-Shot Deconvolution Network (ZS-DeconvNet) was implemented to improve the quality of the Raman Light Sheet Microscopy images by reducing noise and enhancing resolution.

Network Architecture: ZS-DeconvNet

The ZS-DeconvNet is built on a CNN-based encoder-decoder architecture. The encoder compresses the input image into a lower-dimensional latent space through a sequence of Conv2D layers, each followed by batch normalization and max pooling. These layers are designed to extract critical structural features from the image while progressively reducing its dimensionality, enabling the model to focus on the most important information. The decoder mirrors this process, gradually reconstructing the image by applying upsampling and Conv2D layers. Additionally, skip connections between the encoder and decoder allow the model to retain high-resolution details by concatenating features from earlier layers. Finally, a sigmoid-activated Conv2D layer produces the denoised output.

During training, ZS-DeconvNet follows a zero-shot learning approach. Two corrupted versions of the same image are generated—Denoised Image A and Denoised Image B (Figure 4)—by adding and subtracting noise, respectively. The model learns to map Denoised Image A (input) to Denoised Image B (target) without requiring a clean reference image. This is achieved using a Mean Squared Error (MSE) loss function, which minimizes the pixel-wise difference between the predicted and target images. By optimizing the MSE, the network progressively improves its ability to remove noise and restore details from noisy input data. Once the ZS-DeconvNet model is trained, it can be applied to new, unseen noisy images, resulting a denoised version of the input image by leveraging the learned features from the training process(Table 2).

Table 2. Description of Parameters/Hyperparameters used during ZS-DeconvNet Training.

ZS-DeconvNet Parameters/Hyperparameters	Description/Details
Input Image type	tiff/.tif
Input Image Size	1024×1024
Loss Function	Mean Squared Error
Optimizer	Adam
Epochs	100
Batch Size	1
Learning Rate	0.001
Evaluation Metrics	PSNR, SSIM, RMSE, FRC

The ZS-DeconvNet is designed to generalize well to unseen data. The lack of a need for pre-trained datasets enhances its flexibility, allowing it to be applied in image enhancement tasks. This feature is particularly advantageous in applications such as cell imaging and video microscopy, where pre-trained datasets may not be available or applicable (Figure 4).

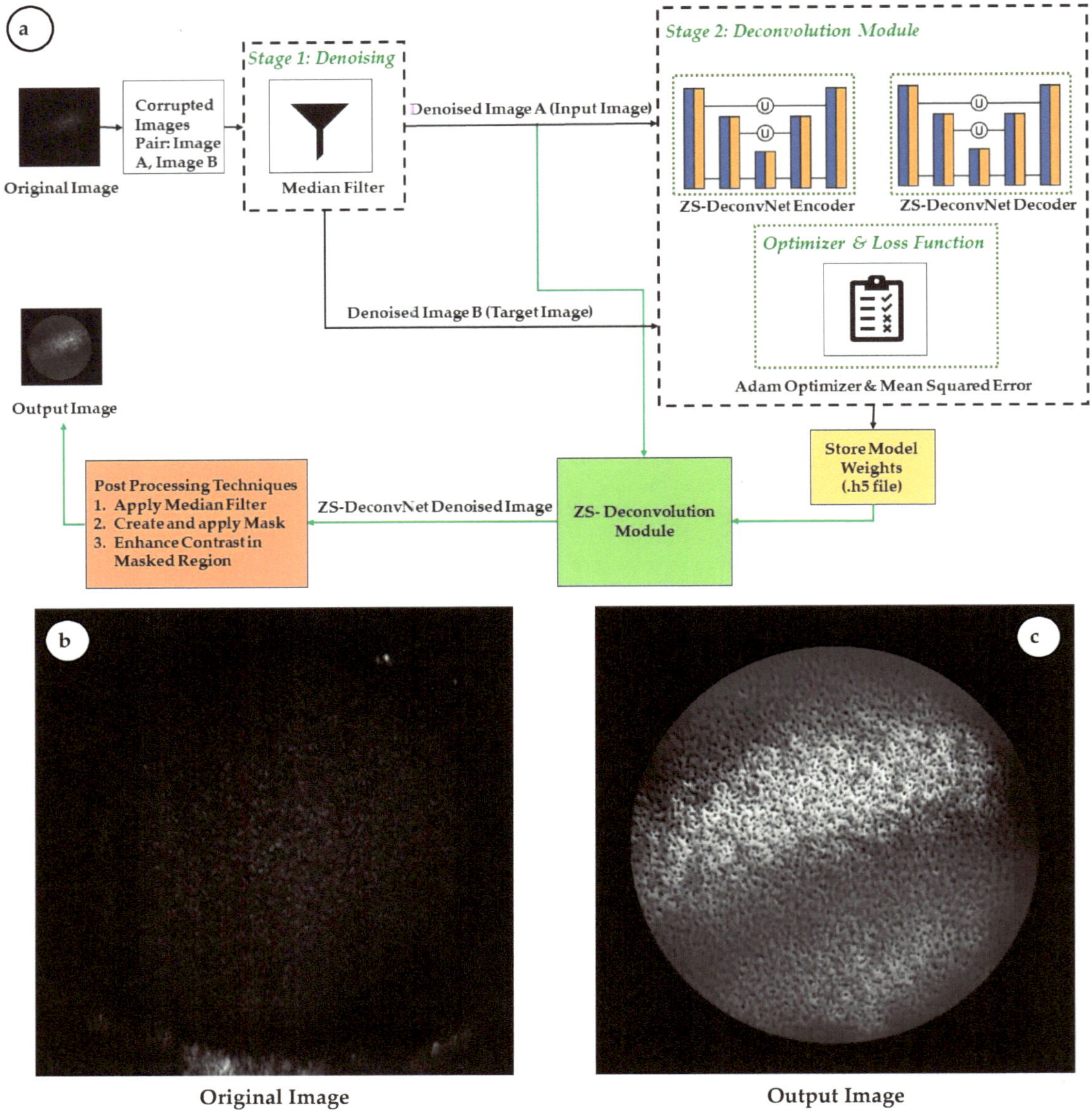

Figure 4. (**a**) The Zero-Shot Deconvolution Network (ZS-DeconvNet) architecture outlines the training workflow, encompassing pre-processing steps—such as corrupted image generation and median filter-based denoising—as well as post-processing techniques, including region-of-interest (ROI) image enhancement and morphological operations. The network's performance is assessed using PSNR, SSIM, and RMSE metrics to achieve enhanced image quality in Raman light sheet microscopy. (**b**,**c**) represent the input (**b**) and output (**c**) of the ZS-DeconvNet architecture, as depicted in (**a**).

Performance Evaluation Metrics

To rigorously assess the performance of the ZS-DeconNet model in image enhancement, we calculated various performance metrics: Peak Signal-to-Noise Ratio (PSNR), Structural Similarity Index (SSIM), Root Mean Square Error and Fourier Ring Correlation

(FRC). PSNR is used to assess the fidelity of a processed image in relation to an uncorrupted reference. Higher PSNR values suggest that the denoised image retains fidelity to the original structural and intensity details, confirming effective noise reduction and minimal loss of crucial information. PSNR thus serves as a direct indicator of ZS-DeconvNet's ability to faithfully restore high-resolution detail, which is essential for accurate imaging in microscopy. SSIM is a perceptual metric that quantifies structural similarity by evaluating luminance, contrast, and spatial composition between the original and processed images. Elevated SSIM scores demonstrate that ZS-DeconvNet not only reduces noise but also maintains the image's inherent structural relationships, essential for preserving context in microscopy where spatial coherence is key to interpretation. SSIM is especially beneficial for validating that enhanced images accurately reflect the morphology and structural integrity of biological samples. RMSE quantifies the average deviation in pixel intensity between the denoised and original images, offering a robust measure of reconstruction accuracy. Lower RMSE values denote minimal divergence from the expected pixel values, highlighting the model's ability to precisely restore image content even under significant noise interference. To objectively evaluate improvements in spatial resolution, we utilize Fourier Ring Correlation (FRC), a frequency-domain metric that quantitatively assesses resolution by comparing spatial frequency content before and after processing. FRC is widely used in super-resolution microscopy and quantifies the effective resolution enhancement achieved by ZS-DeconvNet. Unlike PSNR, SSIM, and RMSE, which primarily address image quality and similarity, FRC directly measures resolution improvements, providing insight into the model's ability to recover or even enhance fine structural details. By achieving a higher FRC resolution threshold, ZS-DeconvNet confirms its utility in super-resolution applications, effectively distinguishing it as a powerful tool for high-resolution image enhancement in multimodal microscopy.

Pre-processing: Before being fed into the ZS-DeconvNet, the raw images (Original Image) undergo a series of crucial preprocessing steps aimed at preparing the data for optimal denoising and model training. In the initial step I, two corrupted image pairs (Image A, Image B) were generated by introducing Gaussian noise to the original image, where Image A had added noise and Image B had inverted noise. After creating a corrupted image pair from the Original Image, median filtering was applied to these images to remove high-intensity "salt-and-pepper" noise, which is commonly seen in images captured through noisy channels, such as biomedical imaging. A median filter with a kernel size of 3 was applied, effectively suppressing noise while preserving edges and fine details within the image. This enhances the network's ability to focus on meaningful structural elements during training and inference, contributing to more precise noise removal. (ref. Figure 4)

These pairs serve as input (Denoised Image A) and target (Denoised Image B) images during the training phase of ZS-DeconvNet. The model learns to predict Denoised Image B from Denoised Image A, simulating a noise-to-noise learning framework.

Post-processing: Following the ZS-DeconvNet denoising process, additional post-processing steps were performed, such as applying a region of interest (ROI) mask, applying Median Filter with Kernal size 3, and enhancing contrast to refine the output image. Here we also calculated PSNR, SSIM, RMSE, and FRC metrics for performance evaluation [19,20]. These enhancements further improved the clarity and usability of the denoised image by focusing on key structures (Figure 4).

These operations help in emphasizing small structures within the spheroids, particularly in the cellular boundaries. If required, edge detection algorithms (such as Canny edge detection) can also be applied to highlight critical boundaries and structural details. This step is essential in the analysis of biological images, where the accurate delineation of subcellular components plays a crucial role in data interpretation.

3. Results

3.1. Denoising Performance and Image Clarity

This study assesses the denoising performance of ZS-DeconvNet on 11B spheroid samples, comparing both treated and untreated conditions following exposure to 50 µM cisplatin for 72 h (refer to the Sample Preparation section for further details). Images were captured using two laser excitations, 660 nm and 785 nm, and processed across multiple imaging modalities. The primary objective was to determine the effectiveness of ZS-DeconvNet in reducing noise while preserving critical image features, and to identify any structural or molecular changes induced by the treatment in the spheroids.

Laser Excitation at 660 nm: For the 660 nm laser, three distinct modalities were used:

(1) Rayleigh Scattering (Power: 1 mW and AOTF: 650 nm): The raw images captured using 660 nm Rayleigh scattering were heavily impacted by noise, making it difficult to discern fine structural details in both treated and untreated spheroids. As demonstrated in Figure 5a,b, the original image (left) contains substantial noise that obscures surface-level information. After applying ZS-DeconvNet, the denoised image (right) exhibited a marked reduction in noise, allowing for the visualization of key features that were previously hidden. The treated spheroids exposed to 50 µM cisplatin for 72 h revealed subtle structural alterations, such as surface roughness and changes in texture, which were not discernible in the noisy image. ZS-DeconvNet's ability to enhance image clarity at such low power (1 mW) demonstrates its robustness in handling noisy datasets without sacrificing the essential information within the image.

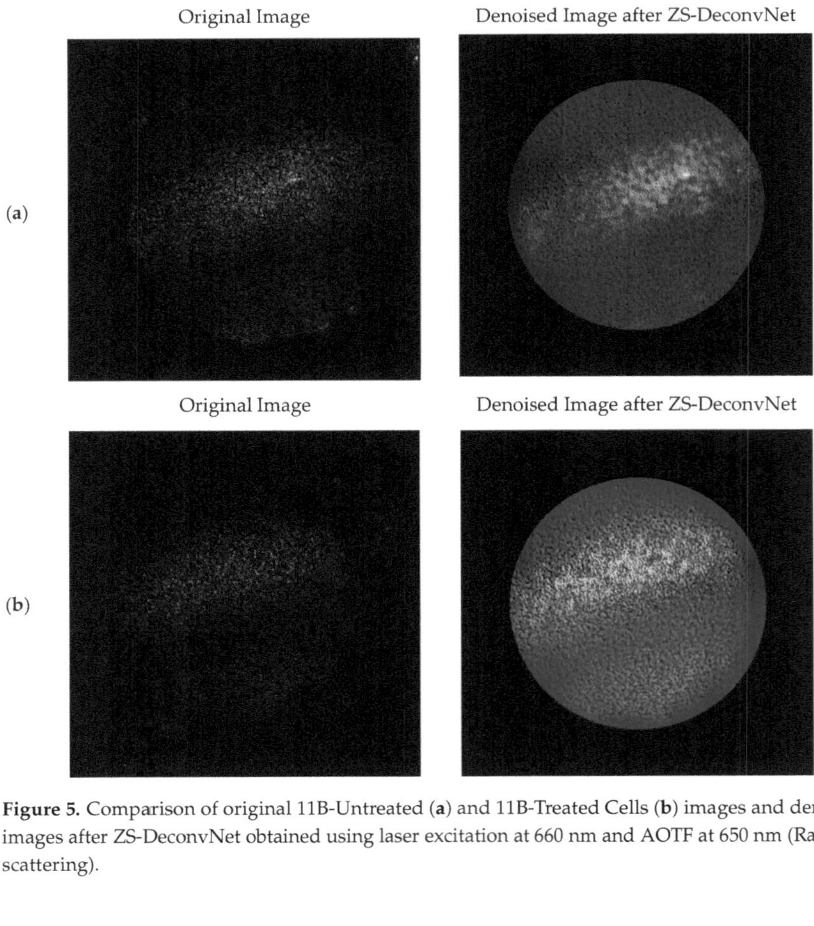

Figure 5. Comparison of original 11B-Untreated (**a**) and 11B-Treated Cells (**b**) images and denoised images after ZS-DeconvNet obtained using laser excitation at 660 nm and AOTF at 650 nm (Rayleigh scattering).

(2) Raman Scattering (Power: 130 mW and AOTF: 817 nm): Denoising significantly enhanced the signal-to-noise ratio (SNR), enabling clearer identification of molecular changes induced by 50 μM cisplatin. Treated spheroids showed distinct Raman shifts and enhanced peaks, while untreated spheroids maintained stable profiles. ZS-DeconvNet preserved these features, improving interpretability (Figure 6).

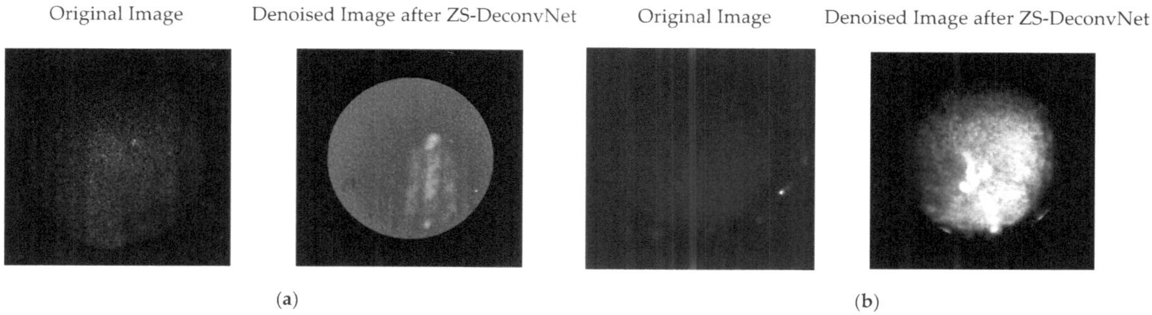

Figure 6. Comparison of original 11B-Untreated Cells (**a**) and 11B-Treated Cells (**b**) images and denoised images after ZS-DeconvNet obtained using laser excitation at 660 nm and AOTF at 817 nm (Raman scattering).

(3) Fluorescence (Power: 130 mW, AOTF: 694 nm): Fluorescence imaging showed substantial improvement after denoising, with noise suppression enhancing signal clarity. Treated spheroids exhibited increased fluorescence intensity, indicating structural or cellular changes, while untreated spheroids displayed more uniform fluorescence. ZS-DeconvNet preserved signal integrity, making the fluorescence data more interpretable (Figure 7).

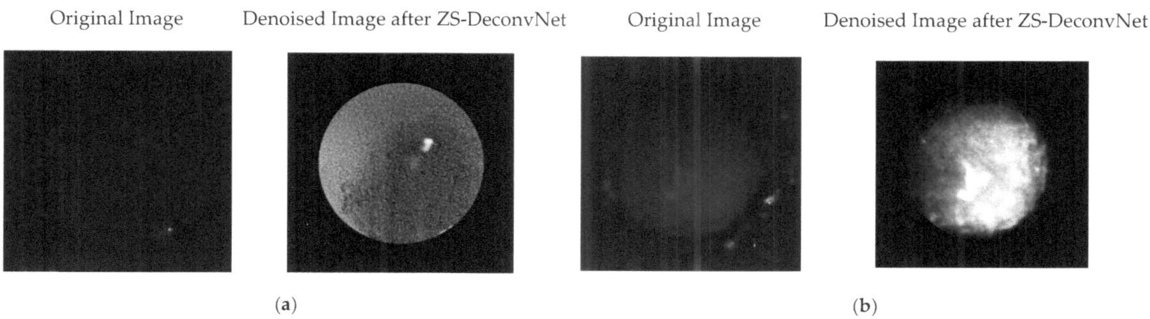

Figure 7. Comparison of original 11B-Untreated Cells (**a**) and 11B-Treated Cells (**b**) images and denoised images after ZS-DeconvNet obtained using laser excitation at 660 nm and AOTF at 694 nm (fluorescence).

Laser Excitation at 785 nm: For the 775 nm laser, three distinct modalities were used:

(1) Rayleigh Scattering (Power: 1 mW and AOTF: 775 nm): In the 785 nm Rayleigh scattering modality, ZS-DeconvNet provided substantial image quality enhancement. The denoised images of cisplatin-treated spheroids revealed previously masked surface irregularities, such as increased roughness and textural changes, that were critical for assessing treatment effects (Figure 8).

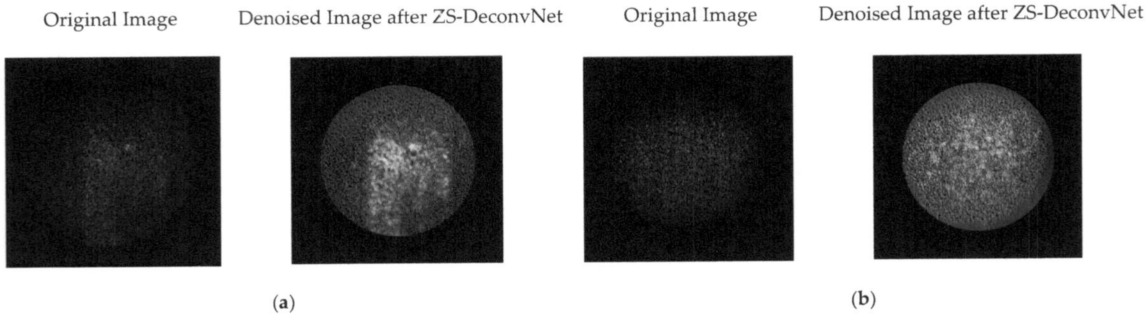

Figure 8. Comparison of original 11B-Untreated Cells (**a**) and 11B-Treated Cells (**b**) images and denoised images after ZS-DeconvNet obtained using laser excitation at 785 nm and AOTF at 775 nm (Rayleigh scattering).

3.2. *Quantitative Evaluation of Image Quality*

The denoising capabilities of ZS-DeconvNet were quantitatively assessed using Peak Signal-to-Noise Ratio (PSNR), Structural Similarity Index (SSIM), and Root Mean Square Error (RMSE), as shown in Figure 9. These metrics provide a comprehensive evaluation of how well the denoised images preserve structural integrity and reduce noise, enhancing the overall quality of the data for further analysis.

PSNR: The denoised image achieved a PSNR value of 16.74 dB, indicating a substantial reduction in noise when compared to the raw image. Although this PSNR value reflects moderate image fidelity, it represents a significant improvement in signal clarity, allowing for better visualization of features previously masked by noise. This improvement underscores the effectiveness of ZS-DeconvNet in restoring image quality in high-noise conditions such as Rayleigh scattering at 660 nm.

SSIM: The SSIM score of 0.168 indicates that some structural information was preserved post-denoising, although there is room for further optimization. Despite the relatively low score, this increase in structural similarity highlights ZS-DeconvNet's ability to recover key features from the noisy input, enabling a more interpretable output. This is particularly relevant in imaging modalities where fine structural details are critical for accurate analysis, such as in the treated spheroids exposed to cisplatin.

RMSE: The RMSE value of 0.146 demonstrates the model's efficiency in minimizing the error between the original noisy image and the denoised output. The reduction in RMSE confirms that ZS-DeconvNet effectively suppresses noise without introducing artifacts, thereby preserving essential structural and molecular details crucial for the accurate evaluation of treatment effects in cisplatin-treated spheroids.

In addition to these metrics, Fourier Ring Correlation (FRC) analysis (Figure 1) was conducted to further evaluate the resolution enhancement. The FRC curve shows improved frequency preservation across multiple shells, indicating that ZS-DeconvNet enhanced the image's spatial resolution while maintaining relevant frequency details. This result highlights the model's capability to improve image quality even in noisy and low-signal environments.

Overall, these quantitative metrics affirm the robustness of ZS-DeconvNet in effectively denoising images, particularly in the context of high-noise biomedical imaging applications, facilitating clearer feature extraction and more reliable data interpretation for both treated and untreated conditions.

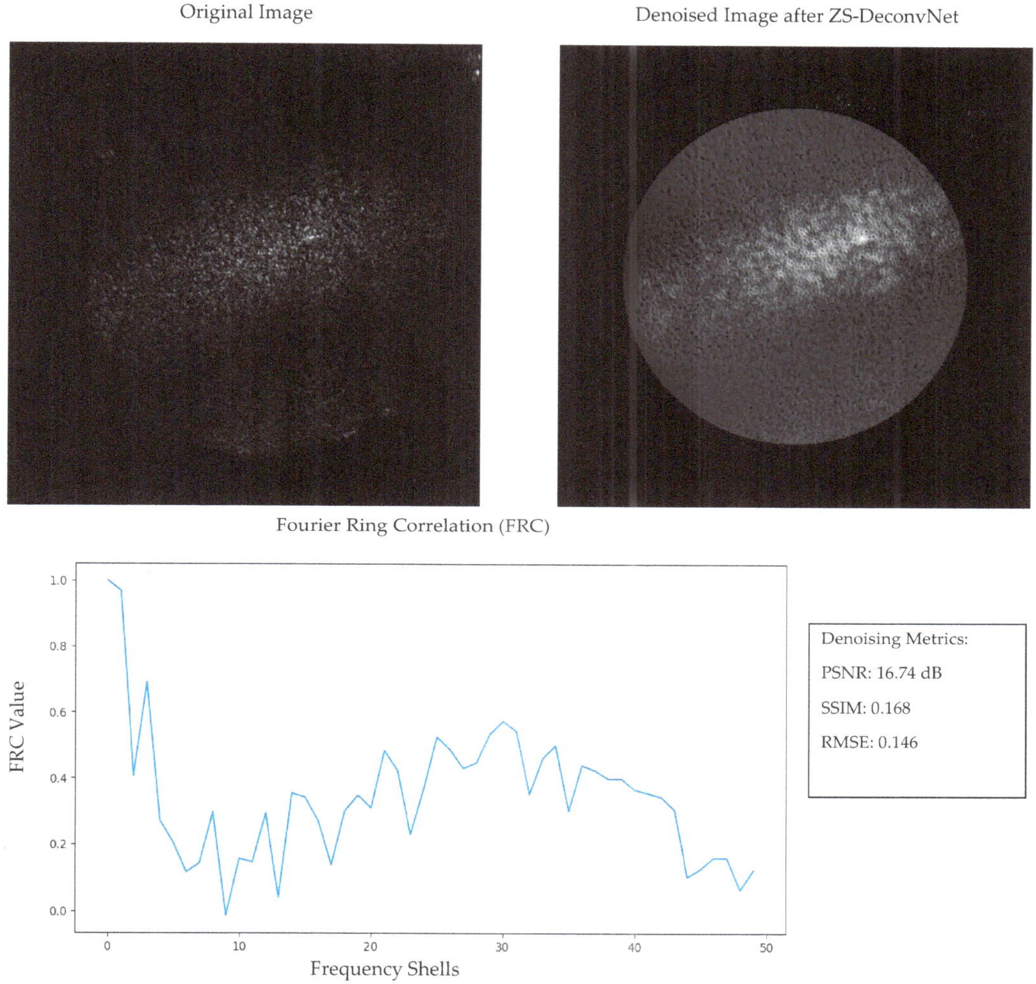

Figure 9. Denoising performance of ZS-DeconvNet on 11B-Untreated cells images using 660 nm laser for Rayleigh spectra: original image, denoised image after ZS-DeconvNet, FRC analysis and denoising metrics (PSNR, SSIM and RMSE).

4. Discussion

This study highlights the effectiveness of ZS-DeconvNet in combination with Multimodal Raman Light Sheet Microscopy for high-resolution imaging of spheroids derived from UMSCC-11B cell lines. The ability to visualize both structural and molecular changes in spheroids exposed to 50 μM cisplatin for 72 h significantly advanced our understanding of treatment-induced effects. By leveraging the denoising capabilities of ZS-DeconvNet, we were able to enhance the clarity of images across multiple imaging modalities, with a particular emphasis on Raman scattering channels [2,10].

Noise Reduction and Image Clarity: The major finding of this work is the significant improvement in signal-to-noise ratio (SNR) achieved by ZS-DeconvNet, which enabled the extraction of valuable structural and molecular details that were otherwise obscured by noise. This was particularly evident in the 660 nm Rayleigh scattering and Raman scattering channels, where noise levels are typically high due to the sensitivity of these modalities to low signal intensities. The PSNR improvements in denoised images reflect

the model's ability to suppress noise without compromising the integrity of critical image features. Additionally, FRC analysis confirmed that the high-frequency components of the images were well-preserved, leading to better resolution and sharper image details, which is essential for studying subtle structural changes in spheroids. The ability to reveal treatment-induced surface irregularities and molecular shifts in cisplatin-treated spheroids demonstrates the utility of ZS-DeconvNet in high-noise imaging environments. For instance, Raman scattering at 660 nm (130 mW, AOTF: 817 nm), which is particularly sensitive to molecular vibrations, revealed distinct shifts in the spectral profiles of treated spheroids after denoising. These shifts were crucial for identifying treatment-induced molecular changes, which were previously masked by noise in the raw images. In contrast, the untreated spheroids maintained stable Raman profiles, further emphasizing the specificity of the cisplatin-induced changes and the model's effectiveness in differentiating between treated and control conditions.

Structural Preservation and Molecular Insights: While ZS-DeconvNet excelled in noise reduction, as evidenced by improved PSNR and RMSE values, the SSIM scores indicate that there is room for further optimization, particularly in preserving intricate structural details. Nevertheless, the overall structural integrity of the denoised images was maintained, as demonstrated by the clear visualization of cisplatin-induced surface irregularities in treated spheroids. This preservation of structural features is critical in biomedical imaging, where even slight distortions can lead to misinterpretation of biological changes. The ability to retain structural fidelity while reducing noise enabled the detection of molecular changes that provide deeper insights into the spheroids' responses to cisplatin treatment. Fluorescence imaging at 660 nm (130 mW, AOTF: 694 nm) benefitted significantly from denoising, with the treated spheroids displaying increased fluorescence intensity, suggesting possible alterations in cell viability or metabolic activity. The preserved fluorescence signals in denoised images allowed for more accurate assessments of these biological processes, facilitating a deeper understanding of treatment-induced cellular changes.

Implications for Biomedical Research: The application of ZS-DeconvNet in this study offers substantial implications for biomedical research, particularly in fields such as cancer biology, drug discovery, and tissue engineering. The ability to visualize real-time molecular changes in 3D spheroids, which are physiologically relevant models for tumor behavior, provides critical insights into how treatments like cisplatin affect cellular architecture and molecular composition. Furthermore, the flexibility of ZS-DeconvNet—which does not require extensive pre-training on specific datasets—makes it a versatile tool for various imaging modalities and experimental setups. Additionally, the integration of post-processing techniques such as image segmentation, contrast enhancement and edge detection can further enhance the usability of the denoised images for downstream analysis. These enhancements ensured that the images were ready for detailed analysis, such as subcellular structural studies or quantitative assessments of spheroid viability. The results of this study suggest that ZS-DeconvNet, when combined with advanced imaging modalities, can significantly improve the quality of data available for quantitative biomedical research.

Future Directions: While ZS-DeconvNet demonstrated strong denoising performance, future research could explore hybrid approaches that integrate the noise suppression capabilities of ZS-DeconvNet with advanced structural preservation techniques. This would ensure even higher SSIM values while maintaining the improvements in PSNR and RMSE. Furthermore, integrating ZS-DeconvNet with deep learning-based segmentation techniques could open new avenues for automated analysis of spheroid morphology and molecular dynamics in response to various treatments. In conclusion, this study demonstrates that ZS-DeconvNet, combined with Multi-modal Raman Light Sheet Microscopy, offers a powerful and flexible framework for imaging 3D spheroids. The model's ability to denoise images in real-time without sacrificing critical structural or molecular information makes it an invaluable tool for biomedical research. By providing high-quality, denoised images that are ready for detailed analysis, ZS-DeconvNet facilitates a more precise understanding

of treatment effects on live cells and tissues, paving the way for new applications in drug discovery, cancer research, and tissue engineering.

5. Conclusions

This study evaluated the performance of ZS-DeconvNet for denoising high-noise biomedical images of 11B spheroids treated with 50 μM cisplatin for 72 h. Metrics like PSNR, SSIM, RMSE, and FRC demonstrated the model's ability to significantly reduce noise while preserving important structural and molecular details.

The model showed a marked improvement in PSNR, confirming its effectiveness in noise suppression and image clarity, while FRC analysis highlighted its ability to retain high-frequency information. This approach enables resolution enhancement beyond the diffraction limit by recovering high-frequency details, allowing for visualization of sub-diffraction structures without additional hardware adjustments. Although SSIM scores indicated some limitations in preserving fine details, ZS-DeconvNet successfully maintained key features, especially in Rayleigh and Raman scattering modalities at both 660 nm and 785 nm. Its zero-shot learning framework further underscores the novelty of this multimodal approach, allowing adaptive enhancement across multiple imaging modalities without the need for pre-trained models. This flexibility addresses unique challenges of multimodal microscopy, providing a unified solution for image enhancement in Raman, fluorescence, and other microscopy techniques.

Denoised images revealed critical treatment-induced changes in cisplatin-treated spheroids, previously masked by noise, enabling more accurate comparisons between treated and untreated samples. This underscores ZS-DeconvNet's effectiveness in high-noise, low-signal imaging environments typical of biomedical applications.

Overall, ZS-DeconvNet provides a powerful tool for real-time image denoising in 3D spheroid imaging and Raman Light Sheet Microscopy, offering faster processing and superior image quality without needing pre-trained datasets. Future research could focus on hybrid approaches to combine its noise reduction capabilities with advanced structural preservation techniques for even better results in biomedical imaging.

Author Contributions: Conceptualization, P.K.; methodology, P.K. and S.K.; software, P.K.; validation, P.K.; formal analysis, P.K.; investigation, P.K.; resources, E.S. and J.K.; data curation, P.K.; writing—original draft preparation, P.K.; writing—review and editing, Matthias Raedle and P.K.; visualization, P.K.; supervision, J.K. and M.R.; project administration, P.K.; funding acquisition, Matthias Raedle. CeMOS Research and Transfer Center designed the entire set of experiments including the setup. CeMOS Research and Transfer Center build the setup and conducted the experiments. Also, it analyzed the obtained images and spectra and drafted this manuscript. All co-authors contributed to discussion, interpretation, and final writing. All authors have read and agreed to the published version of the manuscript.

Funding: We would like to acknowledge funding support from the graduate program Perpharmance (BW6_07) provided by the Ministry of Science, Research, and the Arts (MWK) of Ba-den-Württemberg.

Institutional Review Board Statement: Not Applicable.

Informed Consent Statement: Not Applicable.

Data Availability Statement: Data are contained within the article.

Acknowledgments: In this article, the authors draw on contributions from many members of the CeMOS Research and Transfer Center specially Björn Van Marvick and Johann Strischakov. All images and plots without source were created at the CeMOS Research and Transfer Center, 68163 Mannheim, Germany.

Conflicts of Interest: The authors declare no conflicts of interest.

References

1. Ravi, M.; Paramesh, V.; Kaviya, S.R.; Anuradha, E.; Solomon, F.D. 3D Cell Culture Systems: Advantages and Applications. *J. Cell. Physiol.* **2014**, *230*, 16–26. [CrossRef] [PubMed]
2. Manser, S.; Keck, S.; Vitacolonna, M.; Wühler, F.; Rudolf, R.; Raedle, M. Innovative Imaging Techniques: A Conceptual Exploration of Multi-Modal Raman Light Sheet Microscopy. *Micromachines* **2023**, *14*, 1739. [CrossRef] [PubMed]
3. Park, Y.; Huh, K.; Kang, S.-W. Applications of Biomaterials in 3D Cell Culture and Contribution of 3D Cell Culture to Drug Development and Basic Biomedical Research. *Int. J. Mol. Sci.* **2021**, *22*, 2491. [CrossRef]
4. Oshima, Y.; Sato, H.; Kajiura-Kobayashi, H.; Kimura, T.; Naruse, K.; Nonaka, S. Light sheet-excited spontaneous Raman imaging of a living fish by optical sectioning in a wide field Raman microscope. *Opt. Express* **2012**, *20*, 16195–16204. [CrossRef]
5. Eberhardt, K.; Stiebing, C.; Matthäus, C.; Schmitt, M.; Popp, J. Advantages and limitations of Raman spectroscopy for molecular diagnostics: An update. *Expert Rev. Mol. Diagn.* **2015**, *15*, 773–787. [CrossRef] [PubMed]
6. Koenigstein, J. *Introduction to the Theory of the Raman Effect*, 1st ed.; D. Reidel Publishing Company: Dordrecht-Holland, The Netherlands, 1972; pp. 79–133.
7. Zhao, W.; Zhao, S.; Li, L.; Huang, X.; Xing, S.; Zhang, Y.; Qiu, G.; Han, Z.; Shang, Y.; Sun, D.-E.; et al. Sparse deconvolution improves the resolution of live-cell super-resolution fluorescence microscopy. *Nat. Biotechnol.* **2021**, *40*, 606–617. [CrossRef] [PubMed]
8. Wang, H.; Rivenson, Y.; Jin, Y.; Wei, Z.; Gao, R.; Günaydın, H.; Bentolila, L.A.; Kural, C.; Ozcan, A. Deep learning enables cross-modality super-resolution in fluorescence microscopy. *Nat. Methods* **2019**, *16*, 103–110. [CrossRef] [PubMed]
9. Sage, D.; Donati, L.; Soulez, F.; Fortun, D.; Schmit, G.; Seitz, A.; Guiet, R.; Vonesch, C.; Unser, M. DeconvolutionLab2: An open-source software for deconvolution microscopy. *Methods* **2017**, *115*, 28–41. [CrossRef] [PubMed]
10. Qiao, C.; Zeng, Y.; Meng, Q.; Chen, X.; Chen, H.; Jiang, T.; Wei, R.; Guo, J.; Fu, W.; Lu, H.; et al. Zero-shot learning enables instant denoising and super-resolution in optical fluorescence microscopy. *Nat. Commun.* **2024**, *15*, 4180. [CrossRef] [PubMed]
11. Yanny, K.; Monakhova, K.; Shuai, R.W.; Waller, L. Deep learning for fast spatially varying deconvolution. *Optica* **2022**, *9*, 96–99. [CrossRef]
12. Shah, Z.H.; Müller, M.; Wang, T.-C.; Scheidig, P.M.; Schneider, A.; Schüttpelz, M.; Huser, T.; Schenck, W. Deep-learning based denoising and reconstruction of super-resolution structured illumination microscopy images. *Photon. Res.* **2021**, *9*, B168–B181. [CrossRef]
13. Lehtinen, J.; Munkberg, J.; Hasselgren, J.; Laine, S.; Karras, T.; Aittala, M.; Aila, T. Noise2Noise: Learning Image Restoration without Clean Data. In Proceedings of the International Conference on Machine Learning, Stockholm, Sweden, 10–15 July 2018; pp. 2965–2974.
14. Chen, X.; Zhang, C.; Lin, P.; Huang, K.-C.; Liang, J.; Tian, J.; Cheng, J.-X. Volumetric chemical imaging by stimulated Raman projection microscopy and tomography. *Nat. Commun.* **2017**, *8*, 15117. [CrossRef] [PubMed]
15. Pully, V.V.; Lenferink, A.; Otto, C. Raman-fluorescence hybrid microspectroscopy of cell nuclei. *Vib. Spectrosc.* **2010**, *53*, 2010. [CrossRef]
16. Evans, J.W.; Zawadzki, R.J.; Liu, R.; Chan, J.W.; Lane, S.M.; Werner, J.S. Optical cohe-rence tomography and Raman spectroscopy of the ex-vivo retina. *J. Bio-Photonics* **2009**, *2*, 398–406.
17. Qiao, C.; Li, D.; Guo, Y.; Jiang, T.; Dai, Q.; Li, D. Evaluation and development of deep neural networksfor image super-resolution in optical microscopy. *Nat. Methods* **2021**, *18*, 194–202. [CrossRef] [PubMed]
18. OpenSPIM. Available online: https://openspim.org/ (accessed on 11 June 2024).
19. Nieuwenhuizen, R.P.; A Lidke, K.; Bates, M.; Puig, D.L.; Grünwald, D.; Stallinga, S.; Rieger, B. Measuring image resolution in optical nanoscopy. *Nat. Methods* **2013**, *10*, 557–562. [CrossRef] [PubMed]
20. Diekmann, R.; Deschamps, J.; Li, Y.; Deguchi, T.; Tschanz, A.; Kahnwald, M.; Matti, U.; Ries, J. Photon-free (s)CMOS camera characterization for artifact reduction in high- and super-resolutionmicroscopy. *Nat. Commun.* **2022**, *13*, 3362. [CrossRef] [PubMed]

sensors

MDPI

Article

A Non-Contacted Height Measurement Method in Two-Dimensional Space

Phu Nguyen Trung [1], Nghien Ba Nguyen [1,*], Kien Nguyen Phan [2,*], Ha Pham Van [1], Thao Hoang Van [2], Thien Nguyen [3] and Amir Gandjbakhche [3]

1 Faculty of Information Technology, Hanoi University of Industry, No. 298 Cau Dien, Bac Tu Liem, Hanoi 143510, Vietnam; phunt@haui.edu.vn (P.N.T.)
2 School of Electrical and Electronic Engineering, Hanoi University of Science and Technology, No. 1 Dai Co Viet, Hai Ba Trung, Hanoi 100000, Vietnam
3 *Eunice Kennedy Shriver* National Institute of Child Health and Human Development, National Institutes of Health, 49 Convent Drive, Bethesda, MD 20892-4480, USA; thien.nguyen4@nih.gov (T.N.); gandjbaa@mail.nih.gov (A.G.)
* Correspondence: nguyenbanghien_cntt@haui.edu.vn (N.B.N.); kien.nguyenphan@hust.edu.vn (K.N.P.)

Abstract: Height is an important health parameter employed across domains, including healthcare, aesthetics, and athletics. Numerous non-contact methods for height measurement exist; however, most are limited to assessing height in an upright posture. This study presents a non-contact approach for measuring human height in 2D space across different postures. The proposed method utilizes computer vision techniques, specifically the MediaPipe library and the YOLOv8 model, to analyze images captured with a smartphone camera. The MediaPipe library identifies and marks joint points on the human body, while the YOLOv8 model facilitates the localization of these points. To determine the actual height of an individual, a multivariate linear regression model was trained using the ratios of distances between the identified joint points. Data from 166 subjects across four distinct postures: standing upright, rotated 45 degrees, rotated 90 degrees, and kneeling were used to train and validate the model. Results indicate that the proposed method yields height measurements with a minimal error margin of approximately 1.2%. Future research will extend this approach to accommodate additional positions, such as lying down, cross-legged, and bent-legged. Furthermore, the method will be improved to account for various distances and angles of capture, thereby enhancing the flexibility and accuracy of height measurement in diverse contexts.

Keywords: non-contact; height measurement; MediaPipe

Citation: Nguyen Trung, P.; Nguyen, N.B.; Nguyen Phan, K.; Pham Van, H.; Hoang Van, T.; Nguyen, T.; Gandjbakhche, A. A Non-Contacted Height Measurement Method in Two-Dimensional Space. *Sensors* **2024**, *24*, 6796. https://doi.org/10.3390/s24216796

Academic Editors: Stelios Krinidis and Christos Nikolaos E. Anagnostopoulos

Received: 7 August 2024
Revised: 18 October 2024
Accepted: 21 October 2024
Published: 23 October 2024

1. Introduction

It is critical to have an accurate and convenient method for measuring human height, which is an important variable in healthcare for calculating Body Mass Index (BMI) and determining various treatment-related metrics [1]. BMI enables the classification of individuals as overweight, underweight, or of ideal weight [2]. Moreover, BMI is instrumental in population-based studies due to its widespread acceptance in identifying specific body mass categories that may indicate health or social issues. Recent evidence also suggests that particular BMI ranges are associated with moderate and age-related mortality risks [3].

Height measurement is typically performed in an erect standing posture. For non-critical patients, contact methods such as table scales, standing scales, or medical measuring devices are commonly employed. However, for critically ill patients in intensive care units (ICUs), requiring them to move or assume an upright position for height measurement is often impractical [4]. Additionally, severely ill patients are frequently unconscious or incapacitated, complicating accurate height assessments. Therefore, the development of a non-contact height measurement method for critically ill patients is particularly important in the ICU setting [5,6].

Currently, height measurements for ICU patients in hospitals in Vietnam are frequently conducted by nurses; however, the nurse-to-patient ratio is often insufficient. This situation introduces significant challenges in obtaining accurate measurements. Accurate height data are crucial, as it is integral to calculating treatment parameters such as creatinine indices [7,8]. Thus, the implementation of an automatic, non-contact height measurement method represents a critical step toward ensuring the highest possible accuracy for calculating treatment parameters.

Several non-contact methods have been proposed for measuring height in special populations, such as the elderly, hospitalized individuals, bedridden patients, and those with skeletal deformities. A study conducted at Jimma University demonstrated that height estimates derived from linear body measurements, including arm span, knee height, and half-arm span, serve as useful surrogate measures [9]. However, the study was limited by a narrow age range, including only adults aged 18 to 40 years, which may not adequately represent the broader adult population, especially considering the potential decline in height in older age groups. Furthermore, Haritosh has investigated the use of facial proportions to estimate body height [10]. This method involves calculating height from facial images by extracting facial features through convolutional neural networks and predicting height using artificial neural networks. However, the average error rate in the measurement is approximately 7.3 cm, which constitutes a significant deviation in height assessment.

A common method for estimating human height from images or videos is skeletal extraction [11]. This approach utilizes computer vision and image-processing methodologies to analyze visual data. The accuracy of this method can be affected by various factors, including camera focal length, angle, and ambient-lighting conditions. To enhance the precision of height measurements, we propose a study employing MediaPipe to extract skeletal point coordinates from images capturing both a person and a reference object—a black cardboard of fixed dimensions. These coordinates, represented in a two-dimensional space as X and Y values, are used to calculate the lengths of bone segments, thus facilitating height estimation. Following the extraction of skeletal points, a machine-learning model will be employed to train the input data and estimate human height. We hypothesize that the use of a reference object will improve the accuracy of height measurement.

2. Materials and Methods

2.1. The Proposed Method

Figure 1 illustrates a diagram of the proposed height measurement method. The block diagram consists of six primary blocks. The first block serves as the input, which is an image of a person in a vertical position. The second block identifies and marks human body landmarks (skeleton points) using the OpenCV and MediaPipe libraries. The third block calculates the length of each skeleton using the MediaPipe library. This step involves calculating a centimeter-per-pixel (cm/pixel) ratio using a reference object, counting the number of pixels in each skeleton, and then calculating the skeleton length in centimeters. In the fourth block, the lengths of the skeletons are fed into a multivariate linear regression model to train the model. In the fifth block, human height is predicted using the trained model. Finally, in the sixth block, human height is obtained.

The OpenCV (Open Computer Vision) is a leading open-source library for computer vision, machine learning, and image processing. It is written in C/C++, which enables it to achieve very fast calculation speeds and allows for use in real-time applications [12]. MediaPipe is a series of cross-platform machine-learning solutions used for tasks such as face detection, face mesh, and human pose estimation [13]. It consists of three main parts, namely a framework for inference from sensory data, a set of tools for performance evaluation, and a collection of reusable inference and processing components [14]. YOLOv8 is a computer vision model for object recognition and detection developed by Ultralytics in 2016 [15]. Among different object detection algorithms, the YOLO (You Only Look Once) framework has stood out for its remarkable balance of speed and accuracy, enabling the

rapid and reliable identification of objects in images. Since its inception, the YOLO family has evolved through multiple iterations, each building upon previous versions to address limitations and enhance performance [15].

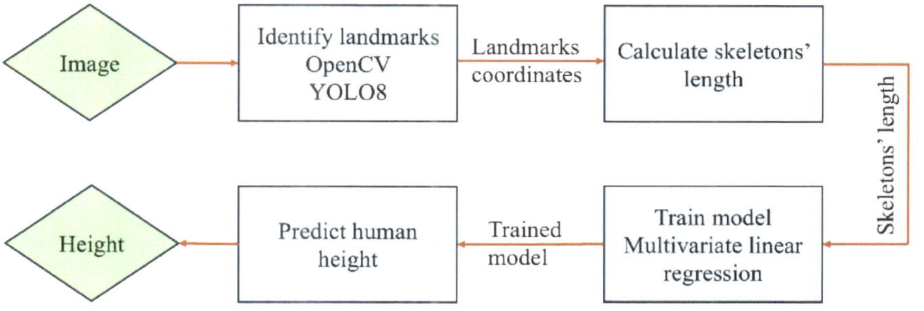

Figure 1. System diagram.

Using the OpenCV and MediaPipe, a total of 501 landmarks (skeleton joints) were identified. These landmarks were then fed to a customized multiclass classification model to understand the relationship between each class and its coordinates for classifying and detecting a body posture [16]. The OpenCV library was first used for image processing. After that, the MediaPipe library was applied to extract x, y, and z coordinates, as well as the number of pixels for each joint. Finally, the YOLOv8 model was employed to identify the black cardboard, calculate the number of pixels in the cardboard, and determine the ratio of cm/pixel according to Equation (1):

$$(\mathrm{k}) = \frac{\mathrm{height(cm)}}{\mathrm{dis(pixel)}} \tag{1}$$

The human body was divided into six segments: h_1 is the distance from the shoulder to the hip, h_2 is the distance from the hip to the knee, h_3 is the distance from the knee to the ankle, h_4 is the distance from the ankle to the sole of the foot, h_5 is the distance from the middle of the shoulder to the middle of the mouth, and h_6 is the distance from the middle of the mouth to the nose. The distance between two points, $A(x_a, y_a)$ and $B(x_b,y_b)$, was calculated. In this project, our calculations were based on the normalized coordinates x_i obtained from MediaPipe y_i. These coordinates were then converted to a pixel coordinate system using Equations (2) and (3).

$$X_i = \mathrm{image\_width} * x_i \tag{2}$$

$$Y_i = \mathrm{image\_height} * y_i \tag{3}$$

The pixel coordinates were then used to calculate the distances between landmarks and the lengths of the skeleton segments in the human body. The coordinates of the midpoint of the shoulder and the coordinates of the midpoint of the hip were used to calculate the distance h_1 (Equation (4)):

$$h_1 = k \times \sqrt{\left(\frac{X_{23}+X_{24}}{2} - \frac{X_{11}+X_{12}}{2}\right)^2 + \left(\frac{Y_{23}+Y_{24}}{2} - \frac{Y_{11}+Y_{12}}{2}\right)^2} \tag{4}$$

The skeletal segment h_2 was calculated as the distance between points 23 and 25 (Equation (5)):

$$h_2 = k \times \sqrt{(X_{25} - X_{23})^2 + (Y_{25} - Y_{23})^2} \tag{5}$$

Similarly, the distance h_3 was calculated as the distance between points 27 and 25 (Equation (6)):

$$h_3 = k \times \sqrt{(X_{27} - X_{25})^2 + (Y_{27} - Y_{25})^2} \tag{6}$$

The distance h_4 from the ankle to the sole of the left foot was calculated as follows:

$$h_4 = k \times \frac{|(Y_{29} - Y_{31})x_{27} + (X_{31} - X_{29})Y_{27} + (Y_{31} - Y_{29})X_{29} - (X_{31} - X_{29})Y_{29}|}{\sqrt{(Y_{29} - Y_{31})^2 + (X_{31} - X_{29})^2}} \tag{7}$$

The distance h_5 was calculated as the distance from the midpoint of the shoulder to the midpoint of the mouth (Equation (8)):

$$h_5 = k \times \sqrt{\left(\left(\frac{X_{11} + X_{12}}{2} - \frac{X_9 + X_{10}}{2}\right)^2 + \left(\frac{Y_{11} + Y_{12}}{2} - \frac{Y_9 + Y_{10}}{2}\right)^2\right.} \tag{8}$$

Finally, the distance h_6 from the midpoint of the mouth to the nose was calculated as follows:

$$h_6 = k \times \sqrt{\left(\left(X_0 - \frac{X_9 + X_{10}}{2}\right)^2 + \left(Y_0 - \frac{Y_9 + Y_{10}}{2}\right)^2\right.} \tag{9}$$

2.2. Predicting Result of Height Measurement

A multivariable linear regression, which is an extension of a single-variable linear regression algorithm, was used to train and predict body height. This algorithm has proven to be highly effective in predicting outcomes based on two or more independent variables.

The multivariate linear regression [17] equation takes the following form:

$$Y = \beta_0 + \beta_1 \times X_1 + \beta_2 \times X_2 + \ldots + \beta_n \times X_n + \varepsilon \tag{10}$$

where Y is the dependent variable that needs to be predicted. $X_1, X_2, \ldots, X_n$ are the independent variables, and $\beta_0, \beta_1, \beta_2, \ldots, \beta_n$ are the relationship coefficients.

After calculating the length of each skeletal segment, we applied a multivariate linear regression equation to predict human height. Equation (10) becomes the following:

$$h = \beta_0 + \beta_1 h_1 + \beta_2 h_2 + \beta_3 h_3 + \beta_4 h_4 + \beta_5 h_5 + \beta_6 h_6 + \varepsilon \tag{11}$$

where h is the predicted height; $h_1, h_2, h_3, \ldots, h_6$ are the calculated distance of skeleton segments; and $\beta_0, \beta_1, \beta_2, \ldots, \beta_6$ are the correlation coefficients obtained during the process of training the multivariate linear regression model.

The multivariate linear regression model is an important tool for investigating relationships between several response variables and multiple predictor variables. The primary focus is on making inferences about the unknown regression coefficient matrices. We propose multivariate bootstrap techniques as a means for drawing inferences about these matrices. A real data example and two simulated data examples that provide finite sample verifications of our theoretical results are presented in [18,19].

2.3. Data Collection

This study was approved by the Hanoi University of Science and Technology. Data were collected from 166 adult subjects who agreed to participate in the study. Photographs of the subjects were taken with a smartphone camera. The smartphone was fixed on a tripod at a height of approximately 115 cm from the ground (Figure 2). The tripod was positioned at distances of 200 cm and 300 cm from the subject. A 20.5 cm × 30.5 cm black cardboard was placed next to the subject on the wall, with the center of the cardboard at a height of approximately 115 cm from the ground. On the opposite side of the subject, a wall height chart was attached.

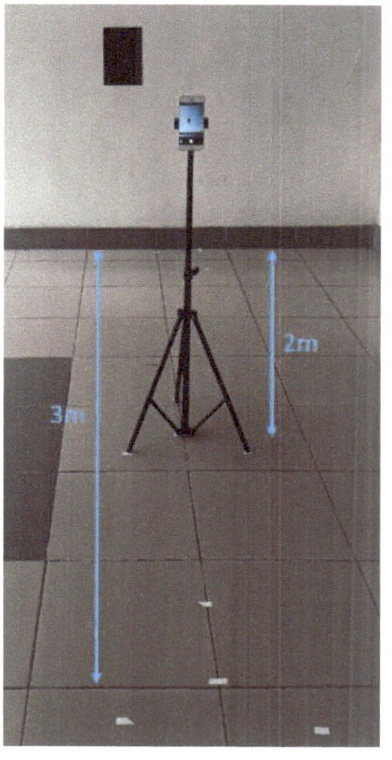

Figure 2. Tripod set up and camera.

Subjects were guided to perform four different postures during the experiment. Firstly, there was the standing-upright position (Figure 3a), where subjects stood straight and looked directly ahead. This position simulates the body in a natural state, with no tilt or rotation. Secondly, in the 45-degree rotation position (Figure 3b), subjects turned their bodies 45 degrees away from the camera while looking straight ahead. This pose simulates the body at a slight angle, which can affect how bone segments appear in the image. Thirdly, in the horizontal 90-degree rotation position (Figure 3c), subjects turned their bodies 90 degrees from the camera and looked straight ahead. This position simulates the body at a greater angle and illustrates patients' positions in a hospital bed. This helps to better understand the differences in measurements of bone segments when the body is in a horizontal state, which is important in medical applications. Finally, in the kneeling position (Figure 3d), subject turned their bodies 90 degrees but bent their knees. This position is especially important for understanding changes in bone segments when the body is in a bent-knee state, simulating situations where the body is not completely upright. In each pose, subjects remained in position throughout the image capturing process to ensure the accuracy of the measurements. Staying steady and immobile during each scan is crucial to ensure that body landmarks are accurately and consistently identified.

2.4. Data Processing

The obtained images were processed using MediaPipe and YOLOv8 to extract the X and Y coordinates of the landmarks on the body, as well as the parameters of the reference object, which served for calculating the lengths of the skeletal segments. After that, the mean and standard deviation (SD) [19] were used to remove outliers to increase model accuracy. Specifically, data outside of the ± 3 SD range were removed. The remaining valid values were used as input for the training model. Finally, a multivariate linear regression model was applied to the skeletal segments to estimate subjects' height. The collected data consisted of 166 samples for each posture, with heights ranging from 148 cm to 184 cm. After eliminating outliners, a new dataset consisting of 162 samples was divided into 80% for training and 20% for testing.

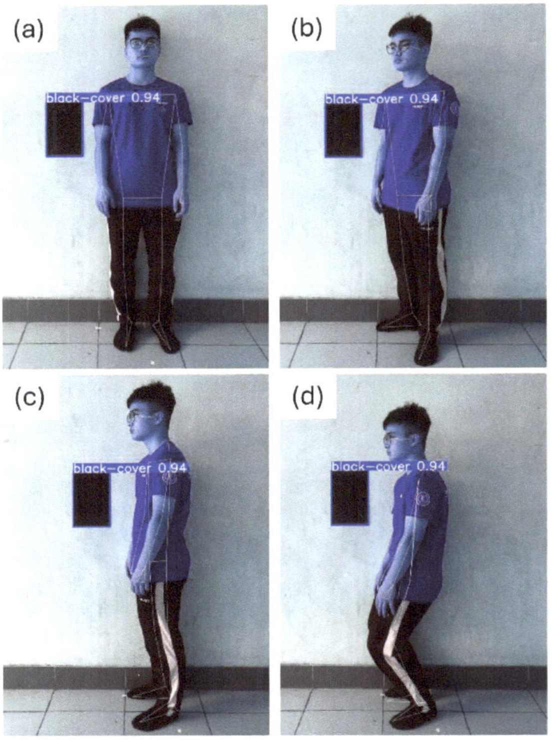

Figure 3. Height measurement in different postures; (**a**) standing-upright position; (**b**) 45-degree rotation position; (**c**) horizontal 90-degree rotation position; and (**d**) Kneeling position. Lines and points in each figure represent segments and joints determined from the OpenCV and the MediaPipe libraries.

3. Results

3.1. Standing Upright Position

After training the model with the training and test samples, we developed Equation (12) to estimate height based on bone segment lengths. Table 1 provides the evaluation results for the standing-upright position. This method has an average error of 1.94 cm (1.14%) across the test data samples. The error is mainly due to a lack of camera calibration and inaccuracies in extracting coordinates from MediaPipe, as well as varying lighting conditions during data collection.

$$H = 1.07533865h_1 + 1.2476316h_2 + 0.59605108h_3 + 0.6496244h_4 + 0.76927537h_5 - 2.13930107h_6 + 58.4779461 \quad (12)$$

Table 1. Prediction results for 17 subjects in the standing-upright posture.

Samples	Actual Height (cm)	Predicted Height (cm)	Error (cm)	Error Rate (%)
1	174	170.9514	3.048575	1.752055
2	177	177.4226	0.422636	0.238778
3	162	167.6994	5.699382	3.518137
4	168	168.3282	0.328223	0.195371
5	172	169.1883	2.811732	1.634728
6	169	172.8558	3.855773	2.281522
7	170	169.2551	0.744909	0.438182
8	170	173.3402	3.340176	1.964809

Table 1. *Cont.*

Samples	Actual Height (cm)	Predicted Height (cm)	Error (cm)	Error Rate (%)
9	180	180.4499	0.449862	0.249923
10	175	173.0343	1.965713	1.123264
11	169	171.7793	2.779289	1.64455
12	175	175.5055	0.505452	0.28883
13	173	175.8168	2.816793	1.628204
14	168	170.0952	2.095163	1.247121
15	171	170.7831	0.216859	0.126818
16	168	167.1057	0.894323	0.532335
17	163	163.9993	0.999287	0.613059
Average	170.8	171.6241	1.939656	1.145746

3.2. 45-Degree Rotation Position

Similarly, Equation (13) was developed to estimate body height for the 45-degree rotation position. Evaluation results for this position are presented in Table 2. The average error is 1.91 cm (1.12%).

$$H = -0.70003005h_1 + 0.98866088h_2 + 0.76985497h_3 + 0.35090296h_4 + 0.68119476h_5 - 0.40682656h_6 + 72.229882 \quad (13)$$

Table 2. Prediction results for 17 subjects in the 45-degree tilted-standing posture.

Samples	Actual Height (cm)	Predicted Height (cm)	Error (cm)	Error Rate (%)
1	174	170.3426	3.657426	2.101969
2	177	178.7488	1.748834	0.988042
3	162	167.0036	5.003625	3.088657
4	168	167.3131	0.686861	0.408846
5	172	174.8213	2.821337	1.640312
6	169	166.2978	2.702194	1.598931
7	170	167.9283	2.071712	1.218654
8	170	168.6685	1.331542	0.78326
9	180	178.4254	1.574556	0.874753
10	175	174.1332	0.866795	0.495312
11	169	171.0258	2.025811	1.198705
12	175	175.7314	0.731434	0.417962
13	173	174.0368	1.03684	0.599329
14	168	164.0762	3.923809	2.335601
15	171	172.5887	1.588712	0.929071
16	168	167.4733	0.526727	0.313528
17	163	162.7955	0.20453	0.125479
Average	170.8	170.6712	1.911926	1.124612

3.3. Horizontal 90-Degree Rotation Position

After training with the dataset for posture 3, Equation (14) was derived to estimate height for the 90-degree turned posture. Table 3 provides the evaluation results for the horizontal 90-degree rotation position. The average error is 2.62 cm (1.54%).

$$H = 0.08162038h_1 + 0.70345667h_2 + 0.67353882h_3 + 0.55258515h_4 + 0.42507086h_5 - 0.20956018h_6 + 73.384567 \quad (14)$$

3.4. Kneeling Position

For the 90-degree sideways bent-knee position, we derived Equation (15) to calculate body height. Results are presented in Table 4. The evaluation results for this position show an average error of 2.45 cm (1.43%).

$$H = 0.3642249\ldots95132h_2 + 0.58705451h_3 + 0.58410078h_4 + 0.35394516h_5 - 0.21066217h_6 + 73.779571 \quad (15)$$

Table 3. Prediction results for 17 subjects in the 90-degree tilted-standing posture.

Samples	Actual Height (cm)	Predicted Height (cm)	Error (cm)	Error Rate (%)
1	177	177.8865	0.886495	0.500845
2	162	169.2808	7.2808	4.494321
3	168	168.0163	0.016319	0.009714
4	172	169.4318	2.568204	1.493142
5	169	168.173	0.826959	0.489325
6	170	176.0739	6.073928	3.572899
7	170	168.528	1.47199	0.865877
8	180	183.2167	3.216669	1.787039
9	175	171.7336	3.266446	1.86654
10	169	174.7692	5.769226	3.413743
11	169	168.7802	0.219849	0.130088
12	175	172.8844	2.115569	1.208896
13	173	175.1489	2.148942	1.242163
14	168	165.562	2.438041	1.451215
15	171	173.9013	2.901262	1.696644
16	168	167.8982	0.10178	0.060583
17	163	166.2274	3.22743	1.980018
Average	170.5	171.6184	2.619406	1.544885

Table 4. Prediction results for 17 subjects in the 90-degree tilted-standing posture with bent knees.

Samples	Actual Height (cm)	Predicted Height (cm)	Error (cm)	Error Rate (%)
1	174	169.3661	4.633906	2.663164
2	177	178.5004	1.50038	0.847672
3	162	163.4223	1.422282	0.877952
4	168	168.3369	0.336886	0.200527
5	172	170.5549	1.445134	0.840194
6	169	168.7444	0.255635	0.151263
7	170	176.0042	6.004188	3.531876
8	170	175.1093	5.109334	3.005491
9	180	182.6475	2.647498	1.470832
10	175	178.3741	3.374108	1.928062
11	169	172.6141	3.614117	2.138531
12	175	172.9836	2.01644	1.152252
13	173	175.8976	2.897555	1.674887
14	168	166.5528	1.44725	0.861458
15	171	169.4226	1.577427	0.922472
16	168	167.0025	0.997496	0.593748
17	163	165.3323	2.332332	1.430879
Average	170.8	171.8156	2.447763	1.428898

4. Discussion

The experimental results demonstrate that the proposed height estimation method can estimate human height relatively accurately, with an average error ranging from 1.91 cm to 2.62 cm (1.12–1.54%). Among the four postures, the height estimation model for the 45-degree rotation position yields the best results, with an average error of 1.91 cm (1.12%). For the other postures, the achieved results are less accurate. The standing-upright position has a result nearly equal to that of the 45-degree rotation posture, with an average error of 1.94 cm (1.14%). However, for the horizontal 90-degree rotation position and the kneeling position, the errors are significantly larger, with average errors of 2.62 cm (1.54%) and 2.45 cm (1.43%), respectively. This is mainly because the MediaPipe model does not

perform as effectively when estimating height in m… plex postures compared to the standing-upright posture. Additionally, other facto… as lighting conditions and camera angles, also affect the accuracy of the measurem…

5. Conclusions

This study presents a non-contact height measurement r… ng the MediaPipe library in conjunction with the YOLOv8 model to extract jo… s and calculate bone lengths, employing a multivariate linear regression fu… cting human height from images. Experimental results indicate that the a… tween the estimated and actual heights range from 1.91 cm to 2.62 cm (1.1… level of accuracy is deemed acceptable for a variety of applications. Futu… us on expanding the methodology to determine the height of individuals… and lying positions. The goal is to develop a flexible and efficient softw… e of measuring height across diverse real-world contexts. The integr… such as MediaPipe and YOLOv8 demonstrates significant potentia… fields such as medicine, sports, and health monitoring, where reliable … measurements from images are essential.

Author Contributions: Conceptualization, P.N.T., N.B.N. and K.N.P.; methodology, … K.N.P., H.P.V. and T.H.V.; software, P.N.T., N.B.N., T.H.V. and K.N.P.; validation, P.N.T., N.B.… H.P.V., T.H.V. and T.N.; formal analysis, P.N.T., N.B.N., T.H.V. and K.N.P.; investigation, P.N.T., N.… T.H.V. and K.N.P.; resources, N.B.N. and K.N.P.; data curation, P.N.T. and T.H.V.; writing, P.N.T., N.B.N., K.N.P. and T.N.; writing—review and editing, P.N.T., N.B.N., K.N.P., H.P.V., T.H.V., T.N. and A.G.; visualization, P.N.T., T.H.V. and T.N.; supervision, N.B.N. and K.N.P.; project administration, N.B.N. and K.N.P.; funding acquisition, N.B.N. and K.N.P. All authors have read and agreed to the published version of the manuscript.

Funding: This research received no external funding.

Institutional Review Board Statement: This study was approved by the Hanoi University of Science and Technology on 28 July 2023.

Informed Consent Statement: Informed consent was obtained from all subjects involved in the study. Written informed consent has been obtained from the patient to publish this paper.

Data Availability Statement: Data will be available from the corresponding authors upon reasonable request.

Conflicts of Interest: The authors declare no conflicts of interest.

References

1. Deaton, A. Height, health, and development. *Proc. Natl. Acad. Sci. USA* **2007**, *104*, 13232–13237. [CrossRef] [PubMed]
2. Obese, H.J.O.R. Body mass index (BMI). *Obes. Res.* **1998**, *6*, 51S–209S.
3. Nuttall, F.Q. Body mass index: Obesity, BMI, and health: A critical review. *Nutr. Today* **2015**, *50*, 117–128. [CrossRef] [PubMed]
4. Phan, K.N.; Anh, V.T.; Manh, H.P.; Thu, H.N.; Thuy, N.T.; Thi, H.N.; Thuy, A.N.; Trung, P.N. The Non-Contact Height Measurement Method Using MediaPipe and OpenCV in a 2D Space. In Proceedings of the 2023 1st International Conference on Health Science and Technology (ICHST), Hanoi, Vietnam, 28–29 December 2023; pp. 1–6. [CrossRef]
5. Dennis, D.M.; Hunt, E.E.; Budgeon, C.A. Measuring height in recumbent critical care patients. *Am. J. Crit. Care* **2015**, *24*, 41–47. [CrossRef] [PubMed]
6. L'her, E.; Martin-Babau, J.; Lellouche, F. Accuracy of height estimation and tidal volume setting using anthropometric formulas in an ICU Caucasian population. *Ann. Intensive Care* **2016**, *6*, 55. [CrossRef] [PubMed]
7. Duc, T.T.; Phan, K.N.; Lan, P.N.; Mai PB, T.; Ngoc, D.C.; Manh, H.N.; Le Hoang, O.; Trung, P.N. Design and Development of Bedside Scale with Embedded Software to Calculate Treatment Parameters for Resuscitated Patients. In Proceedings of the 2023 1st International Conference on Health Science and Technology (ICHST), Hanoi, Vietnam, 28–29 December 2023; pp. 1–6. [CrossRef]
8. Nguyễn, P.K.; Đoàn, B.T.; Lê, H.O. Phần mềm hỗ trợ tính toán các thông số điều trị cho bệnh nhân hồi sức tích hợp với cân bệnh nhân. *Tạp Chí Y Học Việt Nam* **2023**, *529*, 179–183. [CrossRef]
9. Digssie, A.; Argaw, A.; Belachew, T. Developing an equation for estimating body height from linear body measurements of Ethiopian adults. *J. Physiol.-Thropology* **2018**, *37*, 26. [CrossRef] [PubMed]

10. Haritosh, A. A novel method to estimate Height, Weight and Body Mass Index from face images. In Proceedings of the Twelfth International Conference on Contemporary Computing (IC3), Noida, India, 8–10 August 2019; IEEE: Piscataway, NJ, NSA, 2019.
11. Lee, D.S.; Kim, J.S.; Jeong, S.C.; Kwon, S.K. Human height estimation by color deep learning and depth 3D conversion. *Appl. Sci.* **2010**, *10*, 5531. [CrossRef]
12. Bradski, G. The opencv library. *Dr. Dobb's J. Softw. Tools Prof. Program.* **2000**, *25*, 120–123.
13. Lugaresi, C.; Tang, J.; Nash, H.; McClanahan, C.; Uboweja, E.; Hays, M.; Zhang, F.; Chang, C.L.; Yong, M.G.; Lee, J.; et al. Mediapipe: A framework for building perception pipelines. *arXiv* **2019**, arXiv:1906.08172.
14. Lugaresi, C.; Tang, J.; Nash, H.; McClanahan, C.; Uboweja, E.; Hays, M.; Zhang, F.; Chang, C.L.; Yong, M.; Lee, J.; et al. Mediapipe: A framework for perceiving and processing reality. In Proceedings of the Third workshop on computer vision for AR/VR at IEEE computer vision and pattern recognition (CVPR), Long Beach, CA, USA, 17 June 2019; Volume 2019.
15. Terven, J.; Córdova-Esparza, D.M.; Romero-González, J.A. A comprehensive review of yolo architectures in computer vision: From yolov1 to yolov8 and yolo-nas. *Mach. Learn. Knowl. Extr.* **2023**, *5*, 1680–1716. [CrossRef]
16. Singh, A.K.; Kumbhare, V.A.; Arthi, K. Real-time human pose detection and recognition using mediapipe. In Proceedings of the International Conference on Soft Computing and Signal Processing, Hyderabad, India, 18–19 June 2021; Springer Nature: Singapore, 2021; pp. 145–154.
17. Tranmer, M.; Elliot, M. Multiple linear regression. *Cathie Marsh Cent. Census Surv. Res. (CCSR)* **2008**, *5*, 1–5.
18. Eck, D.J. Bootstrapping for multivariate linear regression models. *Stat. Probab. Lett.* **2018**, *134*, 141–149. [CrossRef]
19. Livingston, E.H. The mean and standard deviation: What does it all mean? *J. Surg. Res.* **2004**, *119*, 117–123. [CrossRef] [PubMed]

sensors

MDPI

Article

ARM4CH: A Methodology for Autonomous Reality Modelling for Cultural Heritage

Nikolaos Giakoumidis [1,2] and Christos-Nikolaos Anagnostopoulos [2,*]

[1] KINESIS Lab, Core Technology Platforms, New York University Abu Dhabi, Abu Dhabi P.O. Box 129188, United Arab Emirates; giakoumidis@nyu.edu
[2] Intelligent Systems Lab, Cultural Technology and Communication, University of the Aegean, 811 00 Mitilini, Greece
* Correspondence: canag@aegean.gr; Tel.: +30-2251036624

Abstract: Nowadays, the use of advanced sensors, such as terrestrial, mobile 3D scanners and photogrammetric imaging, has become the prevalent practice for 3D Reality Modeling (RM) and the digitization of large-scale monuments of Cultural Heritage (CH). In practice, this process is heavily related to the expertise of the surveying team handling the laborious planning and time-consuming execution of the 3D scanning process tailored to each site's specific requirements and constraints. To minimize human intervention, this paper proposes a novel methodology for autonomous 3D Reality Modeling of CH monuments by employing autonomous robotic agents equipped with the appropriate sensors. These autonomous robotic agents are able to carry out the 3D RM process in a systematic, repeatable, and accurate approach. The outcomes of this automated process may also find applications in digital twin platforms, facilitating secure monitoring and the management of cultural heritage sites and spaces, in both indoor and outdoor environments. The main purpose of this paper is the initial release of an Industry 4.0-based methodology for reality modeling and the survey of cultural spaces in the scientific community, which will be evaluated in real-life scenarios in future research.

Keywords: reality modeling; autonomous robots; terrestrial laser scanning; LiDAR; UAV; Next Best View

Citation: Giakoumidis, N.; Anagnostopoulos, C.-N. ARM4CH: A Methodology for Autonomous Reality Modelling for Cultural Heritage. *Sensors* **2024**, *24*, 4950. https://doi.org/10.3390/s24154950

Academic Editor: Antonia Spano

Received: 18 June 2024
Revised: 23 July 2024
Accepted: 28 July 2024
Published: 30 July 2024

1. Introduction

In recent years, Reality Modeling (RM) technologies, including cutting-edge sensors and systems such as LiDAR-based 3D scanners, drones, digital twins, augmented reality (AR), and virtual reality (VR), have become increasingly significant in the field of Cultural Heritage (CH) modeling, recording, and management [1–5]. However, the RM of CH remains a significant challenge for surveyors, as the 3D modeling process is largely manual, labor-intensive, and time-consuming. The scanning path and sensor positioning are predominantly reliant on the surveyor's experience, intuition, and perception, as there is currently no standardized automatic procedure [6]. The complexity is compounded by the natural environment surrounding CH sites, the morphological intricacies, and the vulnerability of the monuments. Specifically, to acquire a complete 3D reality model of a large-scale cultural space, multiple manual terrestrial 3D scans (TLS) and aerial surveys with unmanned aerial vehicles (UAVs) are required [1]. This manual approach heavily depends on the operator's expertise to determine the scanning path and identify the optimal scanner positions, a task known in the literature as the Next Best View (NBV) problem. Consequently, optimizing the NBV to efficiently capture large-scale, complex sites or monuments in dynamic environments (e.g., due to growing or changing vegetation) is crucial to reduce surveying time and enhance data quality. Despite its importance, the NBV problem has not been adequately addressed in terms of efficiency and optimality in

existing literature. As a result, the surveying process often takes longer than necessary, with redundant overlaps and additional positions planned as a precaution.

To address these challenges, this paper proposes a technological platform for autonomous 3D Reality Modeling and scanning. The goal is to develop a comprehensive, autonomous, systematic, and optimized 3D scanning procedure that accelerates the overall RM process and enhances data quality. To achieve this, two scientific pillars are essential: (a) a framework consisting of Robotic Agents (RAs) equipped with RM sensors that can navigate and operate autonomously and (b) a methodology to identify the optimal positions and trajectories for scanning, applicable to both terrestrial and aerial surveys, which maximizes area coverage and minimizes the number of scanning positions required (addressing the NBV problem). This approach aims to streamline the 3D RM process, ensuring efficient and high-quality data acquisition for cultural heritage sites.

The contributions of the proposed methodology (ARM4CH) are manifold and may be summarized in the following bullet points with more details available in the table in Section 4:

- Non-invasive and autonomous survey and inspection;
- Scanning operation for hard-to-reach, complex or dangerous areas;
- Reduction of labor costs and time-consuming scanning processes;
- Versatility and an increase in data precision;
- Consistency and optimization of measurements and data acquisition;
- Scanning and survey reproducibility;
- Regular monitoring of a CH site;
- Long-Term Monument Preservation and Management, a fostering of the Digital Twin concept.

2. Robotic Agents and 3D Scanning: A Brief Overview

Three-dimensional scanning using mobile robots has been already applied in recent years, especially in the field of construction, in which 3D scanning and monitoring is required on a regular basis. In most of these scenarios, the robots follow a predefined path or rely on exploration algorithms [7]. Recent years have seen extraordinary progress in the field of robotics, fueled by several key developments. The adoption of advanced control algorithms, such as Model Predictive Control and Deep Reinforcement Learning, advanced the creation of diverse locomotion mobile robots, including bio-inspired quadrupeds capable of navigating through challenging terrains [8,9]. Moreover, the emergence of advanced perception sensors like Depth Cameras, LiDAR, and Global Navigation Satellite Systems (GNSS) and torque-force sensors have revolutionized data acquisition, enabling the capture of extensive, detailed information that offers accurate and comprehensive insights for both environmental conditions and the robots' positions. The integration of artificial intelligence (AI) and machine learning algorithms into robotic systems [10] significantly enhances their autonomy, adaptability, and decision-making capabilities. This is further supported by increased processing power, which facilitates the application of sophisticated perception and AI algorithms directly on the robots, enhancing their efficiency and responsiveness. Moreover, hardware advancements, including batteries with higher energy density [11], more powerful computing units, and more efficient motors, have further advanced the capabilities of robotic systems.

Quadrupedal robots have been utilized for 3D scanning strategies to generate a complete set of point clouds of physical objects through multi-view scanning and data registration [12–14]. Furthermore, the control of quadrupedal robots has seen experimental success in achieving robust and agile locomotion, in the 3D space [15,16]. The utilization of representation-free model predictive control and exact feedback linearization has been implemented on quadrupedal robots, contributing to the stabilization of periodic gaits for quadrupedal locomotion [8]. Additionally, the application of hybrid dynamical systems has achieved physically effective and robust instances of all virtual bipedal gaits on quadrupedal robots [17].

Collectively, all of the above developments tend to transform robots from simple programmable machines into intelligent entities capable of collecting and analyzing complex environmental data, learning from their surroundings, making intricate decisions, and executing autonomous tasks with an unprecedented level of sophistication.

3. Autonomous Reality Modeling for Cultural Heritage (ARM4CH)

The ARM4CH proposed system is designed to automate the 3D Reality Modelling procedures in the field of Cultural Heritage by utilizing both aerial and ground Robotic Agents (RAs). Ground Robots (wheeled or quadrupedal) may navigate terrains with excellent levels of mobility, performing automated operations, tasks, and data capture safely, accurately and frequently. They can enter buildings or confined spaces and capture close-up images or videos at ground level. Since they are not constrained by airspace flying regulations, they can be utilized in areas where drones are not permitted. On the other hand, aerial robots/drones are used when ground scanning is impossible. Each RA (aerial or ground) is equipped with specialized hardware and software to perform autonomous navigation and sensor data acquisition.

A significant feature of ARM4CH is that both ground and aerial robotic agents may be configured to operate cooperatively. The selection of the RA (or a combination of RAs) for the survey is subject to the specifications of the CH site, such as the terrain morphology, necessary regulations to be followed, indoor or outdoor environment, as well as possible requirements set by stakeholders during the survey. For example, for an outdoor, large-scale CH site, with an unpaved trail and with tall artifacts (e.g., large-scale monuments such as the Acropolis of Athens), the best choice would be to employ quadrupedal robots, which have the capability to traverse in complex environments with high mobility, in co-operation with aerial RAs that can capture data from above, to offer an alternative perspective for areas that are inaccessible to the ground robots or when their sensors cannot adequately cover a Point of Interest (POI). Figure 1 depicts an indicative flowchart for the appropriate selection of the group of Robotic Agents for the Reality Modeling task, while a basic description of the RAs configuration is given in the next section.

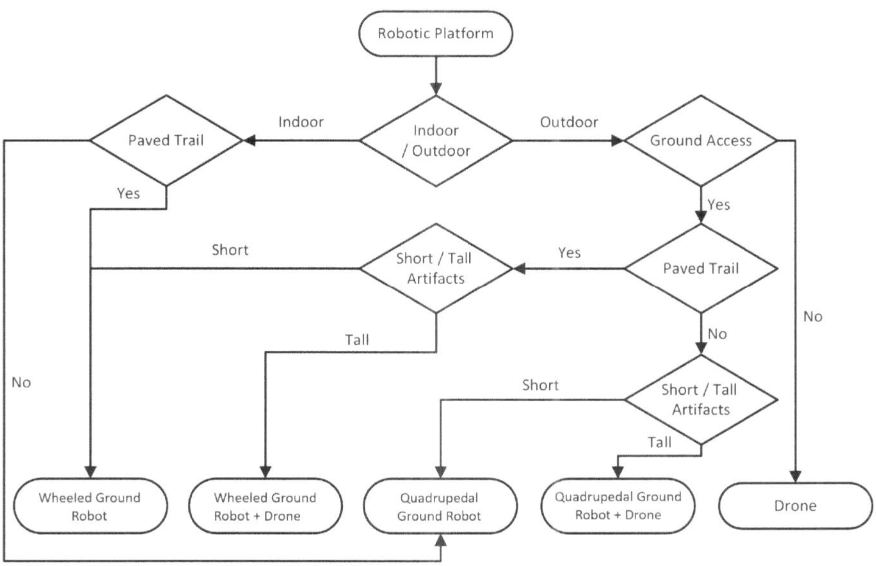

Figure 1. Type of RA decision flowchart.

3.1. Robotic Agent (RA) Architecture

As discussed earlier, the role of the RAs is to navigate autonomously in the CH site to perform the survey. In this section, we present two kinds of agents consisting of five main components (as seen in Figure 2), namely a quadrupedal robot and a drone.

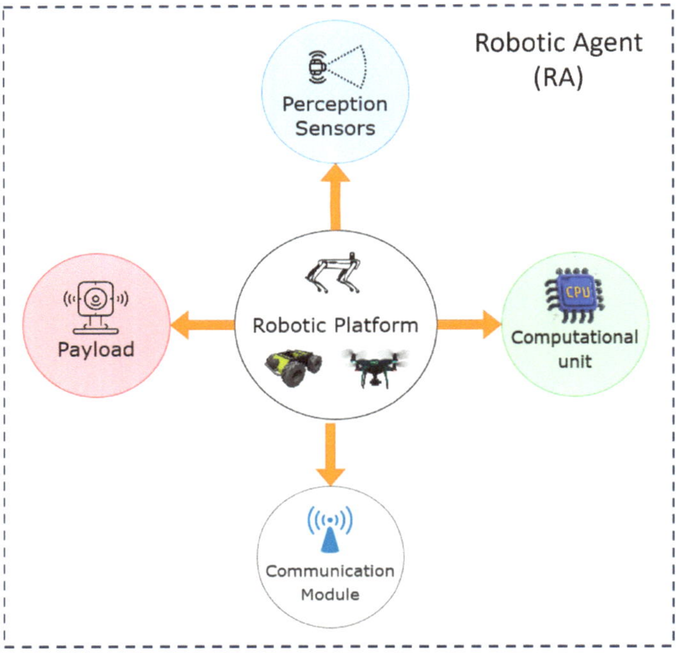

Figure 2. Robotic agent architecture.

The first component is the robotic platform by itself, which is a mobile robot capable of traversing the CH environment. As already mentioned above, there are many different types of mobile robots with different abilities, advantages, and disadvantages such as wheeled robots [18], quadrupedal robots, and aerial robots/drones [19]. The platform is the main core of the mobile agent that carries all of the necessary hardware including the perception sensors, the computation unit, the communication module, and finally, the payload, which is the actual Reality Modelling sensor. The parts of the RAs for the quadrupedal and aerial robots are shown in detail in Figures 3 and 4, respectively.

The perception sensors are responsible for collecting information about the state of the robotic platform and the physical environment around it. These sensors include LiDARs, RGB and depth cameras, motor encoders, Global Navigation Satellite Systems (GNSSs), and Inertial Measurement Units (IMUs), just to name the basics [19,20]. All of the collected data are managed in real time from numerous algorithms to control the robot.

Moreover, the computation unit is crucial for RA functions and operations in order to achieve onboard data processing, data collection, and management for effective navigation. This unit leverages raw data from all of the perception sensors and employs algorithms for odometry, pose estimation, Simultaneous Localization and Mapping (SLAM) [21], obstacle avoidance, motion and path planning, object detection [22], and the exploration of the environment.

The communication module is pivotal for enabling Robotic Agents to interact and co-operate [23]. It facilitates the exchange of information through protocols like TCP-IP, allowing robots to coordinate tasks, share sensor data, and make collective decisions. For instance, in the case of multiple RAs in a collaborative operation mode, it can divide tasks based on RA capabilities or current status, ensuring efficient task completion [24]. Sharing sensor inputs helps in constructing a comprehensive environmental understanding, enhancing decision-making. Additionally, communication is essential for monitoring the system's process by human supervisors.

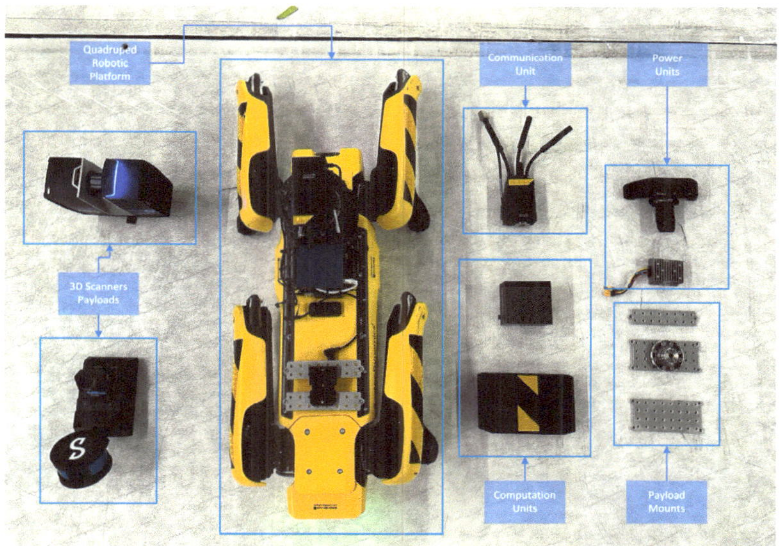

Figure 3. Quadruped RA components.

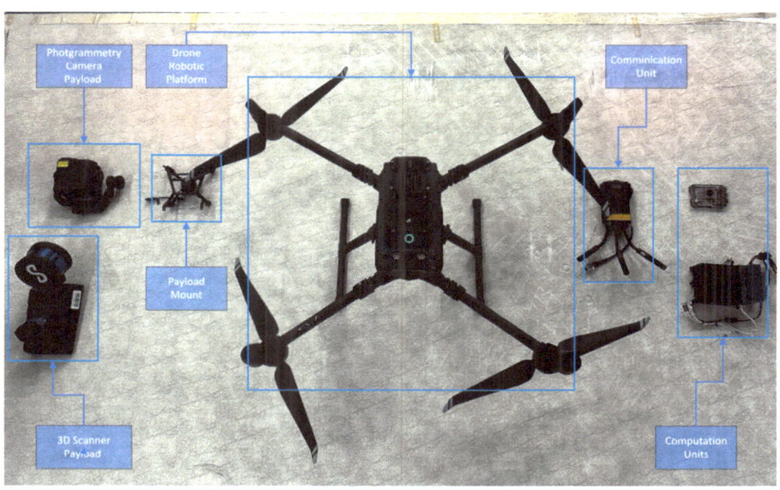

Figure 4. Drone RA components.

Finally, the payload is the main instrument dedicated to the collection of survey data, which in our case is the 3D representation of a CH site. The type of 3D sensor may vary depending on the requirements of the scan, the desired 3D point cloud resolution and accuracy [25,26] and the nature of the artifacts (e.g., shape, size, material, etc.) [27]. Possible payloads may be selected from a group of sensors like 3D Terrestrial Scanners, LiDAR sensors, depth sensors, 360 cameras, and other 2D imaging sensors (RGB or thermal cameras). Table 1 presents a brief comparison with the pros and cons of the three basic 3D point cloud acquisition methods, namely TLS, mobile scanners, and photogrammetry/SfM. In summary, TLS offer very high resolution and accuracy, typically ranging from millimeters to a few centimeters (e.g., 3.5 mm@ 25 m, 1 MPoint/s), while mobile/SLAM scanners offer resolutions around 2–3 cm, 5 mm@10 m, 0.5 Mpoint/s. Hence, the resolution and accuracy of the payloads attached to the RAs specify the resolution and accuracy of every ARM4CH scanning mission.

Table 1. Three-dimensional scanning methods comparison table [27].

Criteria	Terrestrial Laser Scanners (TLS)	SLAM-Based/Mobile Scanners	Photogrammetry/ Structure from Motion (SfM)
Accuracy and Precision	High (millimeter precision)	Moderate (depends on technology)	Variable (high under optimal conditions)
Data Collection Speed	Low (requires setup and multiple stations)	Fast (on-the-go collection)	Medium (depends on the required accuracy and complexity)
Cost	High (expensive hardware)	Moderate (less expensive than Terrestrial Laser)	Low to high (depends on camera equipment)
Operational Complexity	High (requires skilled operation)	Moderate (easier in complex environments)	Moderate (requires photographic expertise)
Environmental Constraints	Sensitive to reflective surfaces	Sensitive to reflective surfaces	Highly dependent on lighting and weather conditions
Post-Processing	Intensive (cleaning, registration, and merging)	Moderate (alignment aids, needs noise reduction)	Automated but long processing and sensitive to image quality
Application Suitability	Ideal for detailed, static environments	Suitable for complex environments	Versatile for various scales under good environment conditions
Common Use Cases	Detailed architectural, archeological, and engineering documentation	Extensive and complex environments like urban areas or large buildings	Large or remote outdoor areas

3.2. Methodology

The ARM4CH methodology comprises of five stages, namely scouting, Point-of-Interest (POI) identification, NBV detection, path planning, and finally, the on-site 3D scanning survey. A flow diagram of ARM4CH is highlighted in Figure 5.

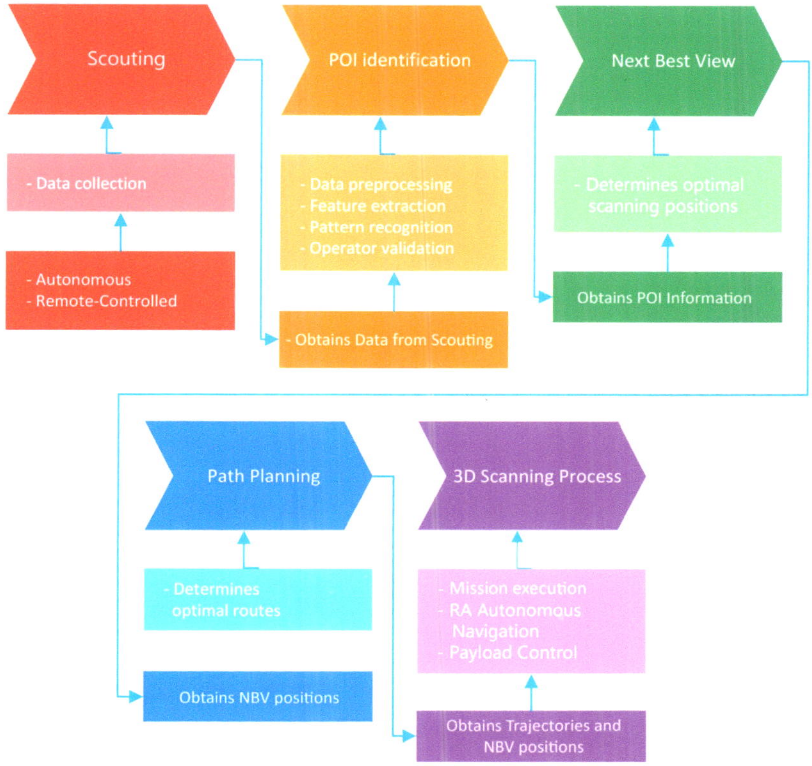

Figure 5. Flow diagram of the ARM4CH methodology.

3.2.1. Scouting

The goal of this step is to collect information related to the heritage site that will be later used for the navigation of the robot during the autonomous scanning process at the final stage. This dynamic operation comes with the limitations of low point cloud resolution, high noise due to motion distortion, and the inability to record an RGB-mapped point cloud. However, as a great advantage, LiDAR sensors calculate rapidly a coarse 3D topological map of the surveyed area, providing occupancy maps for the execution of the next steps.

In the scouting stage, various information may be given by the operator such as general areas of exploration/responsibility and preferable routes for exploration [28], as well as locations of no-go zones (either for ground or aerial units). For successfully achieving the above tasks, fiducial markers (e.g., AprilTags, ArUco markers) [29] and geotagged site images should be incorporated. During scouting, the navigation of the robotic platform can be performed either with autonomous or remote-controlled exploration.

For the former, the robotic agent (RA) autonomously navigates all accessible pathways within a predefined area of responsibility. The primary objective is to maximize coverage of the area, while minimizing the distance traveled. The area of responsibility, along with any designated no-go zones, is the input data to a Frontier-Based Exploration algorithm [7]. This algorithm, in combination with a Simultaneous Localization and Mapping (SLAM) algorithm [21], will then generate an occupancy grid map [30] of the heritage site. For the latter (remote-controlled exploration), the RA is navigated by a remote operator, who manually controls its movement through the predefined area of responsibility [13]. While the operator directs the robot, a SLAM algorithm continuously processes sensor data to

generate an occupancy grid map of the environment. This approach allows for human oversight in navigating complex or sensitive areas, while still benefiting from the automated mapping capabilities of the SLAM algorithm [31].

3.2.2. Point of Interest (POI) Identification

This step involves detecting and recognizing significant locations or objects (Point of Interest—POI) within the heritage environment, which can be accomplished either manually or automatically using Machine Learning.

In manual operation, the operator selects POIs within a visualized multimodal data environment. This environment integrates and displays data collected by the robot's sensors during scouting, including georeferenced images, 3D point clouds, and occupancy grid maps [30]. This digital representation enables the operator to effectively identify areas of interest. Figure 6 displays an example of POI selection in a georeferenced image of a CH site, where ground and drone vehicle POIs are selected along with no-go zones.

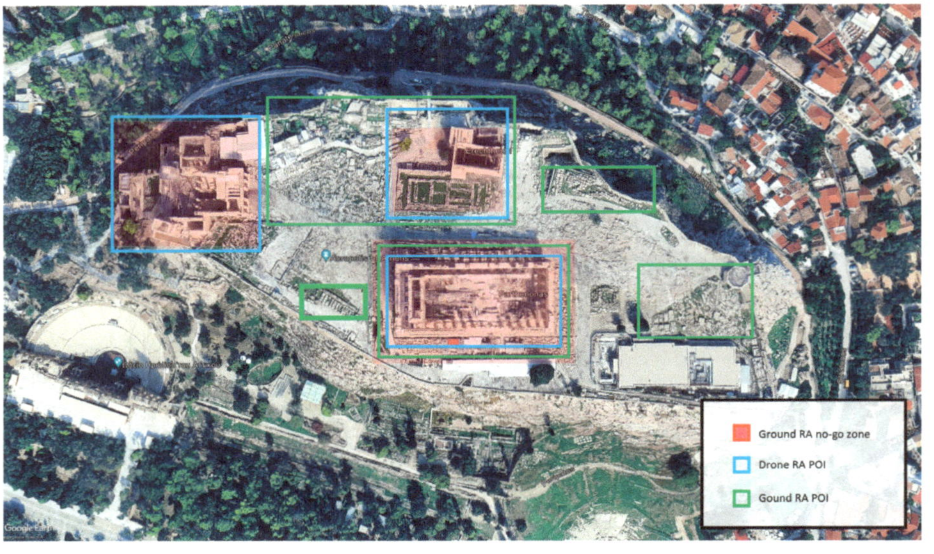

Figure 6. Manual POI annotations.

On the other hand, the automatic POI identification process is a complex Machine Learning classification task that includes several stages [32]. First, data preprocessing is performed to reduce noise. Next, feature extraction is conducted to identify key characteristics within the data [33] and finally, pattern recognition and machine learning techniques are applied to classify and cluster potential POIs [34]. The identified POIs are subsequently mapped onto an occupancy grid, allowing for precise localization and visualization. The final step in automatic POI identification involves operator validation of the identified POIs to ensure accuracy and relevance.

3.2.3. Next Best View Detection

The Next Best View (NBV) process aims to identify the optimal viewpoints for the RA to capture comprehensive 3D scans of the heritage site. Initially, the NBV algorithm evaluates the current state of the environment and determines the next best position and orientation for the robot's sensors. The goal is to maximize the amount of new information captured, reduce redundant scanning, and ensure high-quality, complete 3D models. To this end, estimating the Next Best View (NBV) in 3D environments is a critical aspect of autonomous data acquisition and 3D reconstruction. It involves determining the most

informative viewpoint for a sensor or robotic system to capture data that maximizes the information gain, while considering factors such as occlusions, completeness, and reconstruction quality. Researchers have proposed various approaches for NBV estimation, including probabilistic frameworks [35], volumetric information gain metrics [18,19,36,37], guided NBV for the 3D reconstruction of large complex structures using Unmanned Aerial Vehicles (UAVs) [38], and strategies for selecting the next best view based on ray tracing and already available BIM information [13]. Furthermore, the NBV problem has been addressed in the context of the surface reconstruction of large-scale 3D environments with multiple UAVs [38,39], and effective exploration for Micro Aerial Vehicles (MAVs) based on expected information gain [23,39]. These approaches leverage techniques such as reinforcement learning [24,40], feature tracking, reconstruction for NBV planning [25,41], and history-aware autonomous 3D exploration [26,42]. They aim to address the challenge of selecting the most informative viewpoint for 3D mesh refinement [27,43]. Therefore, NBV is a model-based approach, running within software (virtual environment) on the basis of the prior model obtained from the coarse 3D LiDAR scan of the site or an occupancy map, to defining a planning strategy for the identification of the proper scanning positions. Figure 7 displays optimum positions (blue dots) for Terrestrial Laser Scanning in the Medieval Castle of Chlemoutsi, in Ilia, Greece (https://maps.app.goo.gl/kHMmG7A1DxN8gKLp6, accessed on 1 June 2024), while in the green color are the parts of the Monument that cannot be covered by TLS due to height constraints. These areas will be surveyed using aerial vehicle (drone).

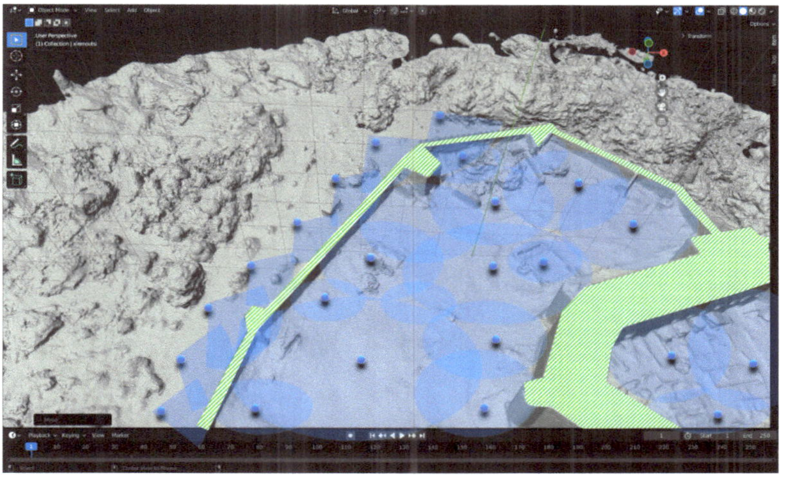

Figure 7. Visualization tool for the calculation of optimum positions (Next Best View) for terrestrial 3D scanning (TLS—blue dots) in Chlemoutsi castle. Uncovered parts are shown in green color. In this figure, Neu-NBV [44] was simulated in Unity and the results are shown in Blender. The 3D model was acquired from a previous manual survey.

3.2.4. Path Planning

Path planning involves determining the optimal routes for the RA to follow, taking into account the NBV recommendations to ensure efficient and comprehensive coverage of the heritage site. Using the data collected during the scouting phase, including sparse cloud points, reference marker positions, and desirable trajectories, the path planning process creates a route that maximizes area coverage while avoiding obstacles and adhering to any specified no-go zones. Path planning algorithms, such as Rapidly-exploring random trees (RRT) and the Probabilistic Roadmap Method (PRM) [45], are employed to compute the most efficient paths. These algorithms consider the occupancy grid map [30] generated by the SLAM [46] algorithm and incorporate the NBV-determined viewpoints, ensuring

that the planned paths are navigable, safe, and optimized for thorough 3D scanning. This integrated approach ensures that the robot can navigate effectively while capturing high-quality data on the heritage site. Figure 8 shows an example of the final trajectories proposed for the robot to follow during the ground and aerial surveys. At this point, Table 2 summarizes possible software packages that may be considered for the former four methodological steps (i.e., Scouting, POI, NBV, and Path Planning).

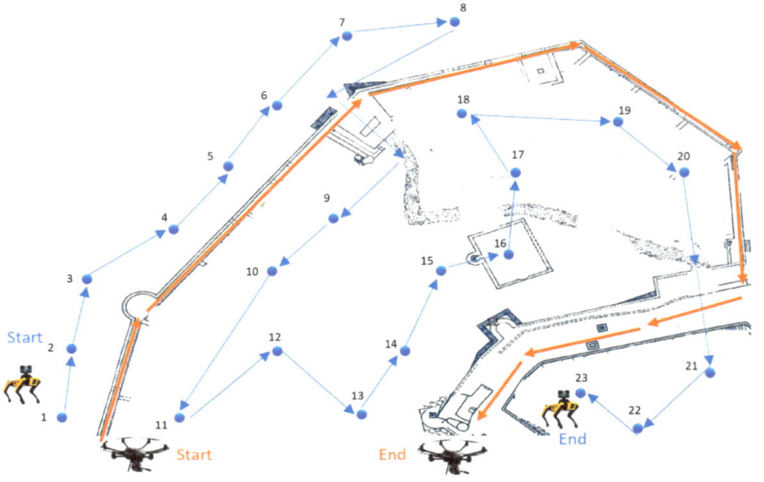

Figure 8. Generated paths/trajectories, with blue for the ground RA and orange arrows for the drone RA in Chlemoutsi castle. The numbers in the blue path indicate the sequence of the proposed positions (stops) for the terrestrial 3D scanning.

Table 2. Potential software algorithms.

Task	Algorithm	Indicative References
Scouting	SLAM-based: ORB-based LIO-SAM ROVIO	[47–49]
	Exploration-based: Frontier-based Graph-based planners SOAR-based space exploration	[50–52]
Point of Interest	Semantic Segmentation: Segment Anything Meta (SAM) Graph-based PSPNet	[53–55]
	Object Detection: YOLO-based R-CNN SSD	[56–58]
Next Best View	Reinforcement Learning NBV (RL-NBV) Point Cloud NBV (PC-NBV) NeU-NBV	[44,59,60]
Path Planning	RRT RPM A* algorithm	[61–63]

3.2.5. Scanning Process

The final stage of the methodology is the 3D scanning process, which builds on the scouting, POI identification, NBV, and path planning stages. In this stage, the RAs follow the pre-determined optimal paths, as outlined by the path planning stage, to conduct comprehensive 3D scanning of the heritage site. Utilizing the optimal viewpoints identified during the NBV process, the RA captures high-resolution 3D data, ensuring that all significant areas and POIs are thoroughly documented. The integration of SLAM ensures continuous localization and mapping accuracy, allowing the RA to adapt in real-time to any changes or obstacles encountered. The resulting 3D scans are then compiled into detailed, high-fidelity models of the heritage site, providing a valuable resource for preservation, analysis, and further research [64]. This systematic approach guarantees that the heritage site is meticulously documented with minimal movement and maximum efficiency. Figure 9 represents a cooperative operation of an aerial and quadrupedal robotic agents (RA) in a CH site, as well as a photo with the respective RAs in the lab.

Figure 9. Robotic agents carrying sensors in the CH site (**left**) and in the KINESIS lab (**right**).

4. Benefits and Barriers of the ARM4CH Methodology

Using the ARM4CH methodology, researchers/surveyors may send ground/quadrupedal robots on autonomous survey missions (both indoors and outdoors) using SLAM and GPS navigation in full co-operation with aerial vehicles (UAV) for analysis, data capture, documentation, and 3D scanning. The great benefits exceed the task of Cultural Heritage 3D scanning, since cooperative autonomous Reality Modelling/inspection features the following advantages:

- The ability to schedule robots remotely on unsupervised data capture and monitoring missions, 24/7, with specific field coverage.
- Ensure accuracy by capturing data from the same locations (viewpoints) multiple times, thus making direct data comparison feasible.
- The ability to create specific schedule plans to capture up-to-date data reliably.
- Reviewing, surveying, and inspecting spaces or places of critical/specific importance or those that pose a level of danger to the human surveyor.
- Complement the advantages of various sensor technologies and boost performance.
- Continuous or periodic monitoring. Thus, once a problem is confirmed, a maintenance team may be sent.

From the above it is evident that a great advantage of ARM4CH methodology is that it may be replicated/executed systematically, as many times as necessary in forthcoming periods, providing the ability to complete follow-up scans of the same place/site. Those follow-up scans introduce the concept of the fourth dimension (4D) in RM, since now the dimension of time is considered. Consecutive follow-up scans facilitate timeline comparison and monitoring of a constantly changing site and thus flag locations that need emergency

actions in times of crisis. To this end, ARM4CH may be extremely valuable during the process of establishing and maintaining a Digital Twin (DT) of a CH site or space. This is due to the fact that, "... a DT is a virtual instance of a physical system that is continually updated with the latter's performance" [65], leveraging the most updated available sensor data, to mirror the corresponding physical counterpart. Figure 10 demonstrates a graphical representation of a Digital Twin, in which ARM4CH may be used as a middleware to maintain updates of the site status.

Figure 10. ARM4CH as a catalyst for continuous model updates in the Digital Twin concept (the case study in this example is the castle of Chalki, Dodecanese).

To summarize the potential benefits and possible barriers of ARM4CH methodology, Table 3 analyses both advantages and disadvantages of embracing this Industry 4.0 methodological framework applied to the field of Cultural Heritage Digitization and Management in general.

Table 3. ARM4CH benefits and barriers.

Benefits	Detailed Analysis
Non-invasive survey and inspection	ARM4CH may carry out detailed Reality Modeling and monitoring without causing any physical disruption to the site, gathering high-resolution images, 3D scans, which are essential for a detailed analysis and documentation of the current state of the CH site.
Access to hard-to-reach areas and complex areas	Human operators are not exposed to missions and roles that might be challenging for them or to conventional surveying equipment. ARM4CH may suit perfectly for the survey or modeling of deteriorating structures of a monument.
Reduce labor costs and survey time	As CH sites often have complex architectures or difficult-to-access areas, ARM4CH may reduce the laborious and time-consuming process of 3D scanning by a human operator. The user now has a supervisory role to extensive surveys or inspections, which can be both time-consuming and expensive.

Table 3. *Cont.*

Benefits	Detailed Analysis
Versatility and precision	ARM4CH can be equipped with different sensors and tools, such as cameras, thermal imaging, and LIDAR, allowing it to perform a wide range of monitoring and surveying tasks with high precision, reducing the chances of human error and ensuring high-quality data collection.
Consistency and optimization	ARM4CH may perform tasks autonomously and systematically, following predefined routes and schedules, ensuring optimized, consistent, and reliable data collection with high precision and consistency.
Survey replication	ARM4CH may be replicated/executed systematically, as many times as necessary, providing the ability to complete follow-up scans of the CH site and update its Digital Twin or Heritage Building Information Model (H-BIM).
Regular monitoring	ARM4CH may lead to regular and consistent monitoring, providing up-to-date information on the CH site (i.e., detecting gradual changes or deterioration over time, damage, structural weaknesses, etc.), facilitating an immediate response to potential problems.
Long-term preservation and management	The detailed and systematic data collected by ARM4CH may assist curators, and conservation experts in prototype planning and executing restoration projects and a holistic CH management strategy.
Cost	The initial cost for creating the ARM4CH core platform including RAs and sensors is high, which might be a barrier for surveying companies or CH stakeholders that would like to operate this methodology.
Training and expertise transfer	The execution of ARM4CH and data management requires the presence of an expert in the survey team, who would supervise the process. This may necessitate additional resources for staff training or hiring skilled personnel.
Data management	If regular surveys are needed, a robust data infrastructure should be available, since large volumes of data collected need to be stored, processed, and managed (e.g., a complete DT platform).
Ethical and cultural concerns	The use of Industry 4.0 equipment in CH sites might raise ethical or cultural concerns among stakeholders who prefer traditional methods or have concerns about the use of frameworks of the latest technology (i.e., malfunctions that may cause unintentional damage to the site).

5. Discussion and Future Work

In this paper, we briefly presented the main steps and stages for a completely new methodology (ARM4CH) to ensure autonomous 3D scanning and digitization for Cultural Heritage spaces. Key enablers of ARM4CH are the following: (a) a technology core platform comprising autonomous ground robots, as well as UAVs, that work cooperatively to navigate and survey large areas using the latest technological sensors and deep learning-based computer vision [66–68], and (b) the operation of a software visualization tool that resolves the Next Best View problem in 3D meshes and identifies the optimum viewpoint position and scanning path for total survey coverage by ground and aerial (drones) robotic agents.

As already mentioned in Section 4, such a methodology could be essential for a "dynamic" DT of a cultural space to actively respond to the urgent need for the efficient management, resilience, and sustainability of CH sites, facilities, buildings, structures (indoors and/or outdoors), and their surrounding environment. This undeniable need is emphasized especially in the light of climate change and the necessity of energy saving.

Therefore, since the preservation and safeguarding of our Cultural Heritage is an urgent responsibility, there is an increased requirement for automated actions and method-

ologies to assist the preservation, data fusion/integration, site monitoring, and holistic management of CH. Using ARM4CH, the "flower" of ARM4CH (see Figure 11) may blossom in critical areas of CH and shift the attention of professionals/experts from a curative towards a more preventive and sustainable approach for CH management.

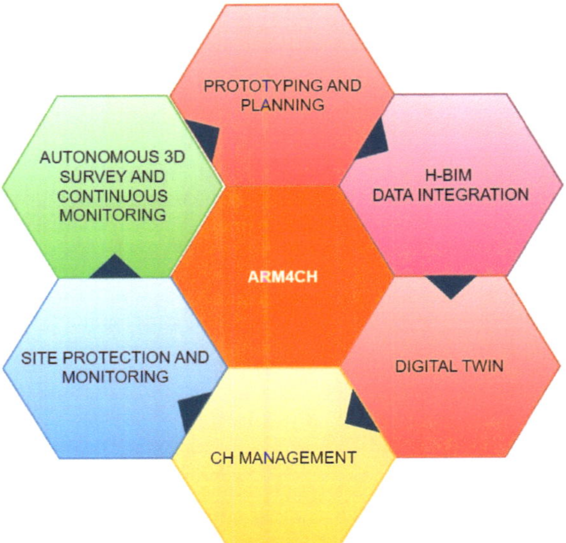

Figure 11. ARMCH as a core platform for various actions related to Cultural Heritage.

As the ARM4CH methodology is not assessed yet, the future actions of our research focus on the provision of a proof-of-concept on an actual CH site. Hence, to validate and verify this framework in a case study, the sequence of necessary steps includes the following:

1. Comparison, evaluation and final selection of the algorithms: This will ensure the seamless operation of the equipment (RAs and payload), as well as those responsible for scouting, POI, NBV, path planning, and scanning.
2. Experimentation and training on a simulated environment: After step 1, this stage involves the training of operational RAs using the latest simulation software platforms, such as the Robot Operation System (ROS) [69], NVIDIA Omniverse [70], and Gazebo [71]. Those platforms are core modules that provide pre-trained models augmented with synthetic data to design, test, and train the autonomous navigation of RAs and deliver scalable and physically accurate virtual environments for high-fidelity simulations.
3. Experimentation in a laboratory environment: This step involves the gradual release of operation in specific scenarios into a controlled environment. Moreover, it will verify that RAs have sufficient control, communication, awareness, and perception, as well as the ability to operate and navigate in dynamic and unpredictable indoor/outdoor environments.
4. Full deployment in a real heritage site: In this final step, ARM4CH will be released and evaluated in a large-scale Cultural Heritage park that includes various infrastructure for public services, protected monuments, and archaeological sites.

Author Contributions: Both authors have equally contributed to the paper. All authors have read and agreed to the published version of the manuscript.

Funding: This research received no external funding.

Institutional Review Board Statement: Not applicable.

Informed Consent Statement: Not applicable.

Data Availability Statement: The original contributions presented in the study are included in the article, further inquiries can be directed to the corresponding author.

Acknowledgments: Research was supported by the Kinesis Core Technology Platform laboratory at New York University Abu Dhabi.

Conflicts of Interest: The authors declare no conflicts of interest.

References

1. Hu, D.; Minner, J. UAVs and 3D City Modeling to Aid Urban Planning and Historic Preservation: A Systematic Review. *Remote Sens.* **2023**, *15*, 5507. [CrossRef]
2. Li, Y.; Zhao, L.; Chen, Y.; Zhang, N.; Fan, H.; Zhang, Z. 3D LiDAR and multi-technology collaboration for preservation of built heritage in China: A review. *Int. J. Appl. Earth Obs. Geoinf.* **2023**, *116*, 103156. [CrossRef]
3. Mitric, J.; Radulovic, I.; Popovic, T.; Scekic, Z.; Tinaj, S. AI and Computer Vision in Cultural Heritage Preservation. In Proceedings of the 2024 28th International Conference on Information Technology (IT), Zabljak, Montenegro, 21–24 February 2024; pp. 1–4. [CrossRef]
4. Caron, G.; Bellon, O.R.P.; Shimshoni, I. Computer Vision and Robotics for Cultural Heritage: Theory and Applications. *J. Imaging* **2023**, *9*, 9. [CrossRef] [PubMed]
5. Aicardi, I.; Chiabrando, F.; Lingua, A.M.; Noardo, F. Recent trends in cultural heritage 3D survey: The photogrammetric computer vision approach. *J. Cult. Herit.* **2018**, *32*, 257–266. [CrossRef]
6. Mahmood, S.; Majid, Z.; Idris, K.M. Terrestrial LiDAR sensor modeling towards optimal scan location and spatial density planning for 3D surveying. *Appl. Geomat.* **2020**, *12*, 467–480. [CrossRef]
7. Prieto, S.A.; Giakoumidis, N.; García De Soto, B. Multiagent robotic systems and exploration algorithms: Applications for data collection in construction sites. *J. Field Robot.* **2024**, *41*, 1187–1203. [CrossRef]
8. Fawcett, R.T.; Pandala, A.; Ames, A.D.; Hamed, K.A. Robust Stabilization of Periodic Gaits for Quadrupedal Locomotion via QP-Based Virtual Constraint Controllers. *IEEE Control. Syst. Lett.* **2022**, *6*, 1736–1741. [CrossRef]
9. Lee, J.; Hwangbo, J.; Wellhausen, L.; Koltun, V.; Hutter, M. Learning quadrupedal locomotion over challenging terrain. *Sci. Robot.* **2020**, *5*, eabc5986. [CrossRef] [PubMed]
10. Soori, M.; Arezoo, B.; Dastres, R. Artificial intelligence, machine learning and deep learning in advanced robotics, a review. *Cogn. Robot.* **2023**, *3*, 54–70. [CrossRef]
11. Mikołajczyk, T.; Mikołajewski, D.; Kłodowski, A.; Łukaszewicz, A.; Mikołajewska, E.; Paczkowski, T.; Skornia, M. Energy Sources of Mobile Robot Power Systems: A Systematic Review and Comparison of Efficiency. *Appl. Sci.* **2023**, *13*, 7547. [CrossRef]
12. Chen, L.; Hoang, D.; Lin, H.; Nguyen, T. Innovative methodology for multi-view point cloud registration in robotic 3d object scanning and reconstruction. *Appl. Sci.* **2016**, *6*, 132. [CrossRef]
13. Park, S.; Yoon, S.; Ju, S.; Heo, J. BIM-based scan planning for scanning with a quadruped walking robot. *Autom. Constr.* **2023**, *152*, 104911. [CrossRef]
14. Kim, P.; Park, J.; Cho, Y. As-is geometric data collection and 3D visualization through the collaboration between UAV and UGV. In Proceedings of the International Symposium on Automation and Robotics in Construction (ISARC), Banff, Canada, 21–24 May 2019; Volume 36, pp. 544–551. [CrossRef]
15. Peers, C.; Motawei, M.; Richardson, R.; Zhou, C. Development of a teleoperative quadrupedal manipulator. In Proceedings of the UKRAS21 Conference: Robotics at Home Proceedings, Online, 2 June 2021; University of Hertfordshire: Hatfield, UK. [CrossRef]
16. Ding, Y.; Pandala, A.; Li, C.; Shin, Y.; Park, H.W. Representation-free model predictive control for dynamic motions in quadrupeds. *IEEE Trans. Robot.* **2021**, *37*, 1154–1171. [CrossRef]
17. Hutter, M.; Gehring, C.; Jud, D.; Lauber, A.; Bellicoso, C.D.; Tsounis, V.; Hwangbo, J.; Bodie, K.; Fankhauser, P.; Bloesch, M.; et al. ANYmal—A highly mobile and dynamic quadrupedal robot. In Proceedings of the 2016 IEEE/RSJ International Conference on Intelligent Robots and Systems (IROS), Daejeon, Republic of Korea, 9–14 October 2016; pp. 38–44. [CrossRef]
18. Borkar, K.K.; Aljrees, T.; Pandey, S.K.; Kumar, A.; Singh, M.K.; Sinha, A.; Sharma, V. Stability Analysis and Navigational Techniques of Wheeled Mobile Robot: A Review. *Processes* **2023**, *11*, 3302. [CrossRef]
19. Rubio, F.; Valero, F.; Llopis-Albert, C. A review of mobile robots: Concepts, methods, theoretical framework, and applications. *Int. J. Adv. Robot. Syst.* **2019**, *16*, 172988141983959. [CrossRef]
20. Camurri, M.; Ramezani, M.; Nobili, S.; Fallon, M. Pronto: A Multi-Sensor State Estimator for Legged Robots in Real-World Scenarios. *Front. Robot. AI* **2020**, *7*, 68. [CrossRef] [PubMed]
21. Macario Barros, A.; Michel, M.; Moline, Y.; Corre, G.; Carrel, F. A Comprehensive Survey of Visual SLAM Algorithms. *Robotics* **2022**, *11*, 24. [CrossRef]
22. Mittal, S. A Survey on optimized implementation of deep learning models on the NVIDIA Jetson platform. *J. Syst. Archit.* **2019**, *97*, 428–442. [CrossRef]
23. Li, Y.; Du, S.; Kim, Y. Robot swarm MANET cooperation based on mobile agent. In Proceedings of the 2009 IEEE International Conference on Robotics and Biomimetics (ROBIO), Guilin, China, 19–23 December 2009.

24. Ivanov, M.; Sergiyenko, O.; Tyrsa, V.; Lindner, L.; Reyes-García, M.; Rodríguez-Quiñonez, J.C.; Hernández-Balbuena, D. *Data Exchange and Task of Navigation for Robotic Group*; Springer International Publishing: Berlin/Heidelberg, Germany, 2020; pp. 389–430.
25. Kalvoda, P.; Nosek, J.; Kuruc, M.; Volarik, T.; Kalvodova, P. Accuracy Evaluation and Comparison of Mobile Laser Scanning and Mobile Photogrammetry Data. *IOP Conf. Ser. Earth Environ. Sci.* **2020**, *609*, 012091. [CrossRef]
26. Dering, G.M.; Micklethwaite, S.; Thiele, S.T.; Vollgger, S.A.; Cruden, A.R. Review of drones, photogrammetry and emerging sensor technology for the study of dykes: Best practises and future potential. *J. Volcanol. Geotherm. Res.* **2019**, *373*, 148–166. [CrossRef]
27. Daneshmand, M.; Helmi, A.; Avots, E.; Noroozi, F.; Alisinanoglu, F.; Arslan, H.S.; Gorbova, J.; Haamer, R.E.; Ozcinar, C.; Anbarjafari, G. 3D Scanning: A Comprehensive Survey. In Proceedings of the Conference on Computer Vision and Pattern Recognition, Salt Lake City, UT, USA, 18–22 June 2018.
28. Chen, K.; Reichard, G.; Akanmu, A.; Xu, X. Geo-registering UAV-captured close-range images to GIS-based spatial model for building façade inspections. *Autom. Constr.* **2021**, *122*, 103503. [CrossRef]
29. Kalaitzakis, M.; Cain, B.; Carroll, S.; Ambrosi, A.; Whitehead, C.; Vitzilaios, N. Fiducial Markers for Pose Estimation. *J. Intell. Robot. Syst.* **2021**, *101*, 71. [CrossRef]
30. Hornung, A.; Wurm, K.M.; Bennewitz, M.; Stachniss, C.; Burgard, W. OctoMap: An efficient probabilistic 3D mapping framework based on octrees. *Auton. Robot.* **2013**, *34*, 189–206. [CrossRef]
31. Wallace, D.; He, Y.H.; Vaz, J.C.; Georgescu, L.; Oh, P.Y. Multimodal Teleoperation of Heterogeneous Robots within a Construction Environment. In Proceedings of the 2020 IEEE/RSJ International Conference on Intelligent Robots and Systems (IROS), Las Vegas, NV, USA, 24 October–24 January 2020.
32. Pierdicca, R.; Paolanti, M.; Matrone, F.; Martini, M.; Morbidoni, C.; Malinverni, E.S.; Lingua, A.M. Point Cloud Semantic Segmentation Using a Deep Learning Framework for Cultural Heritage. *Remote Sens.* **2020**, *12*, 1005. [CrossRef]
33. Câmara, A.; de Almeida, A.; Caçador, D.; Oliveira, J. Automated methods for image detection of cultural heritage: Overviews and perspectives. *Archaeol. Prospect.* **2023**, *30*, 153–169. [CrossRef]
34. Fiorucci, M.; Verschoof-Van Der Vaart, W.B.; Soleni, P.; Le Saux, B.; Traviglia, A. Deep Learning for Archaeological Object Detection on LiDAR: New Evaluation Measures and Insights. *Remote Sens.* **2022**, *14*, 1694. [CrossRef]
35. Potthast, C.; Sukhatme, G.S. A probabilistic framework for next best view estimation in a cluttered environment. *J. Vis. Commun. Image Represent.* **2014**, *25*, 148–164. [CrossRef]
36. Bircher, A.; Kamel, M.; Alexis, K.; Oleynikova, H.; Siegwart, R. Receding horizon path planning for 3D exploration and surface inspection. *Auton. Robot.* **2018**, *42*, 291–306. [CrossRef]
37. Delmerico, J.; Isler, S.; Sabzevari, R.; Scaramuzza, D. A comparison of volumetric information gain metrics for active 3D object reconstruction. *Auton. Robot.* **2018**, *42*, 197–208. [CrossRef]
38. Almadhoun, R.; Abduldayem, A.; Taha, T.; Seneviratne, L.; Zweiri, Y. Guided Next Best View for 3D Reconstruction of Large Complex Structures. *Remote Sens.* **2019**, *11*, 2440. [CrossRef]
39. Palazzolo, E.; Stachniss, C. Effective Exploration for MAVs Based on the Expected Information Gain. *Drones* **2018**, *2*, 9. [CrossRef]
40. Kaba, M.D.; Uzunbas, M.G.; Lim, S.N. A Reinforcement Learning Approach to the View Planning Problem. In Proceedings of the IEEE Conference on Computer Vision and Pattern Recognition (CVPR), Honolulu, HI, USA, 21–26 July 2017.
41. Trummer, M.; Munkelt, C.; Denzler, J. *Combined GKLT Feature Tracking and Reconstruction for Next Best View Planning*; Springer: Berlin/Heidelberg, Germany, 2009; pp. 161–170.
42. Wang, Y.; Del Bue, A. *Where to Explore Next? ExHistCNN for History-Aware Autonomous 3D Exploration*; Springer International Publishing: Berlin/Heidelberg, Germany, 2020; pp. 125–140.
43. Morreale, L.; Romanoni, A.; Matteucci, M. *Predicting the Next Best View for 3D Mesh Refinement*; Springer International Publishing: Berlin/Heidelberg, Germany, 2019; pp. 760–772.
44. Jin, L.; Chen, X.; Rückin, J.; Popović, M. NeU-NBV: Next Best View Planning Using Uncertainty Estimation in Image-Based Neural Rendering. In Proceedings of the 2023 IEEE/RSJ International Conference on Intelligent Robots and Systems (IROS), Detroit, MI, USA, 1–5 October 2023; pp. 11305–11312. [CrossRef]
45. Muhammad, A.; Abdullah, N.R.H.; Ali, M.A.; Shanono, I.H.; Samad, R. Simulation Performance Comparison of A*, GLS, RRT and PRM Path Planning Algorithms. In Proceedings of the 2022 IEEE 12th Symposium on Computer Applications & Industrial Electronics (ISCAIE), Penang, Malaysia, 21–22 May 2022.
46. Bujanca, M.; Shi, X.; Spear, M.; Zhao, P.; Lennox, B.; Luján, M. Robust SLAM Systems: Are We There Yet? In Proceedings of the 2021 IEEE/RSJ International Conference on Intelligent Robots and Systems (IROS), Prague, Czech Republic, 27 September–1 October 2021.
47. Campos, C.; Elvira, R.; Rodríguez JJ, G.; Montiel, J.M.; Tardós, J.D. ORB-SLAM3: An Accurate Open-Source Library for Visual, Visual–Inertial, and Multimap SLAM. *IEEE Trans. Robot.* **2021**, *37*, 1874–1890. [CrossRef]
48. Shan, T.; Englot, B.; Meyers, D.; Wang, W.; Ratti, C.; Rus, D. LIO-SAM: Tightly-coupled Lidar Inertial Odometry via Smoothing and Mapping. In Proceedings of the 2020 IEEE/RSJ International Conference on Intelligent Robots and Systems (IROS), Las Vegas, NV, USA, 25–29 October 2020; pp. 5135–5142. [CrossRef]

49. Bloesch, M.; Omari, S.; Hutter, M.; Siegwart, R. Robust visual inertial odometry using a direct EKF-based approach. In Proceedings of the 2015 IEEE/RSJ International Conference on Intelligent Robots and Systems (IROS), Hamburg, Germany, 28 September–2 October 2015; pp. 298–304. [CrossRef]
50. Yamauchi, B. A frontier-based approach for autonomous exploration. In Proceedings 1997 IEEE International Symposium on Computational Intelligence in Robotics and Automation CIRA'97. "Towards New Computational Principles for Robotics and Automation", Monterey, CA, USA, 10–11 June 1997; pp. 146–151. [CrossRef]
51. Dang, T.; Tranzatto, M.; Khattak, S.; Mascarich, F.; Alexis, K.; Hutter, M. Graph-based subterranean exploration path planning using aerial and legged robots, Special Issue on Field and Service Robotics (FSR). *J. Field Robot.* **2020**, *37*, 1363–1388. [CrossRef]
52. Luo, F.; Zhou, Q.; Fuentes, J.; Ding, W.; Gu, C. A Soar-Based Space Exploration Algorithm for Mobile Robots. *Entropy* **2022**, *24*, 426. [CrossRef] [PubMed]
53. Segment Anything. Available online: https://segment-anything.com/ (accessed on 27 July 2024).
54. Felzenszwalb, P.F.; Huttenlocher, D.P. Efficient Graph-Based Image Segmentation. *Int. J. Comput. Vis.* **2004**, *59*, 167–181. [CrossRef]
55. Zhao, H.; Shi, J.; Qi, X.; Wang, X.; Jia, J. Pyramid Scene Parsing Network. In Proceedings of the 2017 IEEE Conference on Computer Vision and Pattern Recognition (CVPR), Honolulu, HI, USA, 21–26 July 2017; pp. 6230–6239. [CrossRef]
56. Redmon, J.; Divvala, S.; Girshick, R.; Farhadi, A. You Only Look Once: Unified, Real-Time Object Detection. In Proceedings of the 2016 IEEE Conference on Computer Vision and Pattern Recognition (CVPR), Las Vegas, NV, USA, 27–30 June 2016; pp. 779–788. [CrossRef]
57. Girshick, R.; Donahue, J.; Darrell, T.; Malik, J. Rich Feature Hierarchies for Accurate Object Detection and Semantic Segmentation. In Proceedings of the 2014 IEEE Conference on Computer Vision and Pattern Recognition, Columbus, OH, USA, 23–28 June 2014; pp. 580–587. [CrossRef]
58. Liu, W.; Anguelov, D.; Erhan, D.; Szegedy, C.; Reed, S.; Fu, C.Y.; Berg, A.C. SSD: Single Shot MultiBox Detector. In *Computer Vision—ECCV 2016. ECCV 2016. Lecture Notes in Computer Science*; Leibe, B., Matas, J., Sebe, N., Welling, M., Eds.; Springer: Cham, Switzerland, 2016; Volume 9905. [CrossRef]
59. Wang, T.; Xi, W.; Cheng, Y.; Han, H.; Yang, Y. RL-NBV: A deep reinforcement learning based next-best-view method for unknown object reconstruction. *Pattern Recognit. Lett.* **2024**, *184*, 1–6. [CrossRef]
60. Zeng, R.; Zhao, W.; Liu, Y.-J. PC-NBV: A Point Cloud Based Deep Network for Efficient Next Best View Planning. In Proceedings of the IEEE/RSJ International Conference on Intelligent Robots and Systems (IROS), Hangzhou, China, 19–25 October 2020; pp. 7050–7057. [CrossRef]
61. Moon, C.B.; Chung, W. Kinodynamic Planner Dual-Tree RRT (DT-RRT) for Two-Wheeled Mobile Robots Using the Rapidly Exploring Random Tree. *IEEE Trans. Ind. Electron.* **2015**, *62*, 1080–1090. [CrossRef]
62. Kavraki, L.E.; Svestka, P.; Latombe, J.C.; Overmars, M.H. Probabilistic roadmaps for path planning in high-dimensional configuration spaces. *IEEE Trans. Robot. Autom.* **1996**, *12*, 566–580. [CrossRef]
63. Hart, P.E.; Nilsson, N.J.; Raphael, B. A Formal Basis for the Heuristic Determination of Minimum Cost Paths. *IEEE Trans. Syst. Sci. Cybern.* **1968**, *4*, 100–107. [CrossRef]
64. Parrinello, S.; Picchio, F. Digital Strategies to Enhance Cultural Heritage Routes: From Integrated Survey to Digital Twins of Different European Architectural Scenarios. *Drones* **2023**, *7*, 576. [CrossRef]
65. Cimino, C.; Ferretti, G.; Leva, A. Harmonising and integrating the digital twins multiverse: A paradigm and a toolset proposal. *Comput. Ind.* **2021**, *132*, 103501. [CrossRef]
66. Osco, L.P.; Junior, J.M.; Ramos, A.P.M.; de Castro Jorge, L.A.; Fatholahi, S.N.; de Andrade Silva, J.; Matsubara, E.T.; Pistori, H.; Gonçalves, W.N.; Li, J. A review on deep learning in UAV remote sensing. *Int. J. Appl. Earth Obs. Geoinf.* **2021**, *102*, 102456. [CrossRef]
67. Qiu, Q.; Lau, D. Real-time detection of cracks in tiled sidewalks using YOLO-based method applied to unmanned aerial vehicle (UAV) images. *Autom. Constr.* **2023**, *147*, 104745. [CrossRef]
68. Mittal, P.; Singh, R.; Sharma, A. Deep learning-based object detection in low-altitude UAV datasets: A survey. *Image Vis. Comput.* **2020**, *104*, 104046. [CrossRef]
69. Robot Operation System (ROS). Available online: https://www.ros.org/ (accessed on 27 July 2024).
70. NVIDIA Omniverse. Available online: https://www.nvidia.com/en-eu/omniverse/ (accessed on 27 July 2024).
71. Gazebo. Available online: https://gazebosim.org (accessed on 27 July 2024).

sensors

MDPI

Article

Infrared and Visual Image Fusion Based on a Local-Extrema-Driven Image Filter

Wenhao Xiang [1], Jianjun Shen [1], Li Zhang [1] and Yu Zhang [2,*]

1 Department of Electronic Engineering, Tsinghua University, Beijing 100084, China; xiangwh2018@163.com (W.X.); sjj20@mails.tsinghua.edu.cn (J.S.); chinazhangli@tsinghua.edu.cn (L.Z.)
2 School of Astronautics, Beihang University, Beijing 102206, China
* Correspondence: uzeful@163.com

Abstract: The objective of infrared and visual image fusion is to amalgamate the salient and complementary features of the infrared and visual images into a singular informative image. To accomplish this, we introduce a novel local-extrema-driven image filter designed to effectively smooth images by reconstructing pixel intensities based on their local extrema. This filter is iteratively applied to the input infrared and visual images, extracting multiple scales of bright and dark feature maps from the differences between continuously filtered images. Subsequently, the bright and dark feature maps of the infrared and visual images at each scale are fused using elementwise-maximum and elementwise-minimum strategies, respectively. The two base images, representing the final-scale smoothed images of the infrared and visual images, are fused using a novel structural similarity- and intensity-based strategy. Finally, our fusion image can be straightforwardly produced by combining the fused bright feature map, dark feature map, and base image together. Rigorous experimentation conducted on the widely used TNO dataset underscores the superiority of our method in fusing infrared and visual images. Our approach consistently performs on par or surpasses eleven state-of-the-art image-fusion methods, showcasing compelling results in both qualitative and quantitative assessments.

Keywords: infrared and visual image fusion; local-extrema-driven image filter; bright feature map; dark feature map; base image

Citation: Xiang, W.; Shen, J.; Zhang, L.; Zhang, Y. Infrared and Visual Image Fusion Based on a Local-Extrema-Driven Image Filter. *Sensors* **2024**, *24*, 2271. https://doi.org/10.3390/s24072271

Academic Editors: Christos Nikolaos E. Anagnostopoulos and Stelios Krinidis

Received: 2 February 2024
Revised: 30 March 2024
Accepted: 1 April 2024
Published: 2 April 2024

1. Introduction

The need for infrared and visible image fusion arises from the desire to obtain a comprehensive representation of a supervised scenario throughout the day. This technique finds extensive application in both civilian and military surveillance systems, as it can provide valuable information for decision making and situational awareness. Challenges in infrared and visible image fusion include precise segmentation of source images, the integration of salient features without the loss of visual information, and achieving a fusion image with high contrast and visual appeal. Traditional methods, such as spatial-domain and transform-domain approaches, often struggle with these challenges, resulting in suboptimal fusion effects. The motivation for infrared and visible image fusion lies in the complementary nature of the two imaging modalities. Infrared images capture thermal radiation emitted by objects, providing information about their temperature and potentially revealing hidden or camouflaged targets. Visible images, on the other hand, offer high-resolution detail and color information, facilitating the identification and recognition of objects and scenes. By fusing these two types of images, it is possible to achieve a more complete and accurate representation of the supervised scenario.

Various imaging sensors can capture different perspectives of a supervised scenario. The fusion of these multiple images proves invaluable in gaining a comprehensive understanding of the situation at hand [1–3]. For instance, the fusion of multi-modal medical images greatly aids surgeons in accurate disease diagnosis [4–7], while multi-focus image fusion yields a sharp, all-in-focus image [8–12]. In the realm of infrared and visual

image fusion, it results in a composite image that provides a holistic representation of the supervised scenario throughout the day. This technique finds extensive application in both civilian and military surveillance systems [13–17]. Therefore, the development of innovative methods for fusing infrared and visual images is crucial and holds significant utility in both civil and military operations.

In recent years, the field of infrared and visual image fusion has witnessed the emergence of numerous methods, broadly categorized into spatial-domain and transform-domain approaches. Spatial-domain methods involve the initial segmentation of source images into multiple regions, followed by the combination of salient regions to achieve fusion [8,9,11,12,18]. However, these methods often struggle with precise segmentation, leading to suboptimal fusion effects. Transform-domain methods, gaining popularity over the past two decades, mainly include pyramid-based [19,20], wavelet-based [21,22], and sparse-representation-based image-fusion methods [23–25]. These methods extract salient features within a specific domain and integrate them to produce the fusion image, typically visually appealing, but susceptible to blurring or significant information loss.

In recent times, numerous deep learning approaches, particularly those based on convolutional neural networks (CNNs), have been proposed for image fusion [3,6,16,26–30]. Initially, Liu et al. [26] introduced a CNN model to identify the focus decision map of multi-focus images. They refined the focus decision map through post-processing procedures and generated an all-in-focus fusion image by copying focused regions from corresponding partially focused images based on the focus decision map. Subsequently, Li et al. [27] utilized densely connected CNN blocks to construct their image fusion model, achieving significant improvement in fusing infrared and visual images. Afterward, Ma et al. [16] employed a GAN-based model to effectively train their image fusion model for infrared and visual images in an adversarial manner. More recently, Li et al. [29] proposed a representation-learning-based infrared and visual image fusion network, claiming to avoid trial-and-test strategies. Despite their success in image fusion, most of these methods still exhibit low contrast or other types of defects.

In addition to the aforementioned methods, Zhou et al. [18] employed Gaussian and bilateral filters to extract multi-scale feature maps from different input images, subsequently blending them to create their fusion images. Similarly, Zhang et al. [31] devised a multi-scale Bezier filter, utilizing it to extract multiscale bright and dark features from infrared and visual images and integrating these features with the base image to generate their fusion image. Despite these efforts, their proposed image filters did not demonstrate sufficient superiority. Their image-fusion methods primarily concentrated on merging salient features without adequate consideration for the visual effect of the resulting fusion images. Consequently, their fusion images often suffered from low-contrast effects or the loss of visual information, making them unsatisfactory for human visual perception.

To address the limitations of existing methods and integrate the salient features of infrared and visual images while improving the visual quality of the fusion image, in this study, we introduce a simple, yet effective local-extrema-driven image filter. By alternately leveraging local minima and local maxima for image reconstruction, our proposed filter demonstrates exceptional capabilities in extracting both bright and dark features from images. Specifically, the disparities between the filtered and original images reveal these bright and dark features. Additionally, we present a multi-scale local-extrema-filter-based method for fusing infrared and visual images. This method initially extracts multiple scales of bright and dark feature maps and generates corresponding base images from the input infrared and visual images, respectively. It then merges the high-frequency bright and dark feature maps and low-frequency base images using two different fusion rules. Finally, the fusion image is generated by integrating the fused feature maps and the base image. Owing to the exploitation of our advanced local-extrema-driven filter, this method excels in capturing salient dark and bright features from both infrared and visual images, resulting in an informative fusion image. Moreover, the incorporation of our innovative structural similarity- and intensity-based base image fusion scheme enhances

the visual quality of our fusion images, representing a notable improvement over current state-of-the-art image-fusion methods, including deep learning-based approaches.

This paper comprises three primary contributions. Firstly, we introduce an innovative image filter driven by local extrema, which effectively smooths images by removing bright and dark features, thus enabling robust feature extraction for generating salient bright and dark feature maps. Secondly, we propose a novel base image fusion scheme based on structural similarity and intensity considerations. This approach prioritizes obtaining a fused base image that encompasses large-scale structural features and well-distributed intensity, achieved through the generation of a weight map that accounts for these factors within the base images. Consequently, our method consistently produces fusion images with superior visual quality. Lastly, extensive experimental validation demonstrates the effectiveness of our approach, surpassing eleven state-of-the-art transform-domain image-fusion methods and outperforming leading deep learning-based methods. This success underscores the efficacy of our proposed local-extrema image filter and base image-fusion scheme.

The remaining paper is organized as follows. The proposed local-extrema-driven image filter and the proposed image-fusion method based on this filter are elaborated in Section 2. The experimental results and discussions are presented in Section 3. Finally, the conclusions of this paper are drawn in Section 4.

2. Proposed Method

In this study, we present an effective method for fusing infrared and visual images, leveraging our newly developed multi-scale local-extrema-driven image filter. The proposed approach comprises four key steps: Firstly, we apply the local-extrema-driven image filter at varying scales to progressively process the infrared and visual images. Simultaneously, we extract the corresponding bright and dark feature maps from each, while using the resulting filtered images as their base images. Next, we merge the bright and dark feature maps from both the infrared and visual images by selecting their elementwise maximum values, followed by enhancement with a scale-dependent coefficient. Then, we blend the base images of the infrared and visual inputs by a structural similarity-based fusion scheme. Ultimately, the fusion image is generated by integrating the fused bright and dark feature maps with the base image. To facilitate comprehension, we provide a flowchart of our proposed image method in Figure 1. In the following two subsections, the proposed image filter and image-fusion method based on this filter are elaborated, respectively.

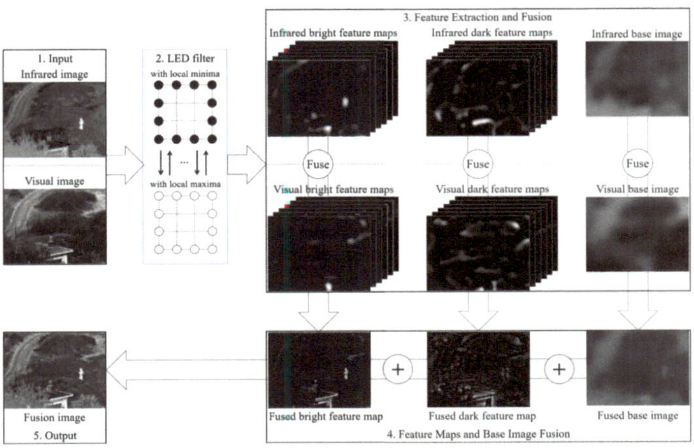

Figure 1. Flowchart of our proposed infrared and visual image-fusion method. Please note that, in order to visualize the dark feature maps (features with negative values), the absolute dark feature maps are presented in this figure. Moreover, the term "LED filter" is short for our proposed local-extrema-driven image filter.

2.1. Local-Extrema-Driven Image Filter

Within an image, bright features, as exemplified by the bright person in the infrared image shown in Figure 1, and dark features, represented by the dark window in the same infrared image, are commonly present. Employing a strategy of smoothing the image and subsequently subtracting the smoothed version from the original has proven to be an effective method for isolating the image's bright and dark features [7,15]. Ideally, the smoothed image should eliminate the bright spots and fill the dark holes in the original, facilitating the extraction of both bright and dark features from the resultant difference image between the original and the smoothed version. To fulfill this objective, our local-extrema-driven image filter is constructed as follows.

Initially, we reconstruct the input image using its local minima, expressed as:

$$F' = H * I_{min}, \tag{1}$$

where $*$ represents the convolution operator. I_{min} represents the local minimum image derived from the input image I, calculated according to Equation (2). Additionally, H represents the convolution kernel, the format of which is defined in Equation (3).

$$I_{min} = imerode(I, se), \tag{2}$$

where *imerode* represents the morphological erosion operator and *se* denotes a disk-shaped structural element with a radius r. Consequently, I_{min} signifies the local minimum image of I with respect to a distance of r.

$$\begin{bmatrix} 1 & 1 & \cdots & 1 & 1 \\ 1 & 0 & \cdots & 0 & 1 \\ \vdots & \vdots & \ddots & \vdots & \vdots \\ 1 & 0 & \cdots & 0 & 1 \\ 1 & 1 & \cdots & 1 & 1 \end{bmatrix}_{(2r+1)\times(2r+1)}. \tag{3}$$

In this manner, every pixel in the original input image is reconstructed based on the local minima of its neighboring pixels, effectively suppressing the bright features present in the original image. Subsequently, the initially filtered image F' undergoes further reconstruction, this time utilizing its local maxima, as follows:

$$F = H * F'_{max}, \tag{4}$$

where F'_{max} represents the local maximum image derived from the initially filtered image F' and can be computed using Equation (5).

$$F'_{max} = imdilate(F', se), \tag{5}$$

where *imdilate* signifies the morphological dilation operator. Consequently, F'_{max} represents the local maximum image of F' with a distance of r.

In contrast to Equation (1), Equation (4) achieves additional removal of salient dark features from the filtered image by reconstructing each pixel in F' based on its local maxima.

To streamline the presentation of the upcoming image-fusion method, we introduce $lextremefilter(\cdot)$ as the function of our devised local-extrema-driven image filter, composed of Equations (1) and (4). The process of smoothing an image with our local-extrema-driven image filter can be succinctly expressed as:

$$F = lextremefilter(I, r), \tag{6}$$

where r denotes the size of the structuring element in Equations (2) and (5).

As is evident, an image comprises both bright and dark features, illustrated by the bright person and the dark window corner in Figure 1. Through the iterative reconstruction of the input image based on the local minima and local maxima, salient bright and dark features can be effectively eliminated, resulting in a well-smoothed image (see the filtered images in the last column of Figure 1). Subsequently, the salient features of the input image can be derived by subtracting the filtered image F from the input image I as per Equation (7). The positive part B captures the bright features (refer to the first column of the Feature Extraction and Fusion Module in Figure 1), while the negative part D corresponds to the dark features (refer to the second column of the Feature Extraction and Fusion Module in Figure 1).

$$\begin{cases} B = \max(I - F, 0) \\ D = \min(I - F, 0) \end{cases}, \tag{7}$$

where B and D represent the bright and dark feature map of I, respectively.

Furthermore, the local-extrema-driven image filter can be scaled to multiple levels through successive applications of the filter driven by local minima and local maxima on the input image I, as outlined in Equation (8).

$$F_i = lextremefilter\left(F_{(i-1)}, r_i\right), \tag{8}$$

where i represents the current scale of the image filter, with i incrementing from 1 to n sequentially. F_i denotes the filtered image at the ith scale, and notably, F_0 corresponds to the original input image I. The parameter r_i denotes the size of the structuring element and convolution kernel at the ith scale. In this study, we designate $r_i = i$ to progressively augment the smoothing degree of our proposed image filter.

Consequently, multiple scales of bright and dark feature maps can be concurrently extracted from the continuously filtered images by

$$\begin{cases} B_i = \max(F_{i-1} - F_i, 0) \\ D_i = \min(F_{i-1} - F_i, 0) \end{cases}. \tag{9}$$

Finally, the last scale of the filtered image is taken as the base image for I:

$$I_{base} = F_n, \tag{10}$$

where n represents the scale number.

2.2. Local-Extrema-Driven Image Fusion

In this study, our objective is to fuse a visual image denoted as I^{vis} and an infrared image denoted as I^{inf}. Utilizing the feature-extraction method outlined in the preceding subsection, multi-scale bright feature maps (represented by B_i^{vis} and B_i^{inf}) and dark feature maps (indicated by D_i^{vis} and D_i^{inf}) are effectively extracted from I^{vis} and I^{inf}. Concurrently, we obtain their respective base images denoted as I_{base}^{vis} and I_{base}^{inf}. The subsequent contents delineate the detailed procedures for fusing a visual image and an infrared image.

Considering that high-frequency bright features usually correspond to sharp and bright features in the image, we combine each scale of bright feature maps from the infrared and visual images by choosing their elementwise maximum values. Likewise, for each scale of dark feature maps, we fuse them using their elementwise minimum values. The mathematical expressions for fusing high-frequency bright and dark features are as follows:

$$\begin{cases} B_i^{fuse} = \max\left(B_i^{vis}, B_i^{inf}\right) \\ D_i^{fuse} = \min\left(D_i^{vis}, D_i^{inf}\right) \end{cases}. \tag{11}$$

Furthermore, the elementwise-fused bright and dark feature maps are individually integrated into single feature maps. As feature maps may contain varied quantities of

features across different scales, potentially leading to redundancy, this study employs a two-step process. Initially, the strengths of these feature maps are dynamically adjusted based on their information content. Subsequently, they are summed together. This adaptation relies on an entropy-based weighting strategy [32], enhancing feature maps with a substantial amount of information while diminishing those with relatively less information. The detailed aggregation of the fused multiple scales of bright and dark feature maps is outlined below.

$$\begin{cases} B^{fuse} = \sum\limits_{i=1}^{n} w_{b,i} \times B_i^{fuse} \\ D^{fuse} = \sum\limits_{i=1}^{n} w_{d,i} \times D_i^{fuse} \end{cases}, \tag{12}$$

where $w_{b,i}$ and $w_{d,i}$ denote the weights of the bright feature map and dark feature map at the ith scale, respectively, and can be calculated as follows:

$$\begin{cases} w_{b,i} = \dfrac{e_{b,i}}{\frac{1}{n}\sum\limits_{j=1}^{n} e_{b,j}} \\ w_{d,i} = \dfrac{e_{d,i}}{\frac{1}{n}\sum\limits_{j=1}^{n} e_{d,j}} \end{cases}, \tag{13}$$

where $e_{b,i}$ and $e_{d,i}$ represent the entropy of B_i^{fuse} and $\left(-D_i^{fuse}\right)$, respectively. This exploited feature aggregation strategy ensures that the fused single bright feature map and dark feature map not only retain the salient high-frequency features, but also eliminate redundant information.

Concerning the low-frequency base images, they commonly contain large-scale structural features, and the intensity distribution of the fused base image plays a crucial role in determining the final appearance of the fusion image. Therefore, in this study, we employed a structural similarity- and intensity-based scheme to fuse the base images of infrared and visual images. Specifically, we initiate the process by averaging the two base images elementwise, yielding an initial base image as follows:

$$I'^{fuse}_{base} = 0.5 \times \left(I^{vis}_{base} + I^{inf}_{base}\right). \tag{14}$$

Subsequently, a provisional fusion image I'^{fuse} is created by combining the fused bright feature map, fused dark feature map, and initially fused base image as follows:

$$I'^{fuse} = B^{fuse} + D^{fuse} + I'^{fuse}_{base}. \tag{15}$$

Afterward, the structural-similarity maps between each base image and the initially fused image are computed, respectively.

$$\begin{cases} S^{vis} = SSIM\left(I'^{fuse}, I^{vis}_{base}\right) \\ S^{inf} = SSIM\left(I'^{fuse}, I^{inf}_{base}\right) \end{cases}, \tag{16}$$

where $SSIM(A, B)$ calculates the structural similarity between image A and image B using the method outlined in [33]. Afterward, we generate a structural similarity-based weight map for fusing base images as follows:

$$w^{struct}_{base,vis} = S^{vis} / \left(S^{vis} + S^{inf}\right). \tag{17}$$

Moreover, the grayscale intensities are closely linked to the appearance of the fusion image. Therefore, we also incorporate an intensity-based weight for fusing base images, which can be computed as follows:

$$w_{base,vis}^{intens} = e^{I^{vis}/\left(I^{vis}+I^{inf}\right)}. \tag{18}$$

To balance the two kinds of weights, we fuse them by

$$w_{base,vis} = G * \left[w_{base,vis}^{struct} \times \left(w_{base,vis}^{intens}\right)^{\alpha}\right], \tag{19}$$

where α serves as a parameter to balance these two weights. G represents a Gaussian kernel employed to smooth the weight distribution map.

Then, the two base images of the infrared and visual images can be fused as follows:

$$I_{base}^{fuse} = w_{base,vis} \times I_{base}^{vis} + (1 - w_{base,vis}) \times I_{base}^{inf}. \tag{20}$$

As depicted in Figure 2, the implementation of our structural similarity- and intensity-based fusion scheme results in a fused base image that not only retains significant large-scale structural features from both base images, but also achieves an advantageous intensity distribution, thereby enhancing visual perception in the final fusion image. Specifically, when compared to exclusively utilizing the structural similarity-based fusion scheme (see Figure 2f), our comprehensive fusion scheme produces a fused base image (see Figure 2h) with a more suitable intensity distribution. Similarly, in contrast to relying solely on an intensity-based fusion scheme (see Figure 2g), our comprehensive fusion approach retains a greater number of structural features in the fused base image (see Figure 2h). Furthermore, compared to simply averaging the two base images (see Figure 2e), our complete base image-fusion scheme generates an intensity-distributed fused base image (see Figure 2h) while preserving richer textures. Additionally, by comparing the fusion images generated from the fused base images in Figure 2e,f, it effectively validates the efficacy of our base image fusion scheme to a significant extent.

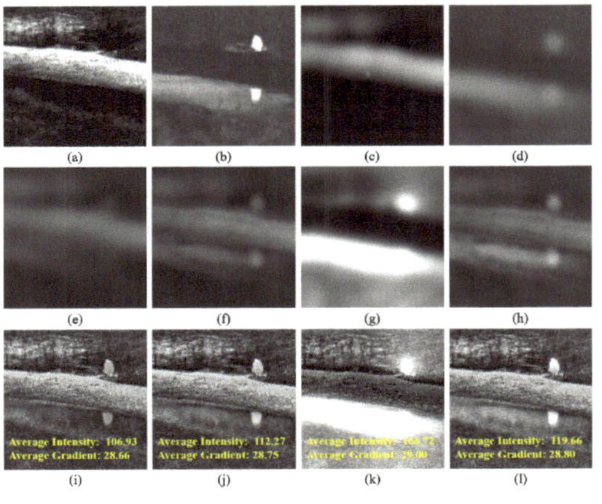

Figure 2. Demonstration example of our base image fusion scheme. (**a**,**b**) present the original visual and infrared images, respectively. (**c**,**d**) depict the base images corresponding to the infrared and visual inputs, respectively. (**e**–**h**) exhibit the resulting fused base images derived from the direct average scheme, structural similarity-based fusion, intensity-based fusion, and our novel structural similarity- and intensity-based fusion approach, respectively. (**i**–**l**) showcase the fusion images generated by combining (**e**–**h**) with our fused high-frequency bright and dark feature maps, respectively. The yellow text in (**i**–**l**) highlights the average grayscale intensity and average absolute gradient of the corresponding fused image.

Finally, our proposed method generates the fusion image by combining the fused bright feature map, dark feature map, and base image together, as expressed in Equation (21). Through this process, our fused image not only retains fundamental information from the infrared and visual images, but also effectively highlights the prominent sharp features present in the infrared and visual images.

$$I^{fuse} = B^{fuse} + D^{fuse} + I^{fuse}_{base}. \tag{21}$$

2.3. Parameter Settings

The proposed method involves two parameters: the scale number n and the parameter α for balancing $w^{struct}_{base,vis}$ and $w^{intens}_{base,vis}$. In this study, we employed the grid search method to find the optimal pair of n (ranging from 1 to 10 in increments of 1) and α (ranging from 0.05 to 1 in increments of 0.05) that maximizes the multi-scale structural similarity metric (MSSIM) [34]. The results show that the MSSIM increases with the increase of the scale number, but the running time of our method increases simultaneously. So, we first set the scale number n to six, so that the performance and time cost of our method will be balanced. Afterwards, when $n = 6$, MSSIM is maximized by setting $\alpha = 0.35$. Therefore, throughout this study, consistent parameter settings ($n = 6$ and $\alpha = 0.35$) were used, and the experimental results in the following section validate the efficacy of these chosen parameters for infrared and visual image fusion.

3. Experimental Results and Discussion

To showcase the merits of our novel infrared and visual image-fusion method, we conducted a thorough comparative analysis against eleven state-of-the-art image-fusion techniques. This evaluation was performed on a widely recognized dataset for infrared and visual images. For comprehensive insights into the experimental settings, results, and discussions, please refer to the subsequent subsections.

3.1. Experimental Settings

The experimental setup for this study is summarized as follows. Initially, we assembled twenty pairs of widely used infrared and visual images from the TNO dataset [35]. Subsequently, we selected eleven state-of-the-art image-fusion methods for comparison. These methods include the guided-filter-based image method (GFF) [36], the hybrid multi-scale-decomposition-based image-fusion method (HMSD) [18], the Laplacian pyramid- and sparse-representation-based image-fusion method (LPSR) [25], the Gaussian of differences-based image-fusion method (GDPSQCV) [37], the relative total variation-decomposition-based image-fusion method (RTVD) [38], the parameter-adaptive unit-linking dual-channel PCNN-based image-fusion method (PAULDCPCNN) [39], the GAN-based image-fusion method (FusionGAN) [16], the unified deep learning-based image-fusion method (U2Fusion) [40], the semantic-aware image-fusion method (SeAFusion) [28], and the representation learning-guided image-fusion method (LRR) [29]. For simplicity, we refer to our proposed local-extrema-driven filter-based image-fusion method as LEDIF. Additionally, we conducted comparisons by excluding the utilization of the structural similarity- and intensity-based base image fusion scheme in our method (denoted as LEDIF_0) to evaluate the effectiveness of this scheme.

Afterwards, the thirteen methods underwent both qualitative and quantitative evaluation. In particular, the qualitative assessment involved a visual comparison of the fusion results across the different methods. For the quantitative evaluation, we employed nine metrics to objectively gauge the quality of the fusion images produced by the various approaches. These metrics include the spatial frequency (SF) [8,41], the average absolute gradient (AG) [42], the linear index of fuzziness (LIF) [43], the blind/referenceless image spatial quality evaluator (BRISQUE) [44], the visual information fidelity (VIF) [45], the multi-scale structural similarity index metric (MSSIM) [34], the edge-dependent structural similarity index metric (ESSIM) [46], the edge-similarity-based metric (QABF) [44] and

the sum of correlation differences metric (SCD) [47]. The superior performance of the corresponding image-fusion method is indicated by smaller values for the BRISQUE metric and larger values for the other eight metrics.

Among these metrics, the SF, AG, and LIF quantify the amount of details preserved in the fusion image, while BRISQUE quantifies the clarity and distortion level of the fusion image. The VIF measures the information fidelity of the fusion image concerning the input images, while the MSSIM, ESSIM, QABF, and SCD gauge the structural similarity between the fusion image and the input images from various perspectives. These metrics collectively provide a comprehensive evaluation framework, capturing different aspects of fusion image quality and fidelity.

3.2. *Qualitative Evaluation Results*

In this subsection, we qualitatively assess the thirteen image-fusion methods by visually comparing their fusion results. To offer visual insight into the quality and effectiveness of each fusion method, we present five comparison examples showcasing the fusion outputs of all thirteen methods in Figures 3–7, respectively.

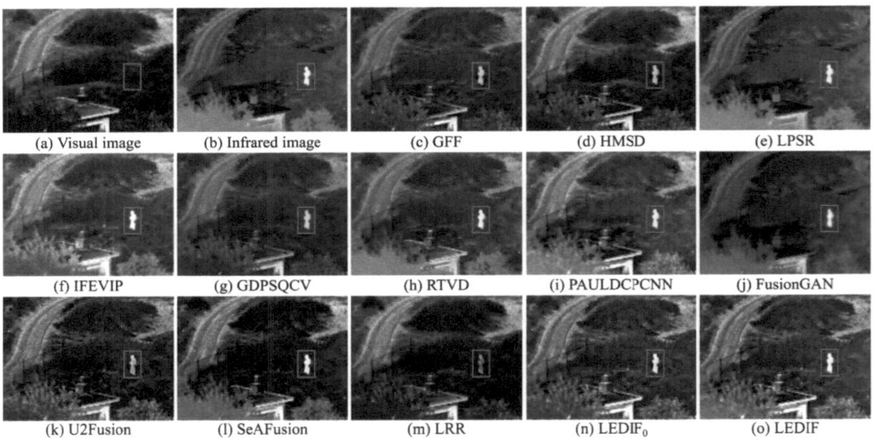

Figure 3. First comparison example of the thirteen image-fusion methods.

In Figure 3, both the infrared and visual images were captured under normal lighting conditions. Notably, a person was standing near the fence, appearing almost invisible in the visual image while prominently visible in the infrared counterpart. Consequently, an ideal fusion image for this image pair should seamlessly integrate the bright person and distinct spots from the infrared image with the intricate textures of the trees and fence from the visual image. It is evident that the areas corresponding to the person in the fusion images produced by the GFF, HMSD, GDPSQCV, U2Fusion, and LRR in (c), (d), (g), (k), and (m) appear dimmer compared to those in other fusion images. Similarly, the tree regions in the fusion images generated by the LPSR, IFEVIP, GDPSQCV, RTVD, and FusionGAN in (e), (f), (g), (h), and (j) exhibit relatively smoother textures than those in other fusion images. Notably, the intensities in the fusion image of PAULDCPCNN, as depicted in (i), are not evenly distributed. Additionally, the background of the fusion image produced by SeAFusion, illustrated in (l), appears noticeably darker compared to others. Finally, (n) and (o) demonstrate that our two fusion images exhibit the most visually appealing results among all fusion images, with the fusion image generated by our complete method in (o) being slightly brighter than that produced by our method without leveraging the proposed base image fusion scheme.

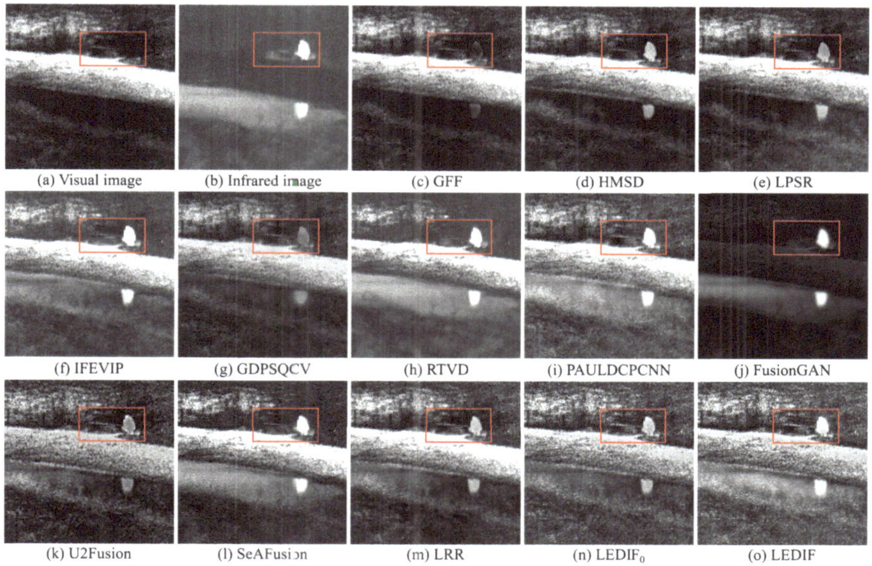

Figure 4. Second comparison example of the thirteen image-fusion methods.

In Figure 4, the infrared and visual images were captured under low-light conditions. The optimal fusion image for this pair should seamlessly integrate distinctive bright features, particularly the two person regions in the infrared image, and the bright textures of the visual image, encompassing the grass and trees, along with the darker features represented by the bench. Among the fusion images depicted in (c), (e), (g), (k), (m), and (n), generated by the GFF, LPSR, GDPSQCV, U2Fusion, LRR, and our LEDIF$_0$, respectively, the intensities of the person regions are notably lower than those in (b), indicating unsatisfactory results in this particular case. Furthermore, the contrast in the fusion results of GDPSQCV and U2Fusion in (g) and (k) is relatively diminished compared to other methods' fusion images. The fusion image of RTVD in (h) is over-exposed, resulting in the loss of many textural details, particularly around the bench. Conversely, the fusion image of FusionGAN in (j) fails to integrate most critical textures of the visual image in (a). While the HMSD, IFEVIP, SeAFusion, and our LEDIF in (d), (f), (l), and (o), respectively, exhibit the most visually appealing results among all fusion images, there are notable observations. IFEVIP's fusion image in (f) appears slightly over-exposed, and the bright infrared features of the HMSD's fusion image in (d) are relatively lower than other methods' results. Additionally, both the IFEVIP and SeAFusion sacrifice some textural details in their fusion images in (f) and (l). In summary, the fusion image generated by our LEDIF in (o) attains the highest visual quality, affirming the effectiveness of our structural similarity- and intensity-based base image fusion scheme in enhancing the overall visual appearance of the final fusion images.

In Figure 5, both the infrared and visual images were captured under normal lighting conditions. The ideal fusion image should effectively combine the various scales of salient bright features from the infrared image with the diverse bright and dark features present in the visual image. It is evident from (c), (f), (g), and (m) that the GFF, IFEVIP, GDPSQCV, and LRR struggle to integrate most of the bright features from the infrared image into their fusion images, as observed in the building area within the red bounding boxes of each image. Among these methods, FusionGAN's fusion image in (f) displays a considerable loss of textures from the visual image, resulting in the poorest visual effect among all thirteen image-fusion methods. U2Fusion manages to integrate the salient features of both the infrared and visual images into its fusion image, as demonstrated in (k). However, the contrast of (k) is relatively low compared to that of the infrared image, the visual image, and

most other fusion images. (l) highlights that the building area of the fusion image generated by SeAFusion is over-exposed, leading to a loss of some building details. Ultimately, the fusion images produced by PAULDCPCNN, our $LEDIF_0$, and our LEDIF in (i), (n), and (o), respectively, exhibit the most favorable visual effects among all fusion images.

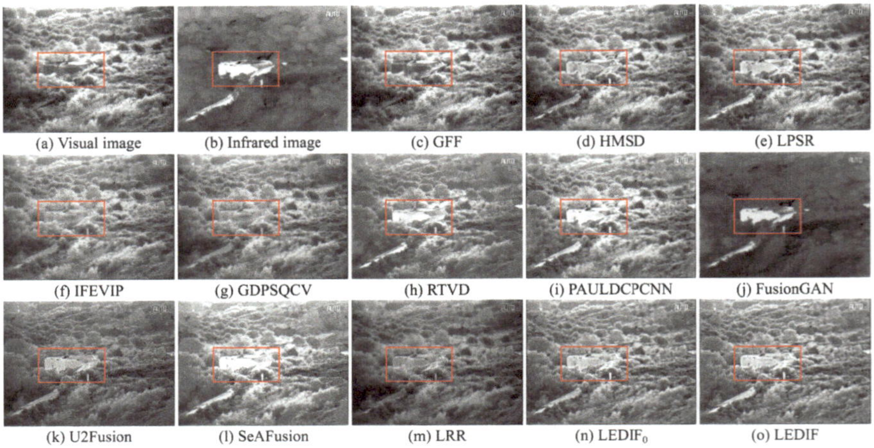

Figure 5. Third comparison example of the thirteen image-fusion methods.

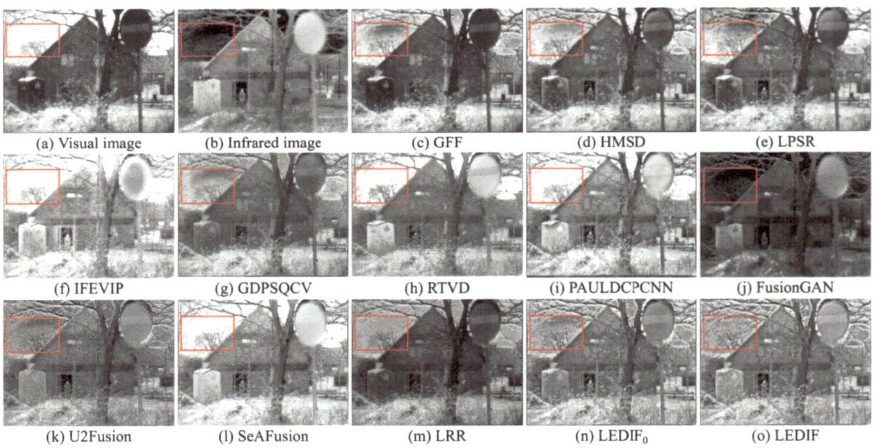

Figure 6. Fourth comparison example of the thirteen image-fusion methods.

In Figure 6, the sky area in the visual image appears over-exposed, necessitating an ideal fusion image for this image pair to accentuate the bright tree features surrounding the sky area from the infrared image. In (c), the GFF demonstrates limitations in incorporating the bright person from the infrared image into its fusion image. While the HMSD and LPSR effectively blend the infrared and visual images in most regions, they struggle to integrate specific bright tree branches from the infrared image, as highlighted in the red bounding boxes of (d) and (e). Moving on to (f), (h), (i), and (l), the IFEVIP, RTVD, PAULDCPCNN, and SeAFusion encounter challenges in including the bright tree branches from the infrared image in their fusion images due to the over-exposed sky area in the visual image. Conversely, the fusion images from the GDPSQCV, FusionGAN, U2Fusion, and LRR in (g), (j), (k), and (m) exhibit the loss of textural details from the visual image, with relatively low contrast compared to other methods. Furthermore, (n) and (o) illustrate that the fusion images generated by the PAULDCPCNN, our $LEDIF_0$, and LEDIF in (i), (n)

and (o) successfully integrate the bright tree branches from the infrared image, displaying good contrast compared to the fusion images from the other methods. Notably, the fusion image from our LEDIF is slightly brighter than that of our LEDIF$_0$, indicating a slight improvement in the visual effect of the fusion image facilitated by the proposed base image fusion scheme in this case.

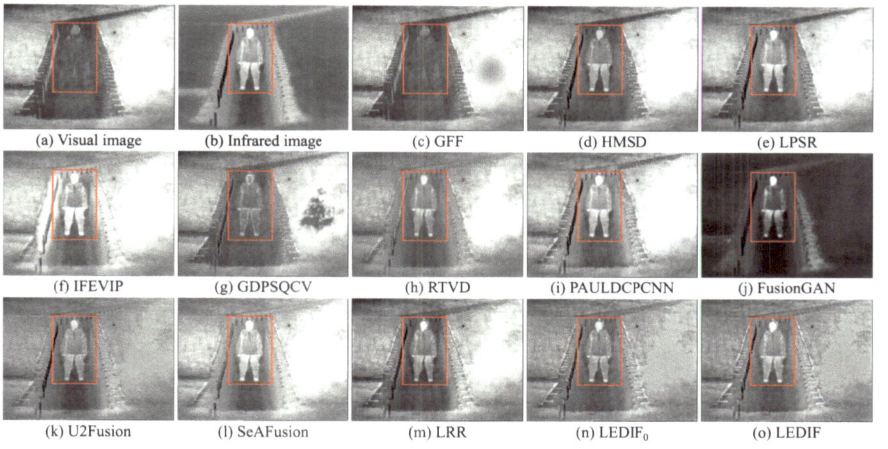

Figure 7. Fifth comparison example of the thirteen image-fusion methods.

In Figure 7, both the infrared and visual images were captured under low-light conditions. The primary goal for this pair was to generate an optimal fusion image that effectively integrates the facial features depicted in the visual image in (a) with the bright person captured in the infrared image in (b). (c) reveals that the GFF fails to effectively integrate the bright person features from the infrared image into its fusion image. Notably, the fusion images of the HMSD, U2Fusion, and our LEDIF$_0$ in (d), (k), and (n), respectively, exhibit relatively low contrast compared to other fusion images. Furthermore, (f), (h), and (i) demonstrate that the fusion images of the IFEVIP, RTVD, and SeAFusion appear over-exposed, resulting in a failure to integrate most facial features from the visual image. In (g), the fusion image generated by the GDPSQCV showcases a significant loss of the person area, while (j) indicates that most background areas of FusionGAN's fusion image fail to integrate from the visual image. Overall, in this scenario, fusion the images obtained from the LPSR, PAULDCPCNN, LRR, and our LEDIF in (e), (i), (m), and (o), respectively, achieve the most favorable visual effects.

The qualitative comparisons across the five examples strongly affirm the efficacy of our proposed method in seamlessly integrating the prominent bright and dark features present in both infrared and visual images, resulting in comprehensive fusion images. Notably, our method consistently performed comparably or even surpassed eleven state-of-the-art image fusion approaches, as evidenced by superior visual observations. Additionally, the visual comparison examples further validate the effectiveness of our proposed base image fusion scheme in enhancing the visual quality of the fusion images.

3.3. Quantitative Evaluation Results

As widely acknowledged, qualitative evaluation heavily depends on subjective observation, potentially resulting in inaccuracies and demanding significant effort. To ensure an objective comparison of the performance of various methods, we additionally utilized nine quantitative metrics, as outlined at the beginning of this section. Subsequently, we provide detailed quantitative evaluation results and discussions.

Table 1 presents the quantitative metrics computed for the thirteen image-fusion methods. Notably, in Table 1, the best, second-best, and third-best values are highlighted in

red, green, and blue, respectively, while the integer in the subscript of each metric value indicates the performance rank among all thirteen image-fusion methods. Additionally, the individual metric values for each fusion image generated by each method are further illustrated in Figure 8.

Table 1. Quantitative evaluation results of different image-fusion methods on the datasets used.

Methods	SF	AG	LIF	BRISQUE	VIF	MSSIM	ESSIM	QABF	SCD
GFF	10.6666_8	9.1331_8	0.4448_9	20.7646_8	0.2521_{11}	0.8558_{10}	0.8418_3	0.6218_1	1.2982_{12}
HMSD	11.7816_4	10.2579_5	0.4493_8	38.3982_{13}	0.3976_7	0.9324_4	0.8319_5	0.5330_4	1.5675_8
LPSR	11.2857_7	9.8681_7	0.4333_{10}	19.0110_4	0.4065_6	0.9289_5	0.8428_2	0.5931_2	1.4170_{12}
IFEVIP	9.5708_9	8.3164_{10}	0.5502_2	21.9260_{10}	0.3231_9	0.8482_{11}	0.7740_{11}	0.4981_8	1.6437_4
GDPSQCV	8.2206_{12}	6.9785_{12}	0.5343_4	20.1511_6	0.2766_{10}	0.8929_7	0.8557_1	0.5078_5	1.5771_7
RTVD	8.4358_{11}	7.2621_{11}	0.5433_3	19.2725_5	0.2122_{12}	0.7893_{12}	0.7878_8	0.4609_{10}	1.5386_9
PAULDCPCNN	11.3139_6	9.9305_6	0.5565_1	18.9329_3	0.4707_4	0.9412_2	0.8330_4	0.5409_3	1.6402_5
FusionGAN	5.7691_{13}	5.0467_{13}	0.2478_{13}	25.4100_{12}	0.1831_{13}	0.7308_{13}	0.6647_{13}	0.2196_{13}	1.0213_{13}
U2Fusion	11.3629_5	10.6915_4	0.4095_{11}	16.9311_2	0.5758_1	0.9250_6	0.7809_{10}	0.4241_{11}	1.6326_6
SeAFusion	11.9697_3	10.7101_3	0.4536_7	10.8181_1	0.4367_5	0.8863_8	0.7862_9	0.4761_9	1.6687_2
LRR	9.4230_{10}	8.4783_9	0.3652_{12}	23.2087_{11}	0.3552_8	0.8709_9	0.7429_{12}	0.3735_{12}	1.4358_{10}
$LEDIF_0$	14.1944_2	12.6043_2	0.4898_6	21.2551_9	0.5468_3	0.9478_1	0.8176_6	0.5015_6	1.6739_1
LEDIF	14.2382_1	12.6777_1	0.5165_5	20.7572_7	0.5661_2	0.9375_3	0.8141_7	0.4986_7	1.6484_3

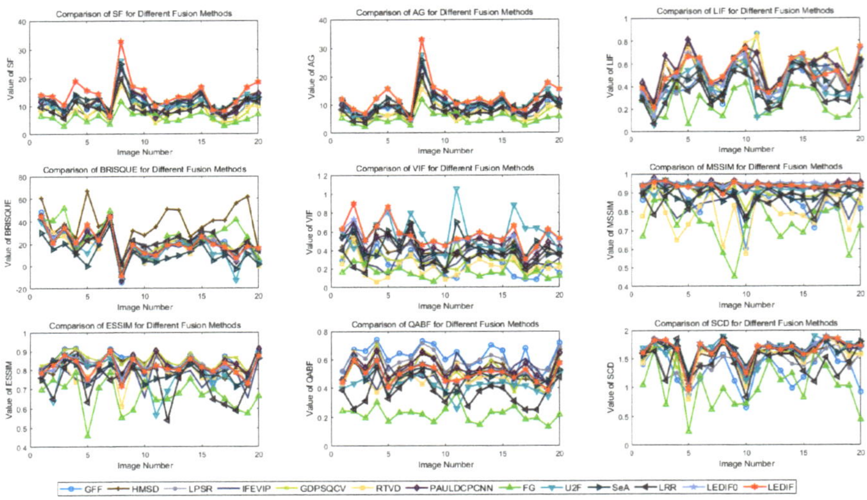

Figure 8. Visual comparison of the quantitative evaluation results.

The analysis of the metrics reveals that our proposed method achieved top performance on two metrics, the SF and AG, while securing the second-best performance on the VIF metric and the third-best performance on the MSSIM and SCD metrics. Furthermore, our method ranked in the top 50% for the other four metrics, including the LIF, BRISQUE, ESSIM, and QABF. Specifically, our method stands out with the largest SF and AG values and the fifth-largest LIF value, indicating superior preservation of textural details compared to the other twelve comparison methods. Regarding BRISQUE, our method ranked seventh, suggesting relatively high-quality image generation with clarity and information retention.

Additionally, our method ranked second on the VIF, indicating high visual information fidelity with respect to the original visual images. In terms of the MSSIM, our $LEDIF_0$ and LEDIF ranked first and third, respectively, on this metric. The MSSIM, being a multi-scale structural similarity measure, is often more robust than other similarity measures like the ESSIM and QABF, where our method ranked seventh. These structural similarity-based metrics validate our method's ability to preserve relatively more structural features from the input infrared and visual images. Similarly, our method ranked third on the SCD metric,

indicating close correlation between the fusion images and the original infrared and visual images, thereby preserving more structural features.

Furthermore, comparing the metric values of our $LEDIF_0$ and LEDIF reveals that the LEDIF preserved more details from the input images in its fusion images compared to $LEDIF_0$, as inferred from the SF, AG, and LIF metrics. The LEDIF also generated fusion images of higher visual quality and fidelity, as indicated by the BRISQUE and VIF metrics. However, incorporating the base image fusion scheme resulted in a slight loss of structural features compared to $LEDIF_0$, evident from metrics like the MSSIM, ESSIM, QABF, and SCD.

The consistency between the average metrics and individual values is further validated by the individual metric values plotted in Figure 8. This consistency reinforces the effectiveness and significance of the quantitative ranks discussed above.

3.4. Further Discussion

When compared to existing or related methods, in particular the approach presented in [31], our method stands out significantly. While both methods rely on a local image filter, the method in [31] is constructed based on the original Bezier interpolation operation, which differs from our construction method. Additionally, the cited method does not address the enhancement of visual quality in the final fusion images. In contrast, our method specifically tackles this issue, particularly addressing the challenge of dim visual effects in fusion images by introducing a novel intensity and structural similarity-based base image fusion scheme. Through both qualitative and quantitative analyses, our newly proposed local-extrema-filter-based image-fusion method and base image fusion scheme prove to be effective for infrared and visual image fusion tasks, performing comparably to or even better than eleven state-of-the-art image-fusion methods.

Furthermore, the efficiency of our image-fusion method is relatively high, requiring approximately 0.21 s to fuse a pair of infrared and visual images. Nevertheless, there exists substantial potential for further efficiency enhancements through the utilization of parallel computing techniques or the optimization of computational operations. Therefore, there is great potential to apply our proposed method to real practical scenarios.

To comprehensively evaluate the generalization ability of our method, we first conducted experiments using the VIFB dataset [48]. The results, depicted in Figure 9, showcase five representative image fusion examples. These examples not only demonstrate our method's capability to fuse images captured under varying lighting conditions, including both daylight and nighttime scenarios, but also its effectiveness in seamlessly integrating salient infrared features with over-exposed visual images.

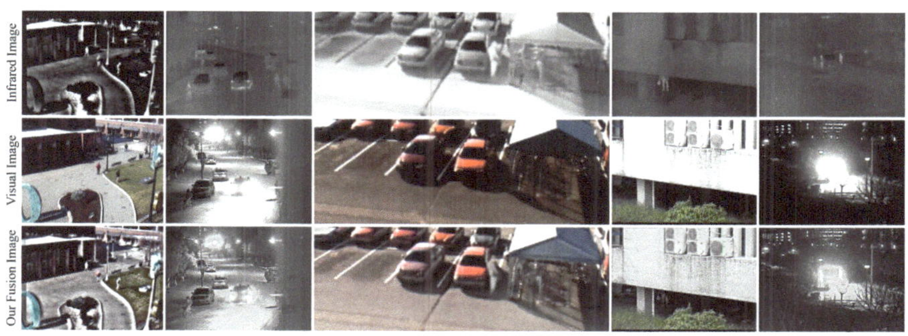

Figure 9. More results of our method for fusing images from other infrared and visual image fusion dataset (i.e., the VIFB dataset [48]).

Expanding beyond infrared–visual fusion, our method was applied to fuse images from diverse modalities, including multi-focus images, multi-exposure images, and multi-modal medical images. As depicted in Figure 10, our approach adeptly integrates salient

features from each pair of source images into the resulting fusion images. This versatility underscores the adaptability and robustness of our method across a wide range of image modalities.

Figure 10. More results of our method for fusing multi-focus, multi-exposure, and multi-modal medical images.

In summary, the positive fusion results observed in both Figures 9 and 10 serve as compelling validation of the robust generalization ability of our method. Its efficient processing time, combined with its effectiveness across varied modalities, positions our approach as a promising solution for real-world image-fusion applications. Through ongoing research and refinement, we remain committed to further advancing the capabilities and applicability of our method in diverse image-fusion scenarios.

Considering both qualitative and quantitative evaluations, our image-fusion method consistently demonstrates performance on par with or superior to the eleven state-of-the-art image-fusion methods.

4. Conclusions

In this study, we have introduced a highly effective local-extrema-driven image filter, meticulously designed for the fusion of infrared and visual images. The proposed filter showcases remarkable capabilities in smoothing images, thereby facilitating the extraction of salient bright and dark features. Through iterative application of this filter, our approach excels at extracting multiple scales of salient textural features from both infrared and visual images. These distinctive features are seamlessly integrated into a single, informative fusion image through two appropriate fusion strategies. Notably, our innovative base image fusion scheme, rooted in structure similarity and intensity, significantly enhances the visual effect of the resulting fusion images.

While our method demonstrates competitive performance against state-of-the-art techniques, several avenues for further research and improvement are apparent. Primarily, the current reliance on grid searching for parameter optimization may not yield the most optimal settings for the infrared and visual image fusion task. To address this limitation, we intend to explore advanced optimization techniques to fine-tune these parameters, ensuring maximal performance and adaptability across diverse datasets and scenarios.

Furthermore, although our method excels in enhancing low-level image features, its current configuration lacks optimization for high-level vision tasks such as image segmentation, object detection, and object tracking. Recognizing the significance of seamlessly integrating these capabilities, our future research endeavors will focus on evolving our framework into a deep learning-driven architecture. By harnessing the power of deep learning, we aim to imbue our method with the capacity to not only preserve critical image features during fusion, but also to facilitate robust performance in subsequent high-level vision tasks, thereby enhancing its utility and applicability in real-world surveillance systems.

Moreover, while our base image fusion scheme yields visually appealing results, we acknowledge its marginal impact on certain quantitative metrics. To address this, we plan to explore novel fusion strategies and evaluation metrics that better capture the holistic quality and utility of fusion images. By refining our approach in this manner, we aim to bridge the gap between subjective visual appeal and objective performance metrics, thereby ensuring a comprehensive assessment of fusion image quality.

In summary, while our method presents a significant advancement in the field of image fusion, we recognize the importance of continuous refinement and adaptation to meet the evolving demands of contemporary surveillance systems. Through targeted research efforts aimed at parameter optimization, the integration of high-level vision tasks, and the refinement of fusion strategies, we are committed to further enhancing the capabilities and applicability of our approach for diverse real-world scenarios.

Author Contributions: Conceptualization, Y.Z. and L.Z.; methodology, W.X. and Y.Z.; investigation, W.X. and J.S.; writing—original draft, W.X., J.S. and Y.Z.; writing—review and editing, L.Z. and Y.Z.; funding acquisition, Y.Z.; resources, Y.Z. and L.Z.; supervision, Y.Z. All authors have read and agreed to the published version of the manuscript.

Funding: This study was supported in part by the National Natural Science Foundation of China under Grant Nos. 62132002 and 62171017.

Institutional Review Board Statement: Not applicable.

Informed Consent Statement: Not applicable.

Data Availability Statement: The code of our image-fusion method and dataset used in this study will be released at https://github.com/uzeful/LEDIF.

Acknowledgments: We sincerely thank the anonymous reviewers for their detailed and constructive comments, which greatly contributed to improving the quality of this paper.

Conflicts of Interest: The authors declare no conflicts of interest.

References

1. Li, S.; Kang, X.; Fang, L.; Hu, J.; Yin, H. Pixel-level image fusion: A survey of the state of the art. *Inf. Fusion* **2017**, *33*, 100–112. [CrossRef]
2. Liu, Y.; Chen, X.; Wang, Z.; Wang, Z.J.; Ward, R.K.; Wang, X. Deep learning for pixel-level image fusion: Recent advances and future prospects. *Inf. Fusion* **2018**, *42*, 158–173. [CrossRef]
3. Zhang, Y.; Liu, Y.; Sun, P.; Yan, H.; Zhao, X.; Zhang, L. IFCNN: A general image fusion framework based on convolutional neural network. *Inf. Fusion* **2020**, *54*, 99–118. [CrossRef]
4. Xu, Z. Medical image fusion using multi-level local extrema. *Inf. Fusion* **2014**, *19*, 38–48. [CrossRef]
5. Liu, Y.; Chen, X.; Cheng, J.; Peng, H. A medical image fusion method based on convolutional neural networks. In Proceedings of the 2017 20th International Conference on Information Fusion (Fusion), Xi'an, China, 10–13 July 2017; pp. 1–7.
6. Wang, K.; Zheng, M.; Wei, H.; Qi, G.; Li, Y. Multi-modality medical image fusion using convolutional neural network and contrast pyramid. *Sensors* **2020**, *20*, 2169. [CrossRef] [PubMed]
7. Zhang, Y.; Xiang, W.; Zhang, S.; Shen, J.; Wei, R.; Bai, X.; Zhang, L.; Zhang, Q. Local extreme map guided multi-modal brain image fusion. *Front. Neurosci.* **2022**, *16*, 1055451. [CrossRef] [PubMed]
8. Huang, W.; Jing, Z. Evaluation of focus measures in multi-focus image fusion. *Pattern Recognit. Lett.* **2007**, *28*, 493–500. [CrossRef]
9. Bai, X.; Zhang, Y.; Zhou, F.; Xue, B. Quadtree-based multi-focus image fusion using a weighted focus-measure. *Inf. Fusion* **2015**, *22*, 105–118. [CrossRef]
10. Zhang, Q.; Levine, M.D. Robust multi-mocus image fusion using multi-task sparse representation and spatial context. *IEEE Trans. Image Process.* **2016**, *26*, 2045–2058. [CrossRef] [PubMed]
11. Zhang, Y.; Bai, X.; Wang, T. Boundary finding based multi-focus image fusion through multi-scale morphological focus-measure. *Inf. Fusion* **2017**, *35*, 81–101. [CrossRef]
12. Liu, Y.; Wang, Z. Dense SIFT for ghost-free multi-exposure fusion. *J. Vis. Commun. Image Represent.* **2015**, *31*, 208–224. [CrossRef]
13. Bai, X. Infrared and visual image fusion through feature extraction by morphological sequential toggle operator. *Infrared Phys. Technol.* **2015**, *71*, 77–86. [CrossRef]
14. Bai, X. Infrared and Visual Image Fusion through Fuzzy Measure and Alternating Operators. *Sensors* **2015**, *15*, 17149–17167. [CrossRef] [PubMed]
15. Zhang, Y.; Zhang, L.; Bai, X.; Zhang, L. Infrared and visual image fusion through infrared feature extraction and visual information preservation. *Infrared Phys. Technol.* **2017**, *83*, 227–237. [CrossRef]

16. Ma, J.; Yu, W.; Liang, P.; Li, C.; Jiang, J. FusionGAN: A generative adversarial network for infrared and visible image fusion. *Inf. Fusion* **2019**, *48*, 11–26. [CrossRef]
17. Ma, J.; Ma, Y.; Li, C. Infrared and visible image fusion methods and applications: A survey. *Inf. Fusion* **2019**, *45*, 153–178. [CrossRef]
18. Zhou, Z.; Li, S.; Wang, B. Multi-scale weighted gradient-based fusion for multi-focus images. *Inf. Fusion* **2014**, *20*, 60–72. [CrossRef]
19. Toet, A. Image fusion by a ratio of low-pass pyramid. *Pattern Recognit. Lett.* **1989**, *9*, 245–253. [CrossRef]
20. Burt, P.J.; Adelson, E.H. The Laplacian pyramid as a compact image code. *IEEE Trans. Commun.* **1983**, *31*, 532–540. [CrossRef]
21. Lewis, J.J.; O'Callaghan, R.J.; Nikolov, S.G.; Bull, D.R.; Canagarajah, N. Pixel- and region-based image fusion with complex wavelets. *Inf. Fusion* **2007**, *8*, 119–130. [CrossRef]
22. Li, H.; Manjunath, B.; Mitra, S.K. Multisensor image fusion using the wavelet transform. *Graph. Model. Image Process.* **1995**, *57*, 235–245. [CrossRef]
23. Yang, B.; Li, S. Multifocus image fusion and restoration with sparse representation. *IEEE Trans. Instrum. Meas.* **2010**, *59*, 884–892. [CrossRef]
24. Li, S.; Yin, H.; Fang, L. Group-sparse representation with dictionary learning for medical image denoising and fusion. *IEEE Trans. Biomed. Eng.* **2012**, *59*, 3450–3459. [CrossRef] [PubMed]
25. Liu, Y.; Liu, S.; Wang, Z. A general framework for image fusion based on multi-scale transform and sparse representation. *Inf. Fusion* **2015**, *24*, 147–164. [CrossRef]
26. Liu, Y.; Chen, X.; Peng, H.; Wang, Z. Multi-focus image fusion with a deep convolutional neural network. *Inf. Fusion* **2017**, *36*, 191–207. [CrossRef]
27. Li, H.; Wu, X.J. DenseFuse: A fusion approach to infrared and visible images. *IEEE Trans. Image Process.* **2018**, *28*, 2614–2623. [CrossRef] [PubMed]
28. Tang, L.; Yuan, J.; Ma, J. Image fusion in the loop of high-level vision tasks: A semantic-aware real-time infrared and visible image fusion network. *Inf. Fusion* **2022**, *82*, 28–42. [CrossRef]
29. Li, H.; Xu, T.; Wu, X.J.; Lu, J.; Kittler, J. LRRNet: A Novel Representation Learning Guided Fusion Network for Infrared and Visible Images. *IEEE Trans. Pattern Anal. Mach. Intell.* **2023**. [CrossRef] [PubMed]
30. Li, F.; Zhou, Y.; Chen, Y.; Li, J.; Dong, Z.; Tan, M. Multi-scale attention-based lightweight network with dilated convolutions for infrared and visible image fusion. *Complex Intell. Syst.* **2023**, *10*, 1–15. [CrossRef]
31. Zhang, Y.; Shen, J.; Guo, S.; Zhong, L.; Zhang, S.; Bai, X. Multi-scale Bézier Filter Based Infrared and Visual Image Fusion. In Proceedings of the Chinese Conference on Image and Graphics Technologies, Beijing, China, 17–19 August 2022; pp. 14–25.
32. Zhang, Y.; Zhang, S.; Bai, X.; Zhang, L. Human chest CT image enhancement based on basic information preservation and detail enhancement. *J. Image Graph.* **2022**, *27*, 774–783.
33. Wang, Z.; Bovik, A.C.; Sheikh, H.R.; Simoncelli, E.P. Image quality assessment: From error visibility to structural similarity. *IEEE Trans. Image Process.* **2004**, *13*, 600–612. [CrossRef]
34. Ma, K.; Zeng, K.; Wang, Z. Perceptual quality assessment for multi-exposure image fusion. *IEEE Trans. Image Process.* **2015**, *24*, 3345–3356. [CrossRef]
35. Toet, A. The TNO Multiband Image Data Collection. *Data Brief* **2017**, *15*, 249–251. [CrossRef]
36. Li, S.; Kang, X.; Hu, J. Image fusion with guided filtering. *IEEE Trans. Image Process.* **2013**, *22*, 2864–2875.
37. Kurban, R. Gaussian of differences: A simple and efficient general image fusion method. *Entropy* **2023**, *25*, 1215. [CrossRef]
38. Chen, J.; Li, X.; Wu, K. Infrared and visible image fusion based on relative total variation decomposition. *Infrared Phys. Technol.* **2022**, *123*, 104112. [CrossRef]
39. Panigrahy, C.; Seal, A.; Mahato, N.K. Parameter adaptive unit-linking dual-channel PCNN based infrared and visible image fusion. *Neurocomputing* **2022**, *514*, 21–38. [CrossRef]
40. Xu, H.; Ma, J.; Jiang, J.; Guo, X.; Ling, H. U2Fusion: A unified unsupervised image fusion network. *IEEE Trans. Pattern Anal. Mach. Intell.* **2020**, *44*, 502–518. [CrossRef]
41. Li, S.; Yang, B. Multifocus image fusion using region segmentation and spatial frequency. *Image Vis. Comput.* **2008**, *26*, 971–979. [CrossRef]
42. Zhao, W.; Wang, D.; Lu, H. Multi-focus image fusion with a natural enhancement via a joint multi-level deeply supervised convolutional neural network. *IEEE Trans. Circuits Syst. Video Technol.* **2018**, *29*, 1102–1115. [CrossRef]
43. Bai, X.; Zhou, F.; Xue, B. Noise-suppressed image enhancement using multiscale top-hat selection transform through region extraction. *Appl. Opt.* **2012**, *51*, 338–347. [CrossRef] [PubMed]
44. Petrovic, V.; Xydeas, C. Objective image fusion performance characterisation. In Proceedings of the 10th IEEE International Conference on Computer Vision (ICCV'05) Volume 1, Beijing, China, 17–21 October 2005; pp. 1866–1871.
45. Han, Y.; Cai, Y.; Cao, Y.; Xu, X. A new image fusion performance metric based on visual information fidelity. *Inf. Fusion* **2013**, *14*, 127–135. [CrossRef]
46. Piella, G.; Heijmans, H. A new quality metric for image fusion. In Proceedings of the 2003 International Conference on Image Processing (Cat. No. 03CH37429), Barcelona, Spain, 14–17 September 2003; Volume 3, pp. III–173–176.

47. Aslantas, V.; Bendes, E. A new image quality metric for image fusion: The sum of the correlations of differences. *AEU Int. J. Electron. Commun.* **2015**, *69*, 1890–1896. [CrossRef]
48. Zhang, X.; Ye, P.; Xiao, G. VIFB: A visible and infrared image fusion benchmark. In Proceedings of the IEEE/CVF Conference on Computer Vision and Pattern Recognition Workshops, Seattle, WA, USA, 14–19 June 2020; pp. 104–105.

sensors

MDPI

Article

The Potential of Diffusion-Based Near-Infrared Image Colorization

Ayk Borstelmann [1,*], Timm Haucke [1,2] and Volker Steinhage [1,*]

[1] Institute of Computer Science IV, University of Bonn, Friedrich-Hirzebruch-Allee 8, 53115 Bonn, Germany; haucke@cs.uni-bonn.de

[2] Computer Science & Artificial Intelligence Laboratory, Massachusetts Institute of Technology, 32 Vassar St., Cambridge, MA 02139, USA

* Correspondence: ayk.borstelmann@uni-bonn.de (A.B.); steinhage@cs.uni-bonn.de (V.S.)

Abstract: Camera traps, an invaluable tool for biodiversity monitoring, capture wildlife activities day and night. In low-light conditions, near-infrared (NIR) imaging is commonly employed to capture images without disturbing animals. However, the reflection properties of NIR light differ from those of visible light in terms of chrominance and luminance, creating a notable gap in human perception. Thus, the objective is to enrich near-infrared images with colors, thereby bridging this domain gap. Conventional colorization techniques are ineffective due to the difference between NIR and visible light. Moreover, regular supervised learning methods cannot be applied because paired training data are rare. Solutions to such unpaired image-to-image translation problems currently commonly involve generative adversarial networks (GANs), but recently, diffusion models gained attention for their superior performance in various tasks. In response to this, we present a novel framework utilizing diffusion models for the colorization of NIR images. This framework allows efficient implementation of various methods for colorizing NIR images. We show NIR colorization is primarily controlled by the translation of the near-infrared intensities to those of visible light. The experimental evaluation of three implementations with increasing complexity shows that even a simple implementation inspired by visible-near-infrared (VIS-NIR) fusion rivals GANs. Moreover, we show that the third implementation is capable of outperforming GANs. With our study, we introduce an intersection field joining the research areas of diffusion models, NIR colorization, and VIS-NIR fusion.

Keywords: near-infrared; diffusion models; camera trapping; unpaired dataset; neural networks; machine learning

Citation: Borstelmann, A.; Haucke, T.; Steinhage, V. The Potential of Diffusion-Based Near-Infrared Image Colorization. *Sensors* **2024**, *24*, 1565. https://doi.org/10.3390/s24051565

Academic Editor: Christos Nikolaos E. Anagnostopoulos

Received: 22 December 2023
Revised: 22 February 2024
Accepted: 26 February 2024
Published: 28 February 2024

1. Introduction

For wildlife monitoring, typically, camera traps are used (see Figure 1). Camera traps show several advantages for wildlife monitoring:

1. Camera traps deliver permanent documentation records of date, location, and species.
2. These camera-trap-based documentation records allow for estimations of animal populations [1] and movements of animals and herds [2].
3. Camera traps can record animal behavior [3].
4. Using invisible infrared flashlights, camera traps work non-invasive and, therefore, have no disturbing effects on animal behavior.
5. Camera traps work efficiently for several weeks [4].
6. Camera trapping allows for synergies between expert and citizen science [5].
7. Images and video clips can be used for education, promotion, and funding acquisition [6,7].

Figure 1. Camera trapping for the Snapshot Serengeti dataset. ©Swanson, Kosmala, Lintott, Simpson, Smith, and Packer [8]. Licensed under a Creative Commons Attribution 4.0 License.

1.1. Problem Statement

During daylight, normal cameras succeed in capturing detailed images. But at dawn or during nighttime, near-infrared (NIR) cameras or normal cameras with incandescent lighting are necessary. Near-infrared light has a wavelength between 750 nm and 1400 nm, which mostly lies outside the visible spectrum (380 nm–780 nm) [9]. Because of that, NIR cameras offer a significant advantage over conventional cameras using incandescent lighting. Incandescent light flashes are visible to animals and may frighten them, leading to animals avoiding the camera location afterward, which in turn would corrupt statistical estimates like population or migration estimations based on the numbers and frequencies of observed animals. Near-infrared light is not visible to animals and thus cannot scare them.

But on the other hand, NIR images appear as grayscale images that do not conform to the human visual spectrum because they lack colors and color textures. Therefore, it can be difficult to perceive the details of observed scenes in NIR images [10]. This discrepancy between NIR and colored images constitutes the domain gap between near-infrared and colored images.

Additionally, in recent years the combination of camera trapping and of artificial intelligence (AI), especially of deep learning approaches, has emerged as a breakthrough in the field of wildlife research and conservation [3,7,11,12]. However, many deep learning approaches are trained for and can benefit from colored images [13] like humans do [10]. This raises the question if deep learning approaches can also benefit from such artificially colored images. We evaluate and discuss if this is the case in Section 4.2.

1.2. Contribution

In this study, we propose the automated conversion of NIR images to colored RGB images

- to derive detail-rich images providing color and texture without scaring animals;
- to gain compatibility with and benefits for existing monitoring systems;
- to improve human comprehension of camera trap data.

1.2.1. Colorizing NIR Images—Luminance and Chrominance

It is important to note that colorizing NIR images is closely related to the colorization of grayscale images, but it differs in one crucial property: the luminance (i.e., the amount of light that is reflected off an object) of grayscale images captured in the visible spectrum is the same as the luminance of colored images. Therefore, only chrominance (i.e., the color component) must be estimated by image colorization systems. However, the objects' reflection properties of NIR light differ from those of visible light. Consequently, due to the differing luminance between RGB and NIR images, conventional image colorization techniques cannot be applied as-is.

1.2.2. Colorizing NIR Images—Paired vs. Unpaired Image Translation

Supervised solutions for this problem exist but require NIR and RGB image pairs with pixel-to-pixel registration and temporal synchronization. For each given NIR image, the corresponding RGB image must have the same information on the pixel locations on both images, and both images must be captured simultaneously to account for motion. However, because many wildlife datasets involve the use of NIR cameras for nighttime and regular cameras for daytime images, it is rare to find paired datasets. This results in unpaired image translation to which unsupervised learning techniques must be applied.

1.2.3. Colorizing NIR Images—GAN-Based Approaches vs. Diffusion Models

Current state-of-the-art approaches leverage generative adversarial networks (GANs) to solve this unpaired image translation problem. For example, Mehri and Sappa [14] proposed a GAN-based approach specifically designed for the task of colorizing NIR images (cf. Section 1.3). However, GANs are known for their unstable training manifesting in mode collapse and hallucinations [15].

Recently, advances in denoising diffusion probabilistic models (short: diffusion models or DDPM) [16] showed superiority over GANs in various other image translation tasks [17,18]. Their training is more stable, while simultaneously a higher sample diversity is observed. This suggests that diffusion models could also influence NIR colorization positively.

1.2.4. Colorizing NIR Images: Refined Contribution

To the best of our knowledge, this study proposes the first automated conversion of NIR images to colored RGB images utilizing diffusion models. The novelty of our approach lies in discovering the key property necessary to let a diffusion model generate realistic images, i.e., the appropriate translation of the NIR image intensities into color intensities.

Thereby, we provide a generic approach in terms of a framework where we first abstract the translation of the intensities to allow for implementations and evaluations of different approaches to intensity translations.

The framework is based on iterative latent variable refinement (ILVR) [19] but comes with the following novel methodical improvements to specialize for near-infrared colorization:

- Replacing the low-pass filter as latent variable refinement technique;
- Differentiating into merging chrominance and merging intensity instead;
- Abstracting the intensity translation.

Based on the abstraction of the intensity translation, we provide and evaluate three different specific implementations. The evaluation of these shows that even the deployment of trivial algorithms inspired by insights gained in the VIS-NIR fusion research field [20] can achieve Fréchet inception distances (FIDs) close to GAN baselines. This employs a connection between the research fields of diffusion models, near-infrared colorization, and visible near-infrared fusion. Finally, we show that our framework is capable of outperforming a GAN baseline, revealing the potential of diffusion-based NIR colorization.

1.3. Related Work

NIR colorization has many applications in addition to wildlife monitoring, e.g., in driver assistance systems or surveillance cameras. As a result, computer vision researchers have studied image colorization during the last decades, and thus, many solutions exist. Solutions for NIR colorization are divided into paired and unpaired image translation problems, and thus, supervised and unsupervised learning techniques are applied. For paired image translation, pixel-to-pixel registration and temporal synchronization are required for each image pair, which adds additional challenges.

Limmer and Lensch [21] used a dataset acquired by a specialized multi-CCD camera that ensures the requirement of the image pair. As a translation mechanism, Limmer and Lensch [21] proposed to use deep multiscale convolutional neural networks (CNN). In pre-processing, a normalized image pyramid is constructed from the NIR input, and

in post-processing, the CNN output is enriched with details from the input image. In later work, Dong et al. [22] introduced an S-shape network consisting of one U-Net-based encoder "ColorNet" and a shallow network that generates an edge loss function "EdgeNet". Dong et al. [22] created a pixel-to-pixel registered dataset using geometric transformations and feature-based correspondence methods.

1.3.1. GAN-Based Approaches

GAN-based methods have established themselves as a powerful approach to unpaired image translation, which can be used for NIR image colorization without the need for paired training data. This is especially useful if such data cannot be obtained, for example, due to dataset limitations. By default, GANs tend to lose the content of the input image. CycleGAN, and an architecture proposed by Zhu et al. [23] uses a cycle consistency loss between the input and the generated image to solve that problem. It uses ResNet [24] as the generator network [23]. Gao et al. [13] trained CycleGAN on a wildlife dataset and showed improved recognition results on the generated images compared to the NIR images. Mehri and Sappa [14] proposed a version of CycleGAN, specifically designed for the task of colorizing NIR images that incorporates enhanced loss functions and utilizes U-Net as a generator. Because of this, we use this CycleGAN of Mehri and Sappa [14] as a GAN baseline for our research.

We use the GAN DeOldify [25] as a second reference method since it is trained on a large dataset.

1.3.2. Diffusion Models

More recently, diffusion models have advanced. First suggested by Sohl-Dickstein et al. [26], diffusion models are neural networks that gradually remove noise from signals. Simultaneously to Sohl-Dickstein et al. [26], Song and Ermon [27] introduced and studied score matching as a way of estimating the given data distribution using its gradients while sampling with Langevin dynamics [28]. Later, Ho et al. [16] first found the connection between diffusion models and score-based models and leveraged this to simplify the training objective of a variational lower bound. They introduced denoising diffusion probabilistic models (DDPM), which is considered a milestone in the development of diffusion models.

Song et al. [18] further analyzed the connection between score matching with Langevin dynamics and diffusion models, proposed a unified framework using the stochastic differential equation, and showed that both DDPM [16] and their previous work [27] can be considered a specialized formulation of it. Further, they introduced deterministic samplers using ordinary differential equations that allow likelihood computation and deterministic latent codes. Most important for us, through their formulation, they derive a conditional sampling method that only uses an unconditional model to control the generation at inference time. This allows applications to image imputation and grayscale colorization, which we base our work on.

Dhariwal and Nichol [17] were the first to outperform GANs in image generation with several architectural improvements. Additionally, they introduced classifier guidance. This sampling method uses unconditional diffusion models and only a classifier during inference to achieve class conditional sampling. With this, they provided a conditional sampling method inspired by Song et al. [18] for DDPMs.

Saharia et al. [29] trained a multi-task image-conditional diffusion model with application to grayscale colorization. In contrast to our approach, they used supervised learning to train a conditional diffusion model and, therefore, required a pair dataset at training time.

Choi et al. [19] leveraged an unconditional diffusion model and iteratively refined the current sample (ILVR). By this, they achieved conditional sampling. Our framework iteratively refines the latent variable by enriching it with information from the near-infrared image. Therefore, we consider our framework heavily based on ILVR's key algorithm. However, ILVR enriches the input image with low-frequency information from the input

image. Binding the low frequencies of the generated image to those of the near-infrared image does not lead to colorization; instead, it just results in grayscale images with similar contours to the given image.

Zhao et al. [30] suggested an energy term for diffusion models, describing the similarity and steering the sampling using its gradient. Their energy is divided into domain-independent energy and domain-specific energy. The domain-independent energy ensures similarity to the input image, while the domain-specific energy ensures realism in the output domain [30]. They employed a low-pass filter as a domain-independent extractor, which NIR colorization does not benefit from. However, it is mentioned that different domain-independent extractors are feasible.

Furthermore, research from the visible-infrared fusion field influences our work. VIS-NIR fusion focuses on enhancing RGB images with NIR images. As our approach can be considered iteratively fusing visible and infrared images, we borrow insights from this field of research. Sharma et al. [20] studied and compared comprehensively multiple visible-infrared fusion techniques. One common similarity between many of the compared methods is that the near-infrared intensities are combined with visible intensities at different scales and with the chrominance of the visible-light image [20,31,32]. We evaluate this principle as a strategy for enriching the latent variable in our framework.

We develop a novel NIR colorization approach to images leveraging the recent advances of diffusion models. Focus is placed on the unpaired image translation because NIR-RGB image pairs are often hard to obtain. We only need to train an unconditional diffusion model in the target domain. Our framework is based on ILVR [19] and abstracts the intensity translation. We present three implementations of this framework. First, we use NIR intensities, which are effectively equivalent to the grayscale colorization of Song et al. [18]. Next, we utilize the connection to VIS-NIR fusion [20] and present an implementation based on fusioning high frequencies of near-infrared images with low frequencies of the colored image. Finally, we show the potential of our method by using CycleGAN itself as an intensity translator.

2. Materials and Methods

Denoising diffusion probabilistic models (DDPMs), as introduced by Ho et al. [16], are recent advances in the field of image generation. We provide a theoretical background for this architecture in Section 2.1 and show how an unconditional diffusion model can be used to sample with inputs. Iterative Seeding, our framework leveraging these diffusion models for colorization, is presented in Section 2.2, and two implementations are presented in Sections 2.2.1 and 2.2.2.

For developing a diffusion near-infrared colorization approach, a dataset containing NIR and colored images from similar settings is required. We choose the Snapshot Serengeti dataset originating in the Serengeti National Park in Tanzania. It consists of 7.1 million images captured over the course of seven seasons of the Snapshot Serengeti Project [8]. There are 61 labeled species, while approximately 76% of the images are labeled as empty. For training and evaluation, we create a subset of the dataset consisting of 10,000 images (5000 NIR and 5000 colored images). We partition this subset into an 8000-image train dataset and two separate datasets for testing and validation, each consisting of 1000 images. Furthermore, night images are chosen only because they align with the application context of near-infrared colorization, and the network does not implicitly learn to translate night images to day images.

2.1. Background

A diffusion process, consisting of T time steps, describes how noise is added to an image. $\mathbf{x}_0$ denotes the original image and $\mathbf{x}_T$ the final noised image. T is chosen, so that $\mathbf{x}_T$ follows an isotropic Gaussian distribution [16].

With $q(\mathbf{x}_t|\mathbf{x}_{t-1})$ we denote the forward process, which describes the distribution of $\mathbf{x}_t$ given a less noised $\mathbf{x}_{t-1}$. The forward process gradually adds Gaussian noise to the image determined by the variance schedule $\beta_1, \ldots, \beta_T$ [16] (Equation (1)).

$$q(\mathbf{x}_t|\mathbf{x}_{t-1}) := \mathcal{N}(\mathbf{x}_t; \sqrt{1-\beta_t}\mathbf{x}_{t-1}, \beta_t\mathbf{I}) \tag{1}$$

To sample $\mathbf{x}_t \sim q(\mathbf{x}_t|\mathbf{x}_0)$, repeated sampling is not necessary because a closed form can be derived (Equation (2)), where $\alpha_t := 1 - \beta_t$ and $\bar{\alpha}_t := \prod_{s=1}^{t} \alpha_s$ [16].

$$q(\mathbf{x}_t|\mathbf{x}_0) = \mathcal{N}(\mathbf{x}_t; \sqrt{\bar{\alpha}_t}\mathbf{x}_0, (1-\bar{\alpha}_t)\mathbf{I}) \tag{2}$$

The reverse process describes how DDPMs operate. Initially, a sample is drawn from the prior distribution $q(\mathbf{x}_T)$, which is nearly an isotropic Gaussian, therefore $\mathbf{x}_T \sim \mathcal{N}(0, \mathbf{I})$. Then we gradually denoise our sample using $q(\mathbf{x}_{t-1}|\mathbf{x}_t)$ until $t = 0$. Because $q(\mathbf{x}_{t-1}|\mathbf{x}_t)$ is not trivially obtainable without knowing the data distribution, we leverage a neural network p_θ to approximate it. θ denotes the parameters of the network. If β_t is small enough, $p_\theta(\mathbf{x}_{t-1}|\mathbf{x}_t)$ will also be Gaussian (Equation (3)) [16]. Note that Ho et al. [16] fix the variance of the reverse process using $\sigma_t^2\mathbf{I}$ with either $\sigma_t^2 = \beta_t$ or $\sigma_t^2 = \tilde{\beta}_t$.

$$p_\theta(\mathbf{x}_{t-1}|\mathbf{x}_t) := \mathcal{N}(\mathbf{x}_{t-1}; \mu_\theta(\mathbf{x}_t, t), \sigma_t^2\mathbf{I}) \tag{3}$$

Ho et al. [16] choose to parameterize $\mu_\theta(\mathbf{x}_t, t)$ as follows, where ϵ_θ is a function to predict ϵ given $\mathbf{x}_t$ (Equation (4)).

$$\mu_\theta(\mathbf{x}_t, t) = \frac{1}{\sqrt{\alpha_t}}\left(\mathbf{x}_t - \frac{\beta_t}{\sqrt{1-\bar{\alpha}_t}}\epsilon_\theta(\mathbf{x}_t, t)\right) \tag{4}$$

Furthermore, Ho et al. [16] suggest a simplified loss (Equation (5)), which uses the μ-parameterization (Equation (4)).

$$L_{simple}(\theta) := \mathbb{E}_{t\sim[1,T],\mathbf{x}_0,\epsilon_t}\left[||\epsilon_t - \epsilon_\theta(\sqrt{\bar{\alpha}_t}\mathbf{x}_0 + \sqrt{1-\bar{\alpha}_t}\epsilon, t)||^2\right] \tag{5}$$

In terms of network architecture, diffusion models usually employ the U-Net architecture for learning the noise $\epsilon_\theta(\mathbf{x}_t, t)$ [14,16,17,33]. The U-Net takes the current noised image $\mathbf{x}_t$ as input and aims to produce the noise that should be removed. We use the refined U-Net architecture from Dhariwal and Nichol [17] which included global attention blocks and embedding of the timestep t.

Our application context of NIR colorization is not able to benefit from the unconditional sampling as derived up until now. We need a method to condition the diffusion model on the given NIR image at inference time (unpaired translation). This is equivalent to sampling from a conditional distribution $p(\mathbf{x}_{t-1}|\mathbf{x}_t, \mathbf{c})$, where $\mathbf{c}$ denotes the NIR image. Similar to Choi et al. [19], we can utilize the unconditional diffusion model, sample $\tilde{\mathbf{x}}_{t-1} \sim p_\theta(\tilde{\mathbf{x}}_t|\mathbf{x}_t)$, and refine $\tilde{\mathbf{x}}_{t-1}$ to be congruent to the condition $\mathbf{c}$ and obtain $\mathbf{x}_t$. Choi et al. [19] use a low-pass filter to maintain similarity to the input image without restricting the sampling procedure. This is not suitable for NIR colorization because the low-pass filter would revert the colorization process performed by the diffusion model.

2.2. Iterative Seeding

Colorization can also be considered as a specialized form of image imputation. As Song et al. [18] showed. Image imputation is the task of restoring lost parts of an image congruent with the known areas of the image. In the case of grayscale colorization, the known part is the intensity, while unknown is the chrominance, which itself can be decomposed into hue and saturation.

As the intensity is not directly known in near-infrared colorization, we take an abstraction approach. In each iteration, we draw a sample $\tilde{\mathbf{x}}_{t-1}$ from the diffusion model given $\mathbf{x}_t$ using $p_\theta(\tilde{\mathbf{x}}_{t-1}|\mathbf{x}_t)$. Simultaneously, we diffuse our input image $\mathbf{y}_0$ to the timestep $t-1$ using $q(\mathbf{y}_{t-1}|\mathbf{y}_0)$. We then decompose both images into their intensity parts $\tilde{\mathbf{x}}_{t-1}^I$, $\mathbf{y}_{t-1}^I$ and

the chrominance parts $\tilde{\mathbf{x}}^C_{t-1}$, $\mathbf{y}^C_{t-1}$ using DECOUPLEINTENSITY without loss of information. A translation function TRANSLATEINTENSITY enriches the intensity $\tilde{\mathbf{x}}^I_{t-1}$ of our current sample with the near-infrared intensities $\mathbf{y}^I_{t-1}$, returning a new visible-light intensity $\mathbf{x}^I_{t-1}$ for the timestep $t-1$. This intensity $\mathbf{x}^I_{t-1}$ is then combined with the chrominance of the sample $\tilde{\mathbf{x}}^C_{t-1}$ and transformed back into the RGB domain using COUPLEINTENSITY to obtain $\mathbf{x}_{t-1}$.

The approach to only sample the intensity for colorization was first proposed by Song et al. [18]. But we do not implement this using a stochastic differential equation. Our procedure is more similar to Choi et al. [19]'s iterative latent variable refinement, and thus, we call this framework Iterative Seeding. In Algorithm 1, we demonstrate the code for our framework.

Algorithm 1 Iterative Seeding

Require: Reference gray-scale image $\mathbf{y}_0$, function TRANSLATEINTENSITY returning intensities of visible light given the current near-infrared intensity and the current visible-light intensity
$\mathbf{x}_T \sim \mathcal{N}(0, \mathbf{I})$
for $t = T, \ldots, 1$ **do**
 $\mathbf{y}_{t-1} \sim q(\mathbf{y}_{t-1}|\mathbf{y}_0)$
 $\tilde{\mathbf{x}}_{t-1} \sim p_\theta(\tilde{\mathbf{x}}_{t-1}|\mathbf{x}_t)$
 $\mathbf{y}^I_{t-1}, \mathbf{y}^C_{t-1} =$ DECOUPLEINTENSITY($\mathbf{y}_{t-1}$)
 $\tilde{\mathbf{x}}^I_{t-1}, \tilde{\mathbf{x}}^C_{t-1} =$ DECOUPLEINTENSITY($\tilde{\mathbf{x}}_{t-1}$)
 $\mathbf{x}^I_{t-1} =$ TRANSLATEINTENSITY($\mathbf{y}^I_{t-1}, \tilde{\mathbf{x}}^I_{t-1}$)
 $\mathbf{x}_{t-1} =$ COUPLEINTENSITY($\mathbf{x}^I_{t-1}, \tilde{\mathbf{x}}^C_{t-1}$)
return $\mathbf{x}_0$

DECOUPLEINTENSITY and COUPLEINTENSITY can theoretically be any invertible transformation where the intensity is decoupled from the color information. Many color spaces fulfill this property, e.g., HSI, HSV, LAB, and YCbCr, but we found empirically that transforming the RGB image using an orthogonal matrix with one dimension in resulting space being the intensity, like Song et al. [18] did, to give the best results.

Therefore, we search for a matrix $C \in \mathbb{R}^{3\times 3}$ such that for any RGB pixel $p = (r, g, b) \in \mathbb{R}^3$ and a fixed scalar $a \in \mathbb{R}$ the requirements of Equation (6) are fulfilled.

$$\begin{aligned} p' &= p \cdot C \Rightarrow p'_1 = a \cdot (r + g + b) \\ p &= (p \cdot C) \cdot C^T \end{aligned} \tag{6}$$

A matrix fulfilling those requirements can be obtained from solving a system of equations or using QR decomposition. We derive the matrix C as Equation (7). Note that Song et al. [18] use a different matrix as solution for this problem.

$$C = \begin{pmatrix} \frac{1}{\sqrt{3}} & -\frac{1}{2} - \frac{1}{2\sqrt{3}} & \frac{1}{2} - \frac{1}{2\sqrt{3}} \\ \frac{1}{\sqrt{3}} & \frac{1}{2} - \frac{1}{2\sqrt{3}} & -\frac{1}{2} - \frac{1}{2\sqrt{3}} \\ \frac{1}{\sqrt{3}} & \frac{1}{\sqrt{3}} & \frac{1}{\sqrt{3}} \end{pmatrix} \approx \begin{pmatrix} 0.577 & -0.789 & 0.211 \\ 0.577 & 0.211 & -0.789 \\ 0.577 & 0.577 & 0.577 \end{pmatrix} \tag{7}$$

TRANSLATEINTENSITY is the central component of our framework as it is an abstract function that is implemented in our study with increasing complex functions (cf. Sections 2.2.1, 2.2.2, 3.2 and 3.3). Thereby, TRANSLATEINTENSITY with its different implementations influences greatly the performance of the colorization. It should integrate information from both the near-infrared intensity $\mathbf{y}^I_{t-1}$ and the approximation of visible-light intensity $\tilde{\mathbf{x}}^I_{t-1}$ and produce an improved approximation $\mathbf{x}^I_{t-1}$ of visible-light intensity. We note that this method is strongly related to the research domain of near-infrared and visible-light fusion (VIS-NIR fusion) since it practically fuses the near-infrared intensity $\mathbf{y}^I_{t-1}$ and the visible-light intensity $\tilde{\mathbf{x}}^I_{t-1}$. So, it is obvious to use approaches from this

domain to implement this method. In Sections 2.2.1 and 2.2.2, we present two different implementations of this method and evaluate them in Section 3.

2.2.1. Near-Infrared Intensities

One simple strategy to implement TRANSLATEINTENSITY is to directly use the near-infrared intensities. In that case, TRANSLATEINTENSITY is the identity function for $\mathbf{y}^I_{t-1}$ (Equation (8)). This implementation of our framework fixates the intensity of the output color to the near-infrared while giving the diffusion model just the freedom to sample the chrominance. Using this method does not reflect any near-infrared properties. It could also be applied to grayscale colorization and is conceptionally equivalent to Song et al. [18] colorization variant.

Leaving the diffusion model no freedom to generate intensity-related changes suggests a weakness of this implementation for near-infrared colorization. In Section 3.1, we evaluate this hypothesis. Compared to other implementations of our framework, this implementation performs the worst in terms of FID, confirming our hypothesis.

$$\textsc{TranslateIntensity}(\mathbf{y}^I_{t-1}, \tilde{\mathbf{x}}^I_{t-1}) := \mathbf{y}^I_{t-1} \tag{8}$$

2.2.2. High-Pass Filtering

A more refined approach to just using the near-infrared intensities is inspired from the VIS-NIR fusion domain. One key insight for fusioning NIR and visible-light images, is to combine the NIR image with intensities from the visible-light image at different scales [20].

A simple implementation of this concept is using the high frequencies of the near-infrared image and combining them with the low frequencies of the visible light's intensity. In practise, a simple Gaussian filter $G \in \mathbb{R}^{k \times k}$ (Equation (9)) can obtain the low frequencies [34], and the high frequencies are obtained by subtracting the low frequencies from the image (Algorithm 2).

$$G_\sigma(u, v) = \frac{1}{2\pi\sigma^2} \exp\left(-\frac{(u^2 + v^2)}{2\sigma^2}\right) \tag{9}$$

Algorithm 2 Implementation using High-Pass Filtering

Require: Gaussian filter G_σ
  procedure $\textsc{TranslateIntensity}(\mathbf{y}^I_{t-1}, \tilde{\mathbf{x}}^I_{t-1})$
    $\tilde{\mathbf{x}}^{I_L}_{t-1} = \tilde{\mathbf{x}}^I_{t-1} * G_\sigma$
    $\tilde{\mathbf{x}}^{I_H}_{t-1} = \tilde{\mathbf{x}}^I_{t-1} - \tilde{\mathbf{x}}^{I_L}_{t-1}$
    $\mathbf{y}^{I_L}_{t-1} = \mathbf{y}^I_{t-1} * G_\sigma$
    $\mathbf{y}^{I_H}_{t-1} = \mathbf{y}^I_{t-1} - \mathbf{y}^{I_L}_{t-1}$
    $\mathbf{x}^I_{t-1} = \tilde{\mathbf{x}}^{I_L}_{t-1} + \mathbf{y}^{I_H}_{t-1}$
    return $\mathbf{x}^I_{t-1}$

With this implementation of our framework, iteratively the generated image is enriched with details of the near-infrared image. Thus, the diffusion model is only restricted to using the high frequencies of the near-infrared image and is free to sample low frequencies and the chrominance. This suggests a better performance can be reached than when only sampling the chrominance. On the other hand, this could also lead to a more difficult generation task, as less information is given.

We apply and evaluate this approach in Section 3.2 and confirm that this implementation performs better, through more freedom for the generator.

Further we investigate the influence of different standard deviations σ controlling the degree of information given in Section 3.2. We discover that this hyperparameter does affect the content preservation and realism of the generated images (Section 3.2).

3. Experimental Results

Our application context of near-infrared colorization lies in closing the gap between NIR and RGB images as inputs for deep learning systems, improving object recognition results by enriching the input with more information, and providing more familiar images to human users.

Since this is an unpaired image translation problem, classic solutions for quantitative evaluation, such as the difference between the absolute intensity values or SSIM [35], cannot be applied. To assess the realism of our results, we calculate the distance between the test dataset, consisting of real images, and our generated images using the Fréchet inception distance (FID) [36]. The FID acts as a distance between two unpaired image sets and is calculated on a classification network's feature abstraction of images [36]. CleanFID is conceptually equivalent to the FID but comes with regularization techniques to make it more robust in terms of distortions, blurring, and compression artifacts [37]. Because of this, we use CleanFID instead of the regular FID. Furthermore, we evaluate our results with two blind/no-reference image quality assessment metrics NIQE [38] and NRQM [39]. Both score the realism images without the reference-image-based properties of the image [38,39].

Initially, an unconditional diffusion model is trained using the improved architecture from Dhariwal and Nichol [17]. All hyperparameters are taken from Dhariwal and Nichol [17] as well but adjusted for less powerful hardware. For both training and inference, an image resolution of 128×128 is used. We use a U-Net with five encode and five decode blocks, where each encode and decode block consists of two residual blocks [17]. Attention blocks are applied at the resolutions of 32×32, 16×16, and 8×8 like Dhariwal and Nichol [17] did. The noise schedule is divided into 1000 linear steps. We train with a batch size of 256 and a learning rate of 10^{-4} for 200 K iterations.

For CycleGAN, we train and evaluate a U-Net for the image resolution of 128×128. We train with a hyperparameter-optimized generator learning rate of 1.5×10^{-5} and 4.5×10^{-5} as the discriminator learning rate. All remaining hyperparameters stay as suggested by Mehri and Sappa [14].

We use DeOldify [25] pretrained from the official GitHub repository (https://github.com/jantic/DeOldify (accessed on 26 February 2024)) because it requires a paired dataset, which is hard to obtain. We consider its training on a larger dataset than ours as beneficial for DeOldify and thus as fair.

First, we evaluate the unconditional sampling of the diffusion model and validate that the results for image synthesis of Dhariwal and Nichol [17] still hold for this dataset:

We observe in Figure 2 that the diffusion model is capable of creating diverse, realistic images. Samples such as the top-left are common for the training and test dataset as well and, therefore, are considered realistic. In Table 1, we see the diffusion network performs strongly in quantitative metrics as well. In comparison with later evaluations of our methods and CycleGAN [14], it performs at least $\sim$20 FID points better. Considering this, we argue this unconditional model is capable of serving as a foundation for effective colorization.

Table 1. Quantitative evaluation of unconditional diffusion sampling. The unconditional diffusion model [17] trained and evaluated on the Snapshot Serengeti dataset [8] containing only night NIR and RGB images. We compare the FID calculated between the test dataset and the generated images.

Model	FID ↓
Unconditional Diffusion Model [17]	55.01

Figure 2. Qualitative evaluation of unconditional diffusion sampling. From left to right, we display sample images from the RGB domain of the Serengeti test dataset [8] and samples produced by the unconditional diffusion model [17].

3.1. The Identity—Using Near-Infrared Intensities

Physically near-infrared light is electromagnetic radiation with wavelengths between 750 nm and 1400 nm, while light from the visible spectrum lies in the range of 380 nm–780 nm [9]. The properties of an object determine which wavelength it reflects and absorbs. Hence, objects might have a strong reflectance of near-infrared light, resulting in high intensities for the observer, while absorbing more of the visible light leads to a lower intensity for the observer. In Figure 3, we show direct comparisons between the intensities of colored images (grayscale) and near-infrared images.

Figure 3. Qualitative comparison between near-infrared and visible-light images. From left to right, we display regular-colored images, intensity/grayscale images by averaging the 3 color channels and near-infrared images. All images are obtained from a dataset introduced by Brown and Süsstrunk [40], which consists of near-infrared and colored image pairs.

As visible, particularly for vegetation (last row), higher reflectance of near-infrared light in comparison to visible light is usual (Figure 3). The primary use case for near-infrared colorization in wildlife monitoring involves the colorization of nighttime images. In night images only a limited cone of illumination is available, resulting in diminished visibility of background elements such as vegetation. Consequently, one could argue in favor of disregarding the physical distinction between near-infrared and visible light and, instead, treating near-infrared images as approximations of the intensity. Song et al. [18] introduced a grayscale colorization method using diffusion models. In the context of our framework, this resolves to the identity function being the TRANSLATEINTENSITY function,

as shown in Section 2.2.1. Note, this is equivalent to the algorithm by Song et al. [18] for grayscale colorization.

In Figure 4, we present samples generated using this approach. We can observe that those generated images are faithful to the input image (Figure 4). Of course, this content-preservation is not a quality learned by the network but induced through our choice of intensity translation function. Most noteworthy is that the colors estimated by the diffusion model appear realistic. In comparison with DeOldify, the colorization is much more advanced. Images generated by DeOldify appear only dully colored. We argue this is because DeOldify has been trained for grayscale colorization and not for near-infrared colorization. Even though our method incorporates properties of grayscale colorization, it is more robust than DeOldify because it can estimate colors to any intensity. Qualitative weaknesses in colorization arise in comparison with CycleGAN: the diversity of images colored through this approach is low, and images appear uniformly colored. This is linked to the limitation of the approximation we made. As the intensity is strictly derived without margin, the chrominance has to be estimated for exactly this intensity. Thus, the choice of color is constrained.

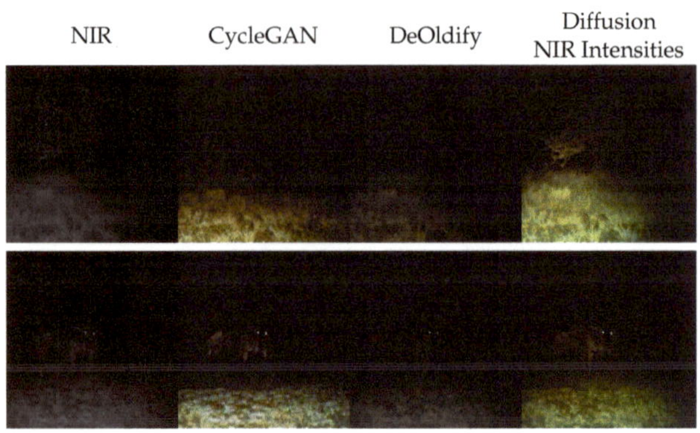

Figure 4. Qualitative evaluation of Iterative Seeding using NIR intensities. From left to right, we present near-infrared images from Snapshot Serengeti dataset [8]. Images generated by CycleGAN [14], by DeOldify [25], and by Iterative Seeding using the NIR implementation from Section 2.2.1.

Despite these minor weaknesses, the results suggest that our approximation of intensity using near-infrared intensity is reasonably accurate. It performs better than DeOldify but worse than CycleGAN, indicating a good result. As previously explained, this can be attributed to the specific condition of night images. In the majority of images, objects that reflect near-infrared light differently than visible light are typically in the background and, therefore, less illuminated. Thus, the near-infrared light approximates the visible light's intensity. CycleGAN, on the other hand, is not restricted to changing the intensity. It is merely trained to produce invertible images and thereby can manipulate the image in favor of realism.

Like the minor qualitative weakness, we also observe the FID of this naive approach to be 12.52 FID points worse than CycleGAN's. For NRQM, this holds too; however, only an irrelevant increase in NIQE is observable (Table 2). Additionally, we see a gap between the unconditional diffusion model and Iterative Seeding of 31.66 FID points. This indicates that this method is too restrictive to allow competitive image generation.

3.2. *Fusing Near-Infrared and Visible Intensity through High-Pass Filtering*

To address the limitation, we can draw inspiration from simple filter-based VIS-NIR fusion methods [41]. Note that the proposed framework can be considered fusioning visible and near-infrared light images in each diffusion step. Thus, it is obvious to apply techniques from the visible near-infrared fusion domain. A more refined approximation is

to use only the high-frequency details of the near-infrared image while the low frequencies can still be sampled by the diffusion model, as explained in Section 2.2.2. This intuitively also provides the diffusion model freedom to sample different illuminations than those provided by the near-infrared image.

Table 2. Quantitative evaluation of Iterative Seeding Using NIR Intensities. Samples of CycleGAN [14], DeOldify [25], and Iterative Seeding on NIR intensities (Section 2.2.1) are generated. The FID [36] is calculated by comparing the test dataset and the set of generated samples.

Model	FID ↓	NIQE ↓	NRQM ↑
CycleGAN [14]	74.15	14.06	5.45
DeOldify [25]	104.07	17.93	4.41
Iterative Seeding Using NIR Intensities	86.67	14.08	4.93

In Figure 5, we evaluate our framework with this translation function. Note that we used $\sigma = 2.3$ for the Gaussian filter employed in Section 2.2.2.

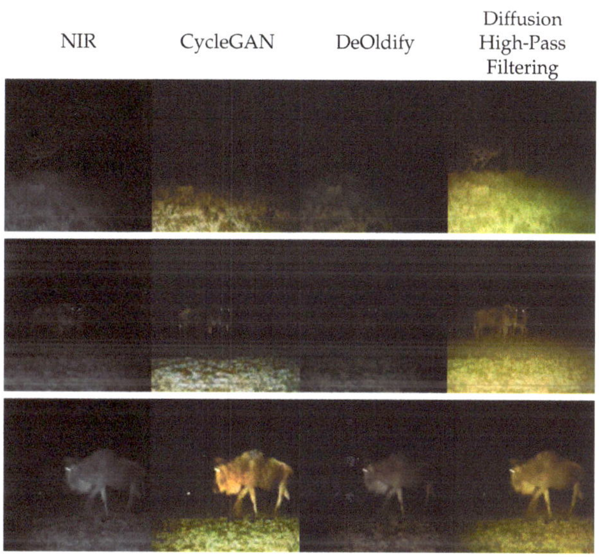

Figure 5. Qualitative evaluation of Iterative Seeding using high-pass filter. From left to right, we present near-infrared image from Snapshot Serengeti dataset [8], samples obtained from CycleGAN given the NIR image [14], from DeOldify [25], and from Iterative Seeding using high-pass filtering (Section 2.2.2).

Samples from the diffusion model appear realistic (Figure 5). Unlike CycleGAN, which modifies smaller regions, the diffusion model modifies the global illumination of the images. In the image of the gnu (middle left), we observe that it can even qualitatively exceed the performance of CycleGAN. In that particular case, the illumination of the scene matches the illumination of the gnu, which is not the case for CycleGAN's colorization. Additionally, Intensity Seeding using high-pass filtering performs better than DeOldify.

Moreover, it is noteworthy that the diffusion model generates a broader diversity of color schemes compared to CycleGAN: incandescent illuminations of the whole image, brown and green grass are all observable and represent the test dataset's distribution of images well (see Figure 2 for some samples from the test dataset). However, in general, we consider CycleGAN's colorization more realistic.

Concerning content preservation, apart from global illuminations of the scene, our framework using high-pass filtering hallucinates in a few instances blue sky, as illustrated in Figure 6.

Figure 6. Examples of hallucinations by the diffusion model sampling. Top row shows given near-infrared images from Snapshot Serengeti dataset [8] and bottom row the samples produced by Iterative Seeding using high-pass filtering (Section 2.2.2).

This can be attributed to a small proportion of the dataset containing images captured during dusk or dawn, where a blue sky is observable. For our test dataset, there were 56 images from 500 images that we consider to have such artifacts, which corresponds to a portion of 11.2%. Therefore, we consider CycleGAN's content preservation stronger.

In Table 3, we provide a quantitative comparison of both methods using the FID [36].

Table 3. Quantitative evaluation of Iterative Seeding using high-pass filter. We compare CycleGAN [14], DeOldify [25], and Iterative Seeding using high-pass filtering (Section 2.2.2) on the Snapshot Serengeti dataset [8]. The FID [36] is calculated between the test dataset and the generated images.

Model	FID ↓	NIQE ↓	NRQM ↑
CycleGAN [14]	74.15	14.06	5.45
DeOldify [25]	104.07	17.93	4.41
Iterative Seeding High-Pass Filtering	83.21	16.28	4.74

CycleGAN outperforms this approach in terms of FID by 9.06 points, as seen in Table 3. Generally, more realistic images most likely contribute to this quantitative difference. However, this result is still 3.46 FID points better than just using the NIR intensities (Table 2). The NIQE score of the approach is 2.22 points worse while NRQM is 0.33 points worse (Table 3). This change is justified by the increase in unrealistic hallucinations, as seen in Figure 6. Thus, by employing high-pass filtering, an outcome closer to CycleGAN can be attained.

While introducing the hyperparameter σ in Section 2.2.2, we further want to discuss its influence on the samples. As σ controls the strength of the Gaussian filter, increasing it reduces information in the extracted low frequencies. The rise of σ leads to more information in the extracted high frequencies. We visualize this effect in Figure 7. Remember, high-frequency intensities of the near-infrared image are combined with the generated low-frequency intensities. Thus, a higher σ also corresponds to more guidance by the near-infrared image, while a lower σ results in more freedom in generation for the model.

In Figure 8, we visualize how σ affects the sampling quality in terms of FID. We sample with the same seed, with the intention to have as few influences of randomness as possible. It is observable that samples with a σ of less than 1 are quantitatively worse (Figure 8). Our hypothesis is that, at this stage, some guidance is provided, but it is insufficient, making sampling within these boundaries too challenging for the model. At σ of 2.3, the minimum

FID is reached. At this point, an adequate amount of guidance is provided for the model to perform reasonable sampling of color, without imposing too many constraints limiting the model's generation freedom.

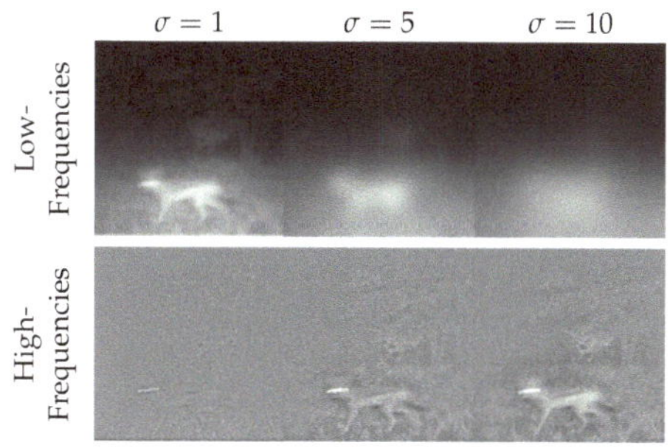

Figure 7. Influence of σ to low and high frequencies. In the top row, we show low frequencies of the image, and in the bottom row, we present high frequencies of the image. From left to right, we display decomposition into low and high frequencies according to Section 2.2.2 for $\sigma = 1$, $\sigma = 5$, and $\sigma = 10$.

For higher σ, the diffusion model receives more guidance, and consequently, it has less freedom for its generation process. Thus, the FID rises. On the other hand, we observe by comparing manually that the proportion of hallucinated blue sky becomes less frequent.

Hence, we regard σ as a hyperparameter that controls the compromise between realism and content preservation.

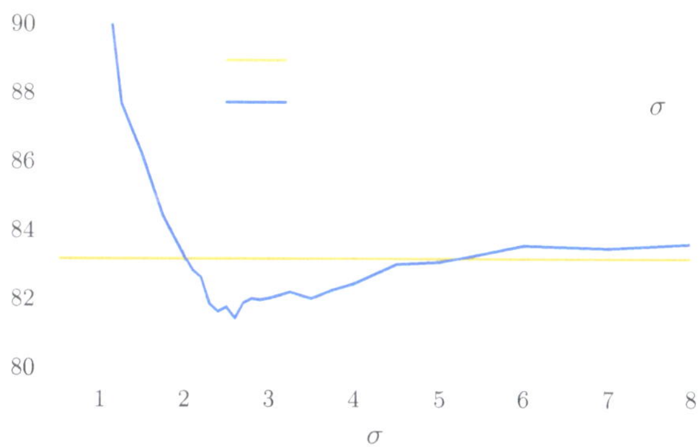

Figure 8. Influence of hyperparameter σ on the FID. On the y-axis, the FID (lower is better) and on the x-axis σ with which was sampled. The blue line shows the FID of Iterative Seeding using high-pass filtering with respect to σ (Section 2.2.2), and the yellow line shows the FID of Iterative Seeding using NIR intensities (Section 2.2.1) for comparison. All FID scores are obtained using the same random seed to reduce outliers.

3.3. The Potential of Diffusion-Based Near-Infrared Colorization—CycleGAN as Intensity Translation Function

We note that our existing translation function implementations do not result in the diffusion model surpassing CycleGAN either quantitatively or qualitatively. However, the translation function we employed was of a trivial nature and did not exhaust the research results from the near-infrared visible fusion field (e.g., see advanced approaches in [20]). Nevertheless, it does generate realistic images (Figure 5), achieves FID scores close to CycleGAN (Table 3), and performs better than the identity (Table 2).

This indicates there is unexhausted potential for this framework. To prove this, we use a translation function known to generate good results: CycleGAN itself. Precisely, this translation function evaluates the trained CycleGAN colorization for the NIR image and uses only the intensity of this image as input for the diffusion model. Therefore, our model still estimates the color but uses the intensity generated from CycleGAN. In Figure 9, we present samples using this translation function.

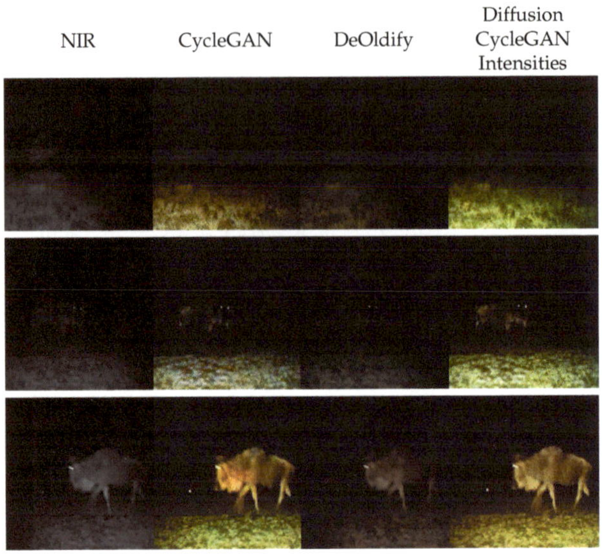

Figure 9. Qualitative evaluation of Iterative Seeding using CycleGAN intensities. From left to right, we present near-infrared images from Snapshot Serengeti dataset [8], images generated by CycleGAN [14], by DeOldify [25], and by Iterative Seeding using CycleGAN intensities.

We observe the diffusion model does not only effectively colorize these images but also surpasses CycleGAN in terms of color selection Figure 9. Unlike CycleGAN, the gnu generated by the diffusion model has matching illumination to the rest of the image.

In quantitative terms, this approach performs 4.47 FID points better than CycleGAN itself (Table 4). The same holds for the NIQE and NRQM metric which indicates a slightly improved realism. Because the model does not only achieve a similar score to CycleGAN using its intensity but also exceeds CycleGAN's score, the potential of diffusion-based colorization is shown. CycleGAN generating the intensity by itself, does not generate colors as good as the diffusion model does.

Table 4. Quantitative evaluation of Iterative Seeding using CycleGAN intensities. Samples are obtained from CycleGAN [14], from DeOldify [25], and from Iterative Seeding using CycleGAN intensities. The FID [36] is calculated by comparing the test dataset and the set of generated samples.

Model	FID ↓	NIQE ↓	NRQM ↑
CycleGAN [14]	74.15	14.06	5.45
DeOldify [25]	104.07	17.93	4.41
Iterative Seeding CycleGAN Intensities	**69.68**	**13.47**	**5.47**

4. Discussion

We introduce a framework for diffusion-based NIR image colorization. It abstracts the translation of intensity and utilizes diffusion models for the effective colorization.

4.1. Implementations of Intensity Translation

We propose and evaluate three variants for the implementation of the intensity translation function.

1. We demonstrate directly using the NIR intensity, effectively representing an identity function. This simple implementation creates a basic colorization, but the diffusion network is too restricted by the NIR intensity, yielding suboptimal realism and FID scores. Even though this colorization is not specialized for near-infrared colorization, it performs more robustly than other grayscale colorization methods, as the comparison with DeOldify suggests.
2. Because the framework can be considered iteratively fusing near-infrared and visible light, we draw inspiration from the VIS-NIR fusion research domain. A key observation made in this field is to use the high-frequency intensities of the near-infrared image and fuse them with the low-frequency intensities of the visible-light image. We apply this insight to our framework in a trivial implementation using a Gaussian filter. An improvement in comparison with just using the NIR images as intensities is observed quantitatively and qualitatively, resulting in an FID score close to our baseline CycleGAN. Additionally, with the ClipFID, a different FID variant not relying on the Inception model and the ImageNet dataset [42], we achieve a score of 7.87 using this translation method compared to CycleGAN 9.15. However, because the ClipFID is not established as a comparison metric, this result has to be treated carefully. Even though this implementation is far from exhausting results from the VIS-NIR fusion domain, it achieves scores close to our baseline, suggesting a more sophisticated implementation can achieve even better results.
3. Finally, we evaluate CycleGAN itself as an intensity translator. Using intensities generated by CycleGAN our framework outperforms CycleGAN quantitatively as well as qualitatively. This indicates the potential of our framework for sophisticated translation functions and diffusion-based NIR colorization in general. Considering this potential, we show that our framework reduces NIR colorization to visible near-infrared fusion, a simpler problem.

4.2. Colorizing NIR Images for Animal Detection

The integration of camera trapping with artificial intelligence (AI), particularly leveraging deep learning methodologies, represents a significant advancement in wildlife research and conservation [3,7,11,12]. Nevertheless, numerous deep learning models are optimized for and perform better with colored images, akin to human perception [10]. To explore this phenomenon, we assess the efficacy of the proposed diffusion-based NIR colorization technique in enhancing image classification within camera trap datasets.

We utilize a subset of randomly selected night images in near-infrared (NIR) from the Snapshot Serengeti dataset [8]. This subset is divided into 4000 images for training and 500 each for the validation and test datasets. Subsequently, all 5000 images are colorized using each method outlined. For every method, we fine-tune a ResNet50 [24] classifier pretrained on ImageNet-21K [43] on the training dataset derived from the respective method. We use a cross-entropy loss with an Adam optimizer ($\beta = (0.9, 0.999)$ and $\text{lr} = 10^{-4}$). Finally, we evaluate the model on the test dataset acquired from each method. This experiment is repeated five times, and the average over all five accuracies is used. Additionally, we repeat this evaluation for a ResNet using the same pretrained weights but with freezing all layers except the final classification layer. We argue this score is a measurement for content preservation, as the network can only classify accurately if the relevant content is translated.

Table 5 displays the classification accuracies of various methods. In both the frozen and unfrozen scenarios, both CycleGAN and our method utilizing its intensities achieve similar accuracies. However, our methods, employing near-infrared intensities or employing high-pass filtering, outperform both CycleGAN and the diffusion approach utilizing CycleGAN's intensities. Specifically, for the unfrozen network, and even more significantly

for the frozen network, our methods achieve an accuracy improvement of approximately 6.83 percentage points.

Table 5. Quantitative evaluation of classification using colorized images Comparison of FID and classification accuracy on images from the NIR dataset, samples from CycleGAN [14] and Iterative Seeding using all presented implementations. The FID [36] is calculated by comparing the test dataset and the set of generated samples. Classification accuracy is obtained by training a ResNet classifier for each method and calculating the accuracy afterward on a test dataset. We either train all layers (non-frozen) or freeze all but the final classification layer (frozen). The accuracy is averaged over 5 runs, and σ displays the corresponding standard deviation.

		Non-Frozen		Frozen	
Model	**FID** ↓	**Accuracy** ↑	σ	**Accuracy** ↑	σ
NIR	-	**0.7276**	0.0124	0.3628	0.0105
CycleGAN [14]	74.15	0.5341	0.0146	0.3447	0.0123
NIR Intensities	86.67	0.6024	0.0102	0.3554	0.0213
High-Pass Filtering	83.21	0.6078	0.0119	**0.3681**	0.0090
CycleGAN Intensities	**69.68**	0.5314	0.0180	0.3367	0.0116

However, when unfrozen, none of the methods used yield improved classification accuracy compared to directly utilizing near-infrared intensities. Conversely, for the frozen network simulating few-shot learning, using high-pass filtering results in accuracies rivaling those achieved by direct NIR intensity utilization. However, the improvement of our variant is only that marginal such that it lies in the standard deviation of both accuracies. This could be attributed to the robustness of ResNet [24], a powerful deep learning approach for object classification, which is not specialized for colored images and can handle various input formats adeptly. When the network's backbone is frozen the feature extraction of the network is settled, and only the classification using those features is trained. Thus, the accuracy benefits from colored images more. The lower accuracies in general result from the fact that the network is also in general more restricted to adjust to the given images.

4.3. *Colorizing NIR Images for Animal Detection Explainability*

We employ the AI-based ResNet [24] approach for visual animal detection. An improvement of the classification accuracy by using colorized images can only marginally be observed.

However, we have to take into account user acceptance and explainability of AI-based approaches to animal detection. Many AI-based systems and especially deep learning (DL) methods (like ResNet) are black-box models that are extremely hard to explain and to understand even by domain experts [44]. Explainable AI (XAI) and explainable machine learning refer to AI approaches that allow users to retain understanding and acceptance. Many AI-based and especially DL-based approaches to object recognition in general and animal recognition in particular are so-called data-centric AI (DCAI) methods. DCAI shifts the focus from hand-crafted model building to curating high-quality, consistently annotated training datasets.

Therefore, understanding and accepting AI-based animal recognition heavily relies on the understanding and acceptance of the employed training datasets, i.e., the training images. Using NIR images for training AI-based systems for animal detection decreases user acceptance because their appearance does not match with human perception [10,21]. Thus, our approach to NIR image colorization improves the utilization of NIR images and NIR video clips for education, promotion, and funding acquisition.

4.4. *Novelty and Scientific Relevance*

The novelty of our approach lies primarily in the iteration step during inference, where we merge the current sample with the given image. Unlike existing methods, we differ-

entiate between merging chrominance and intensity components. Specifically, we extract chrominance from the input image, and we introduce a novel abstraction of intensity merging to suit diffusion models for near-infrared colorization. This innovation is motivated by the three distinct implementations we present, each demonstrating the significance of the performance of this crucial step. From a scientific standpoint, our framework serves as a general framework for diffusion-based near-infrared colorization techniques.

By showcasing the superiority of diffusion-based methods over GAN-based approaches, we contribute to the near-infrared colorization research. Moreover, we present an intersection of near-infrared colorization, near-infrared-visible fusion, and diffusion models, thus contributing to the advancement of these interconnected fields.

4.5. Limitations and Challenges

One limitation of a diffusion model approach is that it requires much more computational resources in training and inference than CycleGAN. This manifests also in the training and sampling durations: although we trained CycleGAN over the course of two days and inference is a matter of mere seconds, the diffusion network required two weeks for training, and generating 500 samples took approximately 40 min on an NVIDIA RTX A5000.

One challenge of our approach is the design of the intensity translation and merging process. It has to incorporate enough information from the given image to preserve the content while it should not use too much to generate a realistic RGB image. We present three example methods for this process and prove its potential; however, future research is needed to find an optimal method. Inspiration can be drawn from further VIS-NIR fusion research; alternatively, different machine learning techniques can also be applied to solve this subproblem (see Section 5.1).

Another challenge faced during this study is the evaluation process. As we focus on unpaired image translation and only have such a dataset, paired evaluation techniques such as mean-squared distances and SSIM can not be used. Instead, we solved this using the unpaired dataset distance FID [36], measures like the classification accuracy and, lastly, no-reference image quality assessment metrics NIQE [38] and NRQM [39]. Although combining all these metrics does provide a robust evaluation, a paired dataset for evaluation and unpaired for training would be optimal (see Section 5.1).

5. Conclusions

This study presents the first framework utilizing diffusion models for the colorization of near-infrared (NIR) images. We show that the effectiveness of colorizing NIR images is primarily controlled by the translation of the intensities of near-infrared light to those of visible light.

1. Iterative Seeding on NIR intensities (ISNIR);
2. Iterative Seeding using high-pass filtering (ISHP);
3. Iterative Seeding using CycleGAN intensities (ISCG).

Inspired by research from visible near-infrared fusion, we have shown that even employing ISHP as a simple algorithm for translating NIR intensities achieves FID scores close to the GAN baseline. Thus, we establish a connection between near-infrared colorization, diffusion models, and visible near-infrared fusion.

Furthermore, our framework is shown to outperform the GAN baseline with the ISCG implementation as indicated by decreasing FID, NIQE, and NRQM values.

In general, our method bridges the domain gap between near-infrared and colored images and addresses challenges of near-infrared colorization including the lack of paired training data, as well as the different reflectance properties of near-infrared and visible light.

5.1. Future Work

For future research, several variations to our proposed framework are feasible. First, recent advances in latent diffusion models (LDMs) [45] could potentially allow

sampling of higher resolutions, such as 1024×1024, increase the sampling speed, and therefore mitigate the drawbacks of our approach in a practical implementation. Deterministic samplers such as DDIM [46] or using a formulation based on ordinary differential equations, as demonstrated by Song et al. [18], could contribute to this.

One potential direction for future research involves exploring more sophisticated approaches from the VIS-NIR fusion field for implementing our framework [20]. Additionally, a hyperparameter scaling the score function, as done by Dhariwal and Nichol [17] for classifier guidance, could allow a built-in approach for controlling the trade-off between realism and content preservation.

Our method, although presented in the context of wildlife monitoring, may not be restricted to it. Therefore, an evaluation of other datasets and application contexts, such as in driver assistance systems could reveal valuable insights. A paired dataset for evaluation and unpaired for training would additionally contribute to an optimal study evaluation.

Finally, using high-frequency details of the near-infrared image to enhance the colored image is a concept not reserved for diffusion models, e.g., introducing a loss between the high-frequency intensities of the generated and given images could potentially benefit CycleGAN, too.

Author Contributions: Conceptualization, A.B., T.H. and V.S.; methodology, A.B., T.H. and V.S.; software, A.B.; validation, A.B., T.H. and V.S.; formal analysis, A.B. and T.H.; investigation, A.B. and T.H.; resources, V.S.; data curation, A.B.; writing—original draft preparation, A.B.; writing—review and editing, T.H. and V.S.; visualization, A.B; supervision, T.H. and V.S.; project administration, V.S.; funding acquisition, V.S. All authors have read and agreed to the published version of the manuscript.

Funding: This research was partially funded by the German Federal Ministry of Education and Research (Bundesministerium für Bildung und Forschung), Bonn, Germany (AMMOD—Automated Multisensor Stations for Monitoring of BioDiversity: FKZ 01LC1903B).

Institutional Review Board Statement: Not applicable.

Informed Consent Statement: Not applicable.

Data Availability Statement: Source code for downloading the evaluation dataset, for our evaluations, and for our proposed framework is available at https://github.com/aykborstelmann/nir-coloring (accessed on 26 February 2024). Our dataset is based on the Snapshot Serengeti dataset [8], which is openly available, as well at https://doi.org/10.5061/dryad.5pt92 (accessed on 26 February 2024).

Acknowledgments: We thank Frank Schindler for proofreading the manuscript.

Conflicts of Interest: The authors declare no conflicts of interest. The funders had no role in the design of the study; in the collection, analyses, or interpretation of data; in the writing of the manuscript; or in the decision to publish the results.

Abbreviations

The following abbreviations are used in this manuscript:

NIR	near-infrared
RGB	red, green, blue (color channels)
VIS-NIR fusion	visible-near-infrared fusion
GAN	generative adversarial network
ILVR	iterative latent variable refinement
FID	Fréchet inception distance
CNN	convolutional neural network

References

1. Haucke, T.; Kühl, H.S.; Hoyer, J.; Steinhage, V. Overcoming the distance estimation bottleneck in estimating animal abundance with camera traps. *Ecol. Inform.* **2022**, *68*, 101536. [CrossRef]
2. Palencia, P.; Fernández-López, J.; Vicente, J.; Acevedo, P. Innovations in movement and behavioural ecology from camera traps: Day range as model parameter. *Methods Ecol. Evol.* **2021**, *12*, 1201–1212. [CrossRef]

3. Schindler, F.; Steinhage, V.; van Beeck Calkoen, S.T.S.; Heurich, M. Action Detection for Wildlife Monitoring with Camera Traps Based on Segmentation with Filtering of Tracklets (SWIFT) and Mask-Guided Action Recognition (MAROON). *Appl. Sci.* **2024**, *14*, 514. [CrossRef]
4. Oliver, R.Y.; Iannarilli, F.; Ahumada, J.; Fegraus, E.; Flores, N.; Kays, R.; Birch, T.; Ranipeta, A.; Rogan, M.S.; Sica, Y.V.; et al. Camera trapping expands the view into global biodiversity and its change. *Philos. Trans. R. Soc. B Biol. Sci.* **2023**, *378*, 20220232. [CrossRef]
5. Green, S.E.; Stephens, P.A.; Whittingham, M.J.; Hill, R.A. Camera trapping with photos and videos: Implications for ecology and citizen science. *Remote Sens. Ecol. Conserv.* **2023**, *9*, 268–283. [CrossRef]
6. Edelman, A.J.; Edelman, J.L. An Inquiry-Based Approach to Engaging Undergraduate Students in On-Campus Conservation Research Using Camera Traps. *Southeast. Nat.* **2017**, *16*, 58–69. [CrossRef]
7. Wägele, J.W.; Bodesheim, P.; Bourlat, S.J.; Denzler, J.; Diepenbroek, M.; Fonseca, V.G.; Frommolt, K.H.; Geiger, M.F.; Gemeinholzer, B.; Glöckner, F.O.; et al. Towards a multisensor station for automated biodiversity monitoring. *Basic Appl. Ecol.* **2022**, *59*, 105–138 [CrossRef]
8. Swanson, A.; Kosmala, M.; Lintott, C.; Simpson, R.; Smith, A.; Packer, C. Data from: Snapshot Serengeti, high-frequency annotated camera trap images of 40 mammalian species in an African savanna. *Sci. Data* **2015**, *2*, 150026. [CrossRef]
9. *ISO 20473:2007*; Optics and Photonics—Spectral Bands. International Organization for Standardization: Geneva, Switzerland, 2007.
10. Toet, A.; Hogervorst, M.A. Progress in color night vision. *Opt. Eng.* **2012**, *51*, 010901. [CrossRef]
11. Adam, M.; Tomášek, P.; Lehejček, J.; Trojan, J.; Jůnek, T. The Role of Citizen Science and Deep Learning in Camera Trapping. *Sustainability* **2021**, *13*, 287. [CrossRef]
12. Simões, F.; Bouveyron, C.; Precioso, F. DeepWILD: Wildlife Identification, Localisation and estimation on camera trap videos using Deep learning. *Ecol. Inform.* **2023**, *75*, 102095. [CrossRef]
13. Gao, R.; Zheng, S.; He, J.; Shen, L. CycleGAN-Based Image Translation for Near-Infrared Camera-Trap Image Recognition. In Proceedings of the Pattern Recognition and Artificial Intelligence: International Conference, ICPRAI 2020, Zhongshan, China, 19–23 October 2020; Springer International Publishing: Cham, Switzerland, 2020; pp. 453–464.
14. Mehri, A.; Sappa, A.D. Colorizing near infrared images through a cyclic adversarial approach of unpaired samples. In Proceedings of the IEEE/CVF Conference on Computer Vision and Pattern Recognition Workshops, Long Beach, CA, USA, 16–17 June 2019.
15. Metz, L.; Poole, B.; Pfau, D.; Sohl-Dickstein, J. Unrolled generative adversarial networks. *arXiv* **2016**, arXiv:1611.02163.
16. Ho, J.; Jain, A.; Abbeel, P. Denoising diffusion probabilistic models. *Adv. Neural Inf. Process. Syst.* **2020**, *33*, 6840–6851.
17. Dhariwal, P.; Nichol, A. Diffusion models beat gans on image synthesis. *Adv. Neural Inf. Process. Syst.* **2021**, *34*, 8780–8794.
18. Song, Y.; Sohl-Dickstein, J.; Kingma, D.P.; Kumar, A.; Ermon, S.; Poole, B. Score-based generative modeling through stochastic differential equations. *arXiv* **2020**, arXiv:2011.13456.
19. Choi, J.; Kim, S.; Jeong, Y.; Gwon, Y.; Yoon, S. Ilvr: Conditioning method for denoising diffusion probabilistic models. *arXiv* **2021**, arXiv:2108.02938.
20. Sharma, A.M.; Dogra, A.; Goyal, B.; Vig, R.; Agrawal, S. From pyramids to state-of-the-art: A study and comprehensive comparison of visible–infrared image fusion techniques. *IET Image Process.* **2020**, *14*, 1671–1689. [CrossRef]
21. Limmer, M.; Lensch, H.P. Infrared colorization using deep convolutional neural networks. In Proceedings of the 2016 15th IEEE International Conference on Machine Learning and Applications (ICMLA), Anaheim, CA, USA, 18–20 December 2016; pp. 61–68.
22. Dong, Z.; Kamata, S.i.; Breckon, T.P. Infrared image colorization using a s-shape network. In Proceedings of the 2018 25th IEEE International Conference on Image Processing (ICIP), Athens, Greece, 7–10 October 2018; pp. 2242–2246.
23. Zhu, J.Y.; Park, T.; Isola, P.; Efros, A.A. Unpaired image-to-image translation using cycle-consistent adversarial networks. In Proceedings of the IEEE International Conference on Computer Vision, Venice, Italy, 22–29 October 2017; pp. 2223–2232.
24. He, K.; Zhang, X.; Ren, S.; Sun, J. Deep residual learning for image recognition. In Proceedings of the IEEE Conference on Computer Vision and Pattern Recognition, Las Vegas, NV, USA, 27–30 June 2016; pp. 770–778.
25. Antic, J. Deoldify. Available online: https://github.com/jantic/DeOldify (accessed on 24 January 2024).
26. Sohl-Dickstein, J.; Weiss, E.; Maheswaranathan, N.; Ganguli, S. Deep unsupervised learning using nonequilibrium thermodynamics. In Proceedings of the International Conference on Machine Learning, PMLR, Lille, France, 7–9 July 2015; pp. 2256–2265.
27. Song, Y.; Ermon, S. Generative modeling by estimating gradients of the data distribution. In Proceedings of the Advances in Neural Information Processing Systems, Vancouver, BC, Canada, 8–14 December 2019; Volume 32.
28. Welling, M.; Teh, Y.W. Bayesian learning via stochastic gradient Langevin dynamics. In Proceedings of the 28th International Conference on Machine Learning (ICML-11), Bellevue, WA, USA, 28 June–2 July 2011; pp. 681–688.
29. Saharia, C.; Chan, W.; Chang, H.; Lee, C.; Ho, J.; Salimans, T.; Fleet, D.; Norouzi, M. Palette: Image-to-image diffusion models. In Proceedings of the ACM SIGGRAPH 2022 Conference Proceedings, Vancouver, BC, Canada, 7–11 August 2022; pp. 1–10.
30. Zhao, M.; Bao, F.; Li, C.; Zhu, J. Egsde: Unpaired image-to-image translation via energy-guided stochastic differential equations. *Adv. Neural Inf. Process. Syst.* **2022**, *35*, 3609–3623.
31. Zhu, Y.; Sun, X.; Zhang, H.; Wang, J.; Fu, X. Near-infrared and visible fusion for image enhancement based on multi-scale decomposition with rolling WLSF. *Infrared Phys. Technol.* **2023**, *128*, 104434. [CrossRef]
32. Bulanon, D.; Burks, T.; Alchanatis, V. Image fusion of visible and thermal images for fruit detection. *Biosyst. Eng.* **2009**, *103*, 12–22. [CrossRef]

33. Ronneberger, O.; Fischer, P.; Brox, T. U-net: Convolutional networks for biomedical image segmentation. In *Medical Image Computing and Computer-Assisted Intervention*; Springer: Berlin/Heidelberg, Germany, 2015; pp. 234–241.
34. Forsyth, D.A.; Ponce, J. *Computer Vision: A Modern Approach*; Prentice Hall: Upper Saddle River, NJ, USA, 2002.
35. Wang, Z.; Bovik, A.C.; Sheikh, H.R.; Simoncelli, E.P. Image quality assessment: From error visibility to structural similarity. *IEEE Trans. Image Process.* **2004**, *13*, 600–612. [CrossRef] [PubMed]
36. Heusel, M.; Ramsauer, H.; Unterthiner, T.; Nessler, B.; Hochreiter, S. Gans trained by a two time-scale update rule converge to a local nash equilibrium. In Proceedings of the Advances in Neural Information Processing Systems, Long Beach, CA, USA, 4–9 December 2017; Volume 30.
37. Parmar, G.; Zhang, R.; Zhu, J.Y. On aliased resizing and surprising subtleties in gan evaluation. In Proceedings of the IEEE/CVF Conference on Computer Vision and Pattern Recognition, New Orleans, LA, USA, 18–24 June 2022; pp. 11410–11420.
38. Mittal, A.; Soundararajan, R.; Bovik, A.C. Making a "completely blind" image quality analyzer. *IEEE Signal Process. Lett.* **2012**, *20*, 209–212. [CrossRef]
39. Ma, C.; Yang, C.Y.; Yang, X.; Yang, M.H. Learning a no-reference quality metric for single-image super-resolution. *Comput. Vis. Image Underst.* **2017**, *158*, 1–16. [CrossRef]
40. Brown, M.; Süsstrunk, S. Multi-spectral SIFT for scene category recognition. In Proceedings of the CVPR 2011, Colorado Springs, CO, USA, 20–25 June 2011; pp. 177–184.
41. Sharma, V.; Hardeberg, J.Y.; George, S. RGB–NIR image enhancement by fusing bilateral and weighted least squares filters. In Proceedings of the Color and Imaging Conference, Society for Imaging Science and Technology, Scottsdale, AZ, USA, 13–17 November 2017; Volume 2017, pp. 330–338.
42. Kynkäänniemi, T.; Karras, T.; Aittala, M.; Aila, T.; Lehtinen, J. The Role of ImageNet Classes in Fréchet Inception Distance. *arXiv* **2022**, arXiv:2203.06026.
43. Deng, J.; Dong, W.; Socher, R.; Li, L.J.; Li, K.; Fei-Fei, L. Imagenet: A large-scale hierarchical image database. In Proceedings of the 2009 IEEE Conference on Computer Vision and Pattern Recognition, Miami, FL, USA, 20–25 June 2009; pp. 248–255.
44. Loyola-González, O. Black-Box vs. White-Box: Understanding Their Advantages and Weaknesses From a Practical Point of View. *IEEE Access* **2019**, *7*, 154096–154113. [CrossRef]
45. Rombach, R.; Blattmann, A.; Lorenz, D.; Esser, P.; Ommer, B. High-resolution image synthesis with latent diffusion models. In Proceedings of the IEEE/CVF Conference on Computer Vision and Pattern Recognition, New Orleans, LA, USA, 18–24 June 2022; pp. 10684–10695.
46. Song, J.; Meng, C.; Ermon, S. Denoising diffusion implicit models. *arXiv* **2020**, arXiv:2010.02502.

sensors

MDPI

Article

Acceleration of Hyperspectral Skin Cancer Image Classification through Parallel Machine-Learning Methods

Bernardo Petracchi, Emanuele Torti, Elisa Marenzi and Francesco Leporati *

Department of Electrical, Computer and Biomedical Engineering, University of Pavia, I-27100 Pavia, Italy; bernardo.petracchi01@universitadipavia.it (B.P.); emanuele.torti@unipv.it (E.T.); elisa.marenzi@unipv.it (E.M.)
* Correspondence: francesco.leporati@unipv.it

Abstract: Hyperspectral imaging (HSI) has become a very compelling technique in different scientific areas; indeed, many researchers use it in the fields of remote sensing, agriculture, forensics, and medicine. In the latter, HSI plays a crucial role as a diagnostic support and for surgery guidance. However, the computational effort in elaborating hyperspectral data is not trivial. Furthermore, the demand for detecting diseases in a short time is undeniable. In this paper, we take up this challenge by parallelizing three machine-learning methods among those that are the most intensively used: Support Vector Machine (SVM), Random Forest (RF), and eXtreme Gradient Boosting (XGB) algorithms using the Compute Unified Device Architecture (CUDA) to accelerate the classification of hyperspectral skin cancer images. They all showed a good performance in HS image classification, in particular when the size of the dataset is limited, as demonstrated in the literature. We illustrate the parallelization techniques adopted for each approach, highlighting the suitability of Graphical Processing Units (GPUs) to this aim. Experimental results show that parallel SVM and XGB algorithms significantly improve the classification times in comparison with their serial counterparts.

Keywords: hyperspectral imaging; machine learning; support vector machine; random forest; eXtreme gradient boosting; GPU

Citation: Petracchi, B.; Torti, E.; Marenzi, E.; Leporati, F. Acceleration of Hyperspectral Skin Cancer Image Classification through Parallel Machine-Learning Methods. *Sensors* **2024**, *24*, 1399. https://doi.org/10.3390/s24051399

Academic Editors: Christos Nikolaos E. Anagnostopoulos, Stelios Krinidis and Jan Cornelis

Received: 6 December 2023
Revised: 29 January 2024
Accepted: 16 February 2024
Published: 21 February 2024

1. Introduction

Skin cancer represents one of the most predominant tumors [1], and in recent years, its occurrence has progressively increased. Such lesions are typically categorized into two main groups: melanoma skin cancer (MSC) and non-melanoma skin cancer (NMSC) [2]. Typically, this cancer type involves three types of cells: squamous, basal, or melanocytic cells.

MSC originates from melanocytes, cells located in the epidermis and responsible for skin color, thanks to melanin production. MSC can be further subdivided into three subtypes: superficial extension, lentigo maligna, and nodular tumor [3]. This is the rarest type of skin cancer, with, if not promptly detected, the highest growth speed and, consequently, is very difficult to treat [4]. Therefore, doctors and surgeons need fast, reliable diagnostic systems for this kind of pathology.

The traditional diagnosis procedure is biopsy, which consists in the removal of a sample of tissue from the living body, followed by histopathological inspection [5,6], representing an onerous and time-consuming process [5–7].

To face these problems, minimally intrusive techniques have been investigated, including hyperspectral imaging (HSI), acquiring information about a scene both in the spatial and in the spectral domain [8]. In fact, a hyperspectral image is represented by a so-called hypercube containing the spectral information of every pixel over a specific wavelength range. HSI allows precise material identification [9] by measuring the fraction of the incident electromagnetic radiation reflected by the surface (reflectance). This is due to the characteristic variation in the reflectance over the wavelength typical of each material, which is called the spectral signature [10]. In contrast with traditional imaging

techniques, HSI allows the acquisition of images with a large number of spectral bands both within the visible and non-visible range. This means that the acquired images contain much more information compared to traditional ones, such as RGB images, and can lead to better performances [11].

However, although the development of accurate tools in the medical field is fundamental, timing requirements should also be taken into consideration when providing a quick diagnosis is necessary. Indeed, the prompt detection of skin lesions facilitates their treatment and increases the probability of survival of the patients.

To achieve this goal, many researchers [12–17] have exploited different kinds of devices suitable for parallel elaboration and computation when the data size is high. Among these, Graphical Processing Units (GPUs), used in different scientific applications [18,19], represent a suitable technology in the field of medical image processing. In addition, compared with other devices such as Field Programmable Gate Arrays (FPGAs), GPUs usually offer a bigger parallel factor due to their high memory bandwidth [20].

Existing works in the literature have focused on the classification of HSI skin cancer images by adopting machine-learning (ML) and deep-learning (DL) methods [11,16,21–31].

In [16], a classification chain based on K-means, Spectral Angle Mapper (SAM), and SVM was considered. The authors also implemented several parallel versions of their classification system exploiting multicore and many-core technologies.

The research in [31] implemented SVM, RF, and XGB, obtaining a mean classification accuracy of 97%, considering only the model's optimization and not the algorithms' parallelization.

Several DL models have been adopted in [32], namely, ResNet-18, ResNet-50, ResNet-101, a ResNet-50 variant, U-Net, and U-Net++ architectures. Since neural networks are time-consuming and computationally expensive, a parallel version of the U-Net++, resulting in the best predictive approach, has been implemented using a low-power NVIDIA Jetson GPU. This parallel version has achieved adequate classification performance satisfying real-time constraints with a low power consumption.

Some works related to ML method parallelization can be found in [16,33], where parallel versions of SVM and XGB have been developed for HSI image classification.

In this paper, we propose the optimization and parallelization of three popular ML methods to accelerate the HSI skin cancer image classification using the Compute Unified Device Architecture (CUDA), a framework for parallel elaboration developed by NVIDIA. More specifically, the considered approaches are SVM, RF, and XGB, which offer a good performance in classifying HSI images when the dimensions of the dataset are limited [31,34]. Furthermore, the works in [16,33,35] showed a great reduction in the classification time developing parallel versions of SVM and XGB, even achieving real-time processing.

This work presents the parallelization techniques implemented on different NVIDIA GPU devices including a GeForce RTX 2080 GPU, a GeForce RTX 4090 GPU, and a cluster composed of five nodes of three Tesla A16 GPUs. Performance differences between the devices in the classification of HSI skin cancer images have also been highlighted. Indeed, GeForce RTX 2080 and 4090 GPUs are optimized for graphics applications, while the cluster is designed for scientific calculations. In particular, the GeForce RTX 4090 is characterized by the latest-generation architecture (Ada Lovelace), while the GeForce RTX 2080 features an older architecture (Turing) and is cheaper than the previous one. Lastly, each Tesla A16 features an Ampere architecture.

Experimental results show a significant improvement of the parallel version of SVM and XGB compared to their serial counterparts, with a speed-up of 130x and 1.4x, respectively, confirming that GPUs represent a valid technology in accelerating the medical diagnosis process.

This manuscript is organized as follows. Section 2 describes the HSI skin cancer dataset and the adopted ML algorithms. Furthermore, the adopted techniques to perform the serial and the parallel inference of the algorithms, and the architectures of the adopted

devices are shown. The obtained results are illustrated in Section 3, while Section 4 presents the discussions, and Section 5 provides conclusions and future developments.

The main contributions of this paper are the following: description of the parallelization of the SVM, RF, and XGB methods targeting GPUs; parallelization on different devices, considering the most recent architectures developed by NVIDIA; and comparison of the results with the state of the art, highlighting the improvement of skin cancer diagnosis through parallel image processing.

2. Materials and Methods

2.1. Hyperspectral Sensors and the Skin Cancer Dataset

The evolution of hyperspectral sensors has resulted in the creation of various platforms, specialized for particular applications and operational needs. The four main sensor types, namely pushbroom, whiskbroom, stereoscopic, and snapshot are fundamental to the hyperspectral imaging landscape [36–38]. Pushbroom sensors function through constant scanning of the scene using a linear or 2D array of detectors. As the platform moves, the sensor captures spectral information for every pixel in the scene, resulting in a continuous spectral image. This technique enhances both spatial and spectral resolution, making pushbroom sensors highly suitable for applications that demand a thorough analysis of specific regions [39].

Whiskbroom sensors operate similarly to pushbroom ones, except for their scanning mechanism. Rather than recording an entire line at once, whiskbroom sensors collect data one point at a time. The sensor sweeps across the scene, gathering spectral information for each point sequentially. Whiskbroom sensors are celebrated for their adaptability and are frequently utilized in airborne and spaceborne reconnaissance [40].

Stereoscopic hyperspectral sensors employ several detectors to capture images from marginally divergent viewpoints. By leveraging stereoscopic vision, these sensors provide not only spectral data but also depth information. This facilitates the creation of 3D models and improves the interpretation of intricate surroundings, such as hilly terrains or urban landscapes [41].

Snapshot sensors, also referred to as snapshot hyperspectral imaging systems, obtain a complete spectral image with a single exposure. This is accomplished through cutting-edge optical designs that record data concurrently for all spectral ranges. Snapshot sensors enable quick data acquisition and are ideal for dynamic scenarios or situations needing promptly available spectral information [42].

A thorough knowledge of the peculiar characteristics of each hyperspectral sensor is crucial to select the most appropriate technology for a particular application. Concerning skin cancer detection, the snapshot sensor is the best choice since it acquires the whole images in a single exposure [25,36].

The HSI skin cancer dataset used is the one considered in [16,21,31,43]; it contains 76 images of skin lesions from 61 subjects, 40 of which are benign and 36 are malignant. They were acquired with a snapshot camera (Cubert UHD, Cubert GmbH, Ulm, Germany) able to cover the 450–950 nm range, distributed over 125 spectral channels [30]. The images were collected in two hospitals of the Canary Islands, Spain: the Hospital Universitario de Gran Canaria Doctor Negrín and the Complejo Hospitalario Universitario Insular-Materno Infantil. The image labelling was led by experts such as dermatologists and pathologists according to the taxonomy described in [32].

The spectral signatures among different patients have been normalized as illustrated in [32] to mitigate the variations in illumination conditions. At the end of preprocessing, the spectral signatures contain 116 bands with values in the range [0, 1].

Figure 1 shows the percentage distributions of the skin lesions that include four possible classes: Benign Epithelial (BE), Benign Melanocytic (BM), Malignant Epithelial (ME), and Malignant Melanocytic (MM).

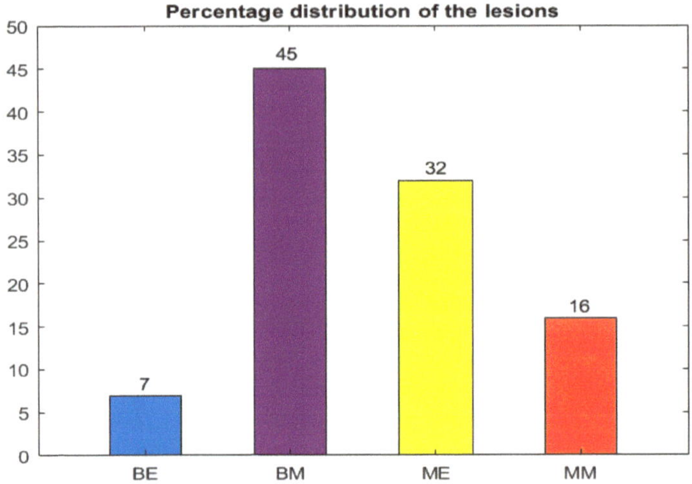

Figure 1. Percentage distribution of each lesion.

Figure 2 shows four images taken from the dataset representing one of the considered lesions, together with the mean spectral signatures of the hyperspectral pixels.

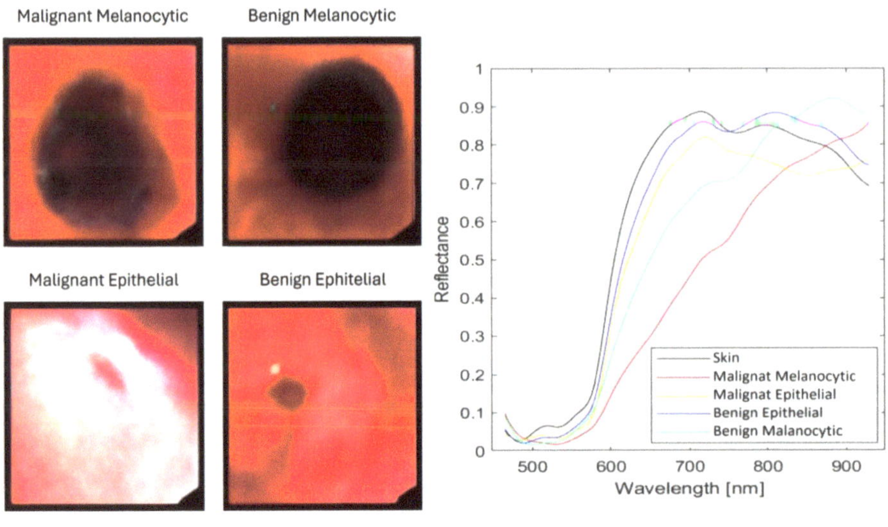

Figure 2. Synthetic RGB images taken from the database to represent each lesion and the mean spectra of the pixels.

2.2. *Machine-Learning Methods*

This section gives a general overview of the SVM, RF, and XGB methods adopted to classify the HSI skin cancer images. Specifically, theoretical aspects of the three algorithms will be presented.

2.2.1. Support Vector Machine

SVM is a supervised machine-learning method proposed by Vapnik and extensively used for classification and regression tasks [44–46]. Originally, SVM performs binary classifications and aims to find the hyperplane which splits the dataset into discrete classes

according to the given training samples [46]. The data points with the minimum distance from the hyperplane are called support vectors (SVs). For multiclass classification, SVM breaks down the multiclass problem into multiple binary classification ones, solving the following equation:

$$\min_{w,b,\zeta} \frac{1}{2} w^T w + C\sum_{i=1}^{n} \zeta_i$$

$$\textit{subject to } y_i\left(w^T x_i + b\right) \geq 1 - \zeta_i, \tag{1}$$

$$\zeta_i \geq 0 \textit{ with } i = 1, \ldots, n$$

where w is the support vectors, C is the penalty term, ζ_i is the distance error from the correct margin, y is the classes, b is the margin, x_i is the training vectors, and n is the number of training samples. Intuitively, the goal is to maximize the margin by minimizing $w^T w$, while incurring a penalty when a sample is misclassified.

The minimization problem described by Equation (1) can be transformed into a dual problem given by Equation (2):

$$\min_{\alpha} \frac{1}{2} \alpha^T Q \alpha - e^T \alpha$$

$$\textit{subject to } y^T \alpha = 0, \tag{2}$$

$$0 \leq \alpha_i \leq C \textit{ with } i = 1, \ldots, n$$

where e is a vector of all ones, and Q is an n by n positive semidefinite matrix whose elements are defined in Equation (3):

$$Q_{ij} = y_i y_j K(x_i x_j) \tag{3}$$

K is the kernel function that maps the data from a low-dimensional space to another space with high dimensions. Once the optimization problem is solved, the output of decision function for a given sample x becomes:

$$\sum_{i \in SV} \alpha_i K(w_i, x) + b \tag{4}$$

where α_i is the dual coefficients. The sign of Equation (4) gives the binary classification, while the multiclass classification is achieved according to the "one-vs.-one" strategy by repeatedly applying Equation (4).

2.2.2. Random Forest

RF was first introduced by Leo Breiman [47]. It is a popular ensemble learning algorithm used for both classification and regression tasks. It combines the predictions of multiple decision trees to improve the predictive accuracy and control over-fitting. Specifically, each tree performs a "partial" prediction, and the class with the most votes becomes the final prediction. Using a random subset of data and features, each decision tree in the RF is built recursively by splitting the data according to various criteria (e.g., Gini impurity or information gain) until a stopping criterion is met. The latter can be a maximum tree depth, a minimum number of samples required to split a node, or a minimum number of samples required in a leaf node.

2.2.3. eXtreme Gradient Boosting

XGB is an ensemble learning algorithm similar to RF. It is based on a generalized gradient boosting method, and is used for classification, regression, and ranking tasks [48–50]. It provides highly accurate classifications by combining the predictions of multiple weak predictive models, typically decision trees. One of the strong points of XGB is the sequential

addition of new models correcting the mistakes made by previous models. Particularly, it optimizes a specific loss function by computing its gradient compared to the predicted values. XGB builds N trees per class; the outputs of the trees belonging to the same class are summed. The soft-max function is then applied to the outputs to obtain the probability values of the class. The class with the biggest value is the final prediction.

2.3. CPU and GPU Technologies

This section describes the architectures and the main features of the CPU and GPU devices employed for the inference implementation of the three algorithms. For the serial inference, we used an Intel Core i9-13900K with a clock frequency of 3 GHz. It is based on the Raptor Lake architecture developed adopting an Intel 7 processor (10 nm), with 24 cores, 32 threads, and 32 MB and 36 MB of L2 and L3 cache memory, respectively. The maximum bandwidth achievable is 89.6 GB/s.

The first two GPU devices considered for the parallel inference were an NVIDIA GeForce RTX 2080 and an NVIDIA GeForce RTX 4090, optimized for graphics applications.

The NVIDIA GeForce RTX 2080 is based on the Turing architecture with 2944 CUDA cores and a clock frequency of 1.5 GHz. Other components of this device include 184 texture units, 64 Render Output Units (ROPs), 368 tensor cores, 46 ray tracing (RT) cores, and 8 GB of GDDR6 modules. The maximum bandwidth achievable is 448 GB/s.

The NVIDIA GeForce RTX 4090 is supported by the Ada Lovelace architecture with 16,384 CUDA cores and a clock frequency of 2.2 GHz. It also contains 512 tensor cores, 176 ROPs, and 128 RT cores. The memory dimension is 24 GB (GDDR6X), and the maximum bandwidth is 1008 GB/s.

The last GPU device considered is a cluster dedicated to the scientific calculation composed of five nodes of three NVIDIA Tesla A16s. Each GPU of the cluster is equipped with four chips and features the Ampere architecture. Every chip of the GPU has 1280 CUDA cores, 40 tensor cores, 16 GB of GDDR6, and a memory bandwidth of 200 GB/s.

2.4. CPU Inference

The inference of the algorithms described in Section 2.2 has been implemented using the best parameters obtained after the training phase as detailed in [31]. Visual Studio 2022 Integrated Development Environment (IDE) was used, adopting the C language.

The serial implementation has been used as a basis for the parallel inference described in Section 2.5.

2.4.1. SVM Inference

The SVM inference consisted in the implementation of Equation (4). The dual coefficients, the margin, the support vectors, and the type of kernel function have been identified after both the training and the parameters tuning described in [31]. The Radial Basis Function (RBF) resulted as the most appropriate kernel function, and it is represented by the following equation:

$$K(w_i, x) = e^{-\gamma||w_i - x||^2} \tag{5}$$

where γ is the kernel parameter, whose best value obtained after the training was 10.

The steps executed to perform the SVM inference can be summarized as follows:

1. Kernel calculation for the sample to classify according to Equation (5);
2. Multiplication between the obtained kernel and the dual coefficients adding the bias b;
3. Pixel classification through the "one-vs.-one" strategy.

The pseudo-code of the SVM inference is reported in Algorithm 1. Lines 2 to 4 perform the kernel calculation by evaluating the squared Euclidean distance between the support vectors and the sample to classify. The second step is executed in lines 6 to 10, where the distance of the sample from the hyperplane is calculated according to Equation (4). Due to the nested loops, the distance is calculated $n_{class} * (n_{class} - 1)/2$ times. With $n_{class} = 5$, 10 values of the distance are obtained. Lines 12 to 21 show the last step that aims to perform

the final prediction by observing the sign of the 10 values of the distance: if d_{ij} is positive (negative), then class i wins (loses) over class j, and the array $\boldsymbol{score_i}$ $(\boldsymbol{score_j})$ is incremented by one. Finally, line 21 finds the index of the maximum value in the array $\boldsymbol{score_i}$, or rather, the class obtaining the greatest number of scores.

Algorithm 1 Serial implementation of Support Vector Machine

Input: $\gamma \rightarrow$ Kernel parameter
$DC_{ij} \rightarrow$ Dual coefficients matrix
$w_i \rightarrow$ Support vectors matrix
$x \rightarrow$ Pixel to classify
$b \rightarrow$ Bias
1: *Step 1 : Kernel calculation*
2: **for** $i = 0$ *to* $n_{sv} - 1$
3: $K(w_i, x) = \exp\left(-\gamma * \|w_i - x\|^2\right);$
4: **end**
5: *Step 2 : Distance of the sample from the hyperplane*
6: **for** $i = 0$ *to* $n_{class} - 1$
7: **for** $j = i + 1$ *to* $n_{class} - 1$
8: $d_{ij} = \sum_{i \in SV} DC_{ij} * K(w_i, x) + b;$
9: **end**
10: **end**
11: *Step 3: "One vs. one" strategy*
12: **for** $i = 0$ *to* $n_{class} - 1$
13: $score_i = 0$
14: **for** $j = i + 1$ *to* $n_{class} - 1$
15: **if** $d_{ij} > 0$
16: $score_i + +;$
17: **else**
18: $score_j + +;$
19: **end**
20: **end**
21: Find ***imax***, index of the $\boldsymbol{score_i}$ maximum
Output: *imax*

2.4.2. RF Inference

The core of serial RF inference is a recursive function representing the tree structure. According to the obtained trained values of the features, the thresholds, as well as the left and right children's nodes of each parent node, the execution follows a specific path in the tree. If the execution ends in a non-leaf node, the function is repeated and drives the execution to the next node depending on the left and right children's values. The recursion stops when the execution ends in a leaf containing the output. The output of this function is an array of 5 elements containing the probability values of the pixel of belonging to each class. Then, a second function was realized with the goal to execute the tree structure N times, where N is the number of decision trees. Therefore, each tree makes its prediction on the pixel, and the class having the greatest number of votes is the final prediction. The number of decision trees used in this work is 425, obtained after the training phase. The pseudo-code of RF inference is shown in Algorithm 2. Line 2 corresponds to the ***tree_structure*** function that outputs the probability array (***prob_array***) exploiting the features, thresholds, and left and right children's node (***input_data***). Lines 4 to 8 perform the forest in which, at each iteration, the ***tree_structure*** function runs and the index of ***prob_array*** maximum is obtained. At the end of the iterations, the array *class* contains the number of votes per each class. The final prediction is the most voted class and is obtained in line 9.

Algorithm 2 Serial implementation of Random Forest

Input: ***input_data*** → Features, thresholds, left and right children's nodes
Step 1: Development of the ***tree_structure*** *function*
The single tree outputs ***prob_array***
Step 2: Building of the forest
for $i = 0$ *to* $n_{trees} - 1$
    tree_structure(***input_data***, ***prob_array***, i);
    Find ***max***, index of ***prob_array*** *maximum*
    $class_{max}++$;
end
Find ***imax***, index of the ***class*** maximum
Output: ***imax***

2.4.3. XGB Inference

XGB is based on the same ***tree_structure*** function of the RF, but in this case, the output is a single value. The forest structure function builds N decision trees for each class; each tree improves the output of the previous tree (belonging to the same class) by considering its prediction mistakes. The optimal number of decision trees obtained after the training was 400, so the forest structure function builds 2000 decision trees overall.

The outputs of the decision trees belonging to the same class are summed. In Algorithm 3, the pseudo-code of the XGB inference is shown. Line 2 is related to the ***tree_structure*** function that outputs the probability value of the single tree. Then, the forest function is described in lines 4 to 8, where the sums of the outputs of the trees belonging to the same class are stored in the Z_i array of 5 elements. Lines 10 to 18 determine the final probability array P_i according to the soft-max function reported in Equation (6). The index of P_i maximum is the final prediction according to line 19.

$$P[i] = \frac{ZE[i]}{\sum_{j=0}^{n_{class}} ZE[j]} \quad (6)$$

Algorithm 3 Serial implementation of eXtreme Gradient Boosting

Input: ***input_data*** → Features, thresholds, left and right children's nodes
1: *Step 1: Development of the* ***tree_structure*** *function*
2: The single tree outputs the probability value of its class
3: *Step 2: Building of the forest*
4: **for** $i = 0$ *to* $n_{class} - 1$
5:     **for** $e = 0$ *to* $n_{trees} - 1$
6:         $Z_i += $ ***tree_structure***(***input_data***, $e * n_{class} + i$);
7:     **end**
8: **end**
9: *Step 3 : Final probability array through soft − max function*
10: **for** $i = 0$ *to* $n_{class} - 1$
11:     $ZE_i = \exp(Z_i)$;
12: **end**
13: **for** $i = 0$ *to* $n_{class} - 1$
14:     $z = \sum_{i \in n_{class}} ZE_i$;
15: **end**
16: **for** $i = 0$ *to* $n_{class} - 1$
17:     $P_i = ZE_i / z$;
18: **end**
19: Find ***imax***, index of the P_i maximum
Output: ***imax***

2.5. GPU Inference

This section describes the parallel inference for the SVM, RF, and XGB algorithms. We adopted the GPU devices described in Section 2.3 and Visual Studio 2022 with CUDA C language.

In the following sections, we will explain some essential terms to define the basic components of the CUDA language. First, we must define the kernel (a CUDA function) that, when called, is executed in parallel by N different CUDA threads. Another important component is the thread block containing a group of threads executed concurrently. The threads belonging to the same block can cooperate through synchronization barriers. A thread block uses the shared memory for inter-thread communication and the data sharing. Finally, a grid is an array of thread blocks executing the same kernel; it reads and writes in the global memory of the GPU. Each thread and block can be identified through the *threadIdx = (threadIdx.x, threadIdx.y, threadIdx.z)* and *blockIdx = (blockIdx.x, blockIdx.y, blockIdx.z)* coordinates, respectively. The dimension of the thread block is defined by the *blockDim = (blockDim.x, blockDim.y, blockDim.z)* array.

2.5.1. Parallel SVM

The most computationally expensive operations in SVM are *Step 1* and *Step 2* of Algorithm 1 in Section 2.4.1. *Step 1* involves the SV matrix (116 × 47,220) and the image to classify (2500 × 116), while *Step 2* performs the product between the obtained kernel (2500 × 47,220) and the dual coefficients matrix (47,220 × 4).

Step 2 was performed through a CUDA kernel using a number of blocks equal to $(N + n_{threads} - 1)/n_{threads}$ with $n_{threads} = 32$ and N being the number of SVs. The choice to use 32 as the number of threads is because the basic unit of execution in an NVIDIA GPU is the warp, a collection of 32 threads executed simultaneously by a Streaming Multiprocessor (SM) of the GPU. Therefore, the resulting number of blocks was 1476. The pseudo-code of Algorithm 4 below represents the kernel calculation through the CUDA syntax.

Algorithm 4 Kernel calculation

Input: $\gamma \rightarrow$ Kernel parameter
$w_i \rightarrow$ Support vector matrix
$x \rightarrow$ Pixel to classify
1: i = *blockIdx.x * blockDim.x + threadIdx.x*
2: if $i < n_{sv}$
3: **for** $i = 0$ *to* $n_{bands} - 1$
4: $d_i = \|w_i - x\|^2$
5: **end**
6: $K(w_i, x) = \exp(-\gamma * d_i)$
Output: $K(w_i, x)$

In line 1, the variables *blockIdx.x* and *threadIdx.x* indicate the current block and thread identifier, while *blockDim.x* is the block dimension along the *x*-axis as described in Section 2.5. In line 4, the squared Euclidean distance d_i is shown; each thread performs the difference between an element of the SV matrix w_i and an element of the sample to classify x in parallel. Finally, in line 6, the kernel $K(w_i, x)$ is obtained.

Then, *Step 2* was implemented by adopting the *cublasSgemm* and the *cublasSaxpy* functions (from the cuBLAS library) explicitly designed for matrix operations: the first has been used to perform the multiplication between the kernel and the dual coefficients matrix, the second to sum the obtained result and b. The result of this step was a vector of 10 elements containing the outputs of the decision function (see Equation (4)). *Step 3* was performed employing 1 block of 5 threads (1 per class), whose task was to apply the "one-vs.-one" strategy. Finally, the *cublasIsamax* function has been used to determine the final prediction.

2.5.2. Parallel RF

For the parallel version of RF, the intrinsic nature of decision trees that is based on sequences of *if–else* statements causes threads divergence, representing a challenge that did not allow the parallelization of the ***tree_structure*** function. Therefore, such function has been declared as a device function using the CUDA keyword ***__device__***, meaning that the function is called by the GPU.

The forest structure was realized with a CUDA kernel composed of 425 blocks of 1 thread, with one block for each decision tree and every block having only one thread in order to avoid the potential thread divergence in the ***tree_structure*** function.

The pseudo-code in Algorithm 5 represents the parallel RF inference. Line 2 refers to the serial RF ***tree_structure*** with the addition of the ***__device__*** declaration, as mentioned above. Lines 4 to 6 perform the forest where each block builds a decision tree and outputs the prediction (***max***) for that same tree. Furthermore, to prevent race conditions in filling the ***class*** array, line 6 performs the *atomicAdd* operation to add the value 1 to all the elements of the array. In line 7, the final prediction ***imax*** is obtained through the *cublasIsamax* function.

Figure 3 shows the flow diagram of the RF classifier and how it is divided between host and device. The input data, stored in the host, are transferred in the device memory through the ***cudaMemcpy*** function, thus representing the input to the forest structure device function, where each block implements a decision tree by calling the ***tree_structure*** function. After that, the ***cublasIsamax*** function has been used to make the prediction for each specific pixel. Since the device output vector contains the predictions of every pixel of the image, its dimension is 2500. At last, the device output vector is transferred to the host memory.

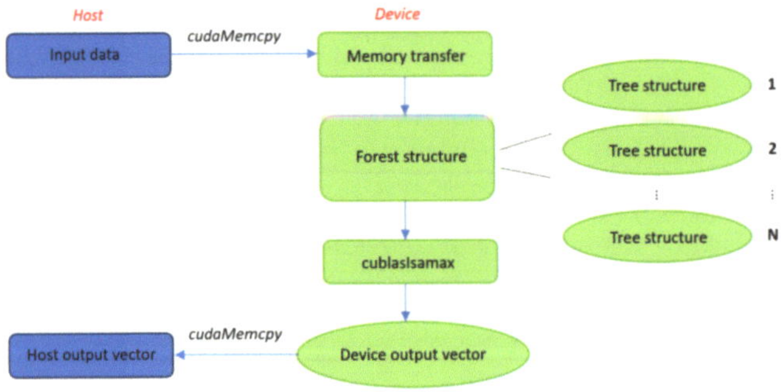

Figure 3. Flow diagram of parallel RF classifier.

2.5.3. Parallel XGB

To perform the parallelized version of the XGB, the forest structure function has been designed similarly to the parallelized RF: 2000 blocks have been adopted, each including 1 thread, and launching the tree structure function. The values obtained for each block have been stored in the vector **Z**. Then, the reduction technique has been used to sum the elements of **Z** related to the same class. To perform this task, the "sequential addressing" strategy has been implemented. The code below shows the sequential addressing reduction technique.

In Code 1, for each class, 400 elements (*n_estimators)* of **Z** are transferred to the GPU shared memory through the array **S**. Then, the ***for*** loop reduces the entire upper portion of the array **S** to the entire lower portion of **S**. With 512 values, the upper 256 values are reduced into the lower 256 values. Then, the upper 128 values of the lower 256 values from before are reduced with the lower 128 values. The loop ends when the sum of all the elements of the array is obtained and stored in the first element of **S**.

The reduction was executed using a 2D grid composed of 1 block of 512 (512 being the first power of 2 greater than 400) threads for the x-axis, and 5 blocks of 1 thread for the y-axis. Each thread of the x-axis transfers one element of **Z** to the shared mem-

ory and sums two elements of $\mathbf{Z}$, while the 5 blocks of the y-axis iterate over the classes. Algorithms 4 and 5, related to SVM and RF, respectively, involve a single index in performing their kernels; therefore, the use of a 1D grid was considered sufficient. In the reduction process, XGB involves two independent indexes, $\boldsymbol{e}$ and $\boldsymbol{b}$, related to the elements of the $\mathbf{S}$ array and to the classes, respectively; as a consequence, a 2D grid has been identified as more suitable compared to a 1D grid.

Code 1 Sequential Addressing Reduction

Input: *tid, e, b* → indexes of the threads and blocks
ncl → number of classes

```
int tid = threadIdx.x;
__shared__ float S[512];
int e = blockIdx.x * blockDim.x + threadIdx.x;
int b = blockIdx.y;
if (tid < n_estimators)
    S[tid] = Z[e * ncl + b];
__syncthreads();
for (s = blockDim.x/2; s > 0; s >>= 1){
    if (tid < s)
        S[tid] += S[tid + s];
    __syncthreads();
}
```

Output: *S*

Algorithm 5 Parallel Random Forest

Input: *input_data* → Features, thresholds, left and right children's nodes

```
Step 1: Development of the device tree_structure function
The single tree outputs max, the prob_array maximum index
Step 2: Building of the forest
i = blockIdx.x;
max = tree_structure(input_data, prob_array, i);
atomicAdd(&class_max, 1.0);
Find imax, index of the class maximum
```

Output: *imax*

The sequential addressing approach solves the warp's divergence and shared memory bank conflict problems of the interleaved addressing reduction. Figure 4 exemplifies the concept of sequential addressing reduction.

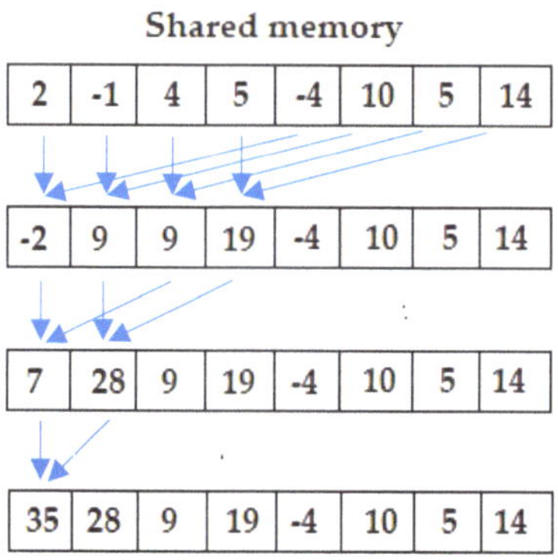

Figure 4. Example of sequential addressing reduction technique.

To conclude, the final probability array P of Equation (6) was obtained using a CUDA kernel composed by 5 blocks of 1 thread.

3. Results

The inference part of SVM, RF, and XGB methods has been implemented in a serial and a parallelized version using C and CUDA languages, respectively. The programs have been developed with the Microsoft Visual Studio 2022 IDE and the CUDA 11.7 toolkit for the NVIDIA GeForce RTX 2080 GPU and the CUDA 12.0 toolkit for the NVIDIA Tesla A16 and NVIDIA GeForce RTX 4090 GPUs. The serial version was compiled with the v143 compiler of Visual Studio, while the parallel code was compiled with the NVCC compiler included in the toolkit. The compiler configuration has been set to release mode, meaning that the optimizations are enabled, and that the full debugging information is not included. Furthermore, we have set the code generation option of the CUDA compiler to 7.5, 8.6, and 8.9 values corresponding to the compute capability of the NVIDIA GeForce RTX 2080, NVIDIA Tesla A16, and NVIDIA GeForce RTX 4090 GPUs. This option allowed us to fully exploit the architectures of the respective GPUs.

The SVM, RF, and XGB inference has been tested using 10 HSI skin cancer images, all having dimensions of 50 × 50 pixels and 116 bands; this dataset contains all the possible skin lesions.

Specifically, the average classification time of such images has been measured for each algorithm and for all the adopted technologies. All the average classification times with the standard deviations and the speed-up (in brackets) are reported in Table 1.

Table 1. Average classification times for SVM, RF, and XGB for all the CPU and GPU devices.

	SVM [s]	RF [s]	XGB [s]
i9-13900K	445.90 ± 105.72	0.51 ± 0.01	1.17 ± 0.02
RTX 2080	14.10 ± 0.09 (32x)	0.77 ± 0.00 (0.66x)	0.98 ± 0.00 (1.19x)
Tesla A16	40.80 ± 0.00 (11x)	1.07 ± 0.00 (0.48x)	1.43 ± 0.00 (0.82x)
RTX 4090	3.44 ± 0.00 (130x)	0.76 ± 0.00 (0.67x)	0.84 ± 0.00 (1.39x)

It is worth noting that the parallel SVM features the greatest speed-up. In fact, all GPU devices have obtained valid results for this algorithm: a speed-up of 32x, 11x, and 130x turned out for the GeForce RTX 2080, Tesla A16, and GeForce RTX 4090, respectively. This confirms that parallelizing SVM is an appropriate solution for the acceleration of skin lesions' detection.

Parallel XGB has outperformed its serial counterpart when using both the GeForce RTX 2080 and GeForce RTX 4090 GPUs, achieving a speed-up of 1.19x for the first and 1.39x for the second device conversely. The cluster has not accelerated the serial version, its average execution time being 1.17 s, whereas 1.43 s is the average execution time of the parallelized version.

Finally, RF is the only algorithm that has not shown improvements; however, some observations should be made: the intrinsic nature of RF did not allow the tree structure to be parallelized since it is based on *if–else* sequences. Hence, this algorithm is not fully parallelizable. Moreover, the number of decision trees used in this work was 425, which is not as big as it should be to adequately exploit the benefits of parallel computing.

NVIDIA GeForce RTX 4090 GPU resulted as the most performant among the GPUs, due to its high number of CUDA cores (16,384) and to its latest-generation architecture, the Ada Lovelace.

As already said, the university cluster achieved the worst performance for all algorithms, probably because the code developed for the parallel inference has not exploited the full computational power of the cluster. Indeed, the cluster is composed of five nodes of three Tesla A16 GPUs, while our code employed the use of one out of four chips equipped on each single GPU.

4. Discussion

To compare the results of our methods with the state of the art, the works proposed in [16,33] can be considered. The authors of [16] have developed a hybrid classification system based on K-means, SAM, and SVM using the same dataset here described. In particular, they implemented several parallel versions of their system using an NVIDIA GeForce RTX 2080 GPU (the same employed in this work) and an NVIDIA Tesla K40 GPU. The best performance was achieved through the version performing the K-means in CUDA using the NVIDIA GeForce RTX 2080 GPU and the SVM in OpenMP. To evaluate the performance, the authors considered nine images and measured the classification times of each image as the mean of five executions. They reported a diagram showing that the classification times of their system were approximately 1 s. However, the SVM implementation in [16] had to classify only a limited number of pixels of the images; namely, the pixels clustered as pigmented skin lesions from the K-means stage. In contrast, this work's SVM classified all the 2500 pixels of the images, discriminating between five different classes. Indeed, the computational complexity of the SVM adopted in [16] is lower than the one described in this work. Not only the number of elements to classify is lower, but also the hyperparameters are different, since a higher number of support vectors is needed by the SVM adopted in this paper.

In [33], a parallel XGB version was developed using an NVIDIA Quadro P4000 to classify the Pavia University (PU), GRSS-DFC2013 Houston (GH13), and GRSS-DFC2018 Houston (GH18) datasets. All three datasets are based on a single HSI image. The PU image features a dimension of 610 $\times$ 340 pixels and 103 channels, while the GH13 image is a cube of dimensions 349 $\times$ 1905 $\times$ 144. Finally, the GH18 Houston image has 4172 $\times$ 1202 pixels and 48 bands. The times taken to classify these images were 6.67 s, 31.05 s, and 347.30 s for the PU, GH13, and GH18 datasets, respectively. Given the big difference between the number of samples and features considered in the datasets of [33] and the one of this work, a quasi-linear relation between the images size and the processing times is observed. Indeed, the structure of XGB is poorly parallelizable, and the performances are strictly related to the number of features and trees. In the proposed work, since the data dimensionality is lower than that of [33], the number of features and trees is small. Moreover, as described in Section 2.5.3, the parallelization is based on assigning each tree to a block, whilst instead, [33] uses a standard approach.

To the best of the authors' knowledge, no prior parallel version of RF has been developed in the HSI field.

Table 2 summarizes the prediction times of this work and the results obtained in the literature.

Table 2. Comparison between classification times of our work with the state of the art.

	K-Means + SAM + SVM [16]	SVM (This Work)	XGB PU [33]	XGB GH13 [33]	XGB GH18 [33]	XGB (This Work)
Time [s]	~1	3.44	6.67	31.05	347.30	0.84
# pixels	From 300 to 1700	2500	207,400	664,845	5,014,744	2500
# channels	116	116	103	144	48	116

5. Conclusions

In this work, a serial and a parallel inference of the SVM, RF, and XGB algorithms to classify a dataset of HS skin cancer images have been proposed. The serial inference has been implemented employing the CPU Intel Core i9-13900K, and to accelerate the serial classification, three different GPUs have been employed: the NVIDIA GeForce RTX 2080, the NVIDIA Tesla A16, and the NVIDIA GeForce RTX 4090.

The results show that our work can significantly accelerate medical diagnosis through image processing techniques. In fact, the parallel versions of both SVM and XGB lead to an acceleration very significant in the case of the most complex SVM and minor but

not neglectable in the case of the less challenging XGB. In any case, this experimentation confirms the validity of the approach used in [16] and in [38] even in case of a problem featuring a low parallelizable algorithm applied to a small dataset with a low number of trees. Again, it is possible to say that hyperspectral image processing can support doctors in timely detecting skin lesions, planning an opportune therapy, and helping surgeons during interventions.

Future works will focus on multi-GPU programming to exploit the full computational power of the cluster, since we only used one out of four GPUs of one NVIDIA Tesla A16. Furthermore, integrated GPU solutions will be explored, such as the NVIDIA Jetson, that is a System on Module (SoM) that features small dimensions, high performance, and embedded CPU, GPU, and memory in a single board. Lastly, datasets with a higher number of patients will be considered to better validate the proposed approach.

Author Contributions: Conceptualization, B.P. and E.T.; methodology, B.P. and E.M.; software, B.P.; validation, B.P., E.T. and E.M.; investigation, B.P., E.T., E.M. and F.L.; writing—original draft preparation, B.P.; writing—review and editing, E.T., E.M. and F.L.; supervision, F.L. All authors have read and agreed to the published version of the manuscript.

Funding: This research received no external funding.

Institutional Review Board Statement: Not applicable.

Informed Consent Statement: Not applicable.

Data Availability Statement: Data available upon request to the corresponding author.

Conflicts of Interest: The authors declare no conflict of interest.

References

1. Ferlay, J.; Colombet, M.; Soerjomataram, I.; Parkin, D.M.; Piñeros, M.; Znaor, A.; Bray, F. Cancer Statistics for the Year 2020: An Overview. *Int. J. Cancer* **2021**, *149*, 778–789. [CrossRef] [PubMed]
2. Abdlaty, R.; Doerwald-Munoz, L.; Farrell, T.J.; Hayward, J.E.; Fang, Q. Hyperspectral Imaging Assessment for Radiotherapy Induced Skin-Erythema: Pilot Study. *Photodiagn. Photodyn. Ther.* **2021**, *33*, 102195. [CrossRef] [PubMed]
3. Scolyer, R.A.; Long, G.V.; Thompson, J.F. Evolving Concepts in Melanoma Classification and Their Relevance to Multidisciplinary Melanoma Patient Care. *Mol. Oncol.* **2011**, *5*, 124–136. [CrossRef]
4. Krensel, M.; Petersen, J.; Stephan, B.; Katalinic, A.; Augustin, J. Comparison of Patient Pathways in the Early Detection of Skin Cancer—A Claims Data Analysis. *JDDG J. Der Dtsch. Dermatol. Ges.* **2021**, *19*, 389–398. [CrossRef] [PubMed]
5. Rey-Barroso, L.; Peña-Gutiérrez, S.; Yáñez, C.; Burgos-Fernández, F.J.; Vilaseca, M.; Royo, S. Optical Technologies for the Improvement of Skin Cancer Diagnosis: A Review. *Sensors* **2021**, *21*, 252. [CrossRef] [PubMed]
6. Jiang, S.; Li, H.; Jin, Z. A Visually Interpretable Deep Learning Framework for Histopathological Image-Based Skin Cancer Diagnosis. *IEEE J. Biomed. Health Inform.* **2021**, *25*, 1483–1494. [CrossRef] [PubMed]
7. Dildar, M.; Akram, S.; Irfan, M.; Khan, H.U.; Ramzan, M.; Mahmood, A.R.; Alsaiari, S.A.; Saeed, A.H.M.; Alraddadi, M.O.; Mahnashi, M.H. Skin Cancer Detection: A Review Using Deep Learning Techniques. *Int. J. Environ. Res. Public. Health* **2021**, *18*, 5479. [CrossRef]
8. Abdlaty, R.; Fang, Q. Skin Erythema Assessment Techniques. *Clin. Dermatol.* **2021**, *39*, 591–604. [CrossRef]
9. Kamruzzaman, M.; Sun, D.-W. Introduction to Hyperspectral Imaging Technology. In *Computer Vision Technology for Food Quality Evaluation*; Elsevier: Amsterdam, The Netherlands, 2016; pp. 111–139.
10. Meyer, J.M.; Kokaly, R.F.; Holley, E. Hyperspectral Remote Sensing of White Mica: A Review of Imaging and Point-Based Spectrometer Studies for Mineral Resources, with Spectrometer Design Considerations. *Remote Sens. Environ.* **2022**, *275*, 113000. [CrossRef]
11. Johansen, T.H.; Møllersen, K.; Ortega, S.; Fabelo, H.; Garcia, A.; Callico, G.M.; Godtliebsen, F. Recent Advances in Hyperspectral Imaging for Melanoma Detection. *WIREs Comput. Stat.* **2020**, *12*, e1456. [CrossRef]
12. Zhang, Q.; Bai, C.; Liu, Z.; Yang, L.T.; Yu, H.; Zhao, J.; Yuan, H. A GPU-Based Residual Network for Medical Image Classification in Smart Medicine. *Inf. Sci.* **2020**, *536*, 91–100. [CrossRef]
13. Pandey, M.; Fernandez, M.; Gentile, F.; Isayev, O.; Tropsha, A.; Stern, A.C.; Cherkasov, A. The Transformational Role of GPU Computing and Deep Learning in Drug Discovery. *Nat. Mach. Intell.* **2022**, *4*, 211–221. [CrossRef]
14. Wang, H.; Peng, H.; Chang, Y.; Liang, D. A Survey of GPU-Based Acceleration Techniques in MRI Reconstructions. *Quant. Imaging Med. Surg.* **2018**, *8*, 196–208. [CrossRef] [PubMed]
15. Kalaiselvi, T.; Sriramakrishnan, P.; Somasundaram, K. Survey of Using GPU CUDA Programming Model in Medical Image Analysis. *Inform. Med. Unlocked* **2017**, *9*, 133–144. [CrossRef]

16. Torti, E.; Leon, R.; La Salvia, M.; Florimbi, G.; Martinez-Vega, B.; Fabelo, H.; Ortega, S.; Callicó, G.M.; Leporati, F. Parallel Classification Pipelines for Skin Cancer Detection Exploiting Hyperspectral Imaging on Hybrid Systems. *Electronics* **2020**, *9*, 1503. [CrossRef]
17. Shi, L.; Liu, W.; Zhang, H.; Xie, Y.; Wang, D. A Survey of GPU-Based Medical Image Computing Techniques. *Quant. Imaging Med. Surg.* **2012**, *2*, 188–206. [CrossRef]
18. Jimenez, L.I.; Sanchez, S.; Martan, G.; Plaza, J.; Plaza, A.J. Parallel Implementation of Spatial–Spectral Endmember Extraction on Graphic Processing Units. *IEEE J. Sel. Top. Appl. Earth Obs. Remote Sens.* **2017**, *10*, 1247–1255. [CrossRef]
19. Marenzi, E.; Torti, E.; Leporati, F.; Quevedo, E.; Callicò, G.M. Block Matching Super-Resolution Parallel GPU Implementation for Computational Imaging. *IEEE Trans. Consum. Electron.* **2017**, *63*, 368–376. [CrossRef]
20. Cong, J.; Fang, Z.; Lo, M.; Wang, H.; Xu, J.; Zhang, S. Understanding Performance Differences of FPGAs and GPUs. In Proceedings of the 2018 IEEE 26th Annual International Symposium on Field-Programmable Custom Computing Machines (FCCM), Boulder, CO, UAS, 29 April–1 May 2018; IEEE: New York, NY, USA, 2018; pp. 93–96.
21. Leon, R.; Martinez-Vega, B.; Fabelo, H.; Ortega, S.; Melian, V.; Castaño, I.; Carretero, G.; Almeida, P.; Garcia, A.; Quevedo, E.; et al. Non-Invasive Skin Cancer Diagnosis Using Hyperspectral Imaging for In-Situ Clinical Support. *J. Clin. Med.* **2020**, *9*, 1662. [CrossRef]
22. Tian, C.; Xu, Y.; Zhang, Y.; Zhang, Z.; An, H.; Liu, Y.; Chen, Y.; Zhao, H.; Zhang, Z.; Zhao, Q.; et al. Combining Hyperspectral Imaging Techniques with Deep Learning to Aid in Early Pathological Diagnosis of Melanoma. *Photodiagn. Photodyn. Ther.* **2023**, *43*, 103708. [CrossRef]
23. Kazianka, H.; Leitner, R.; Pilz, J. Segmentation and Classification of Hyper-Spectral Skin Data. In *Data Analysis, Machine Learning and Applications*; Springer: Berlin/Heidelberg, Germany, 2008; pp. 245–252.
24. Vinokurov, V.; Khristoforova, Y.; Myakinin, O.; Bratchenko, I.; Moryatov, A.; Machikhin, A.; Zakharov, V. Neural Network Classifier for Hyperspectral Images of Skin Pathologies. *J. Phys. Conf. Ser.* **2021**, *2127*, 012026. [CrossRef]
25. Pardo, A.; Gutiérrez-Gutiérrez, J.A.; Lihacova, I.; López-Higuera, J.M.; Conde, O.M. On the Spectral Signature of Melanoma: A Non-Parametric Classification Framework for Cancer Detection in Hyperspectral Imaging of Melanocytic Lesions. *Biomed. Opt. Express* **2018**, *9*, 6283. [CrossRef] [PubMed]
26. Räsänen, J.; Salmivuori, M.; Pölönen, I.; Grönroos, M.; Neittaanmäki, N. Hyperspectral Imaging Reveals Spectral Differences and Can Distinguish Malignant Melanoma from Pigmented Basal Cell Carcinomas: A Pilot Study. *Acta Derm. Venereol.* **2021**, *101*, adv00405. [CrossRef] [PubMed]
27. Liu, L.; Qi, M.; Li, Y.; Liu, Y.; Liu, X.; Zhang, Z.; Qu, J. Staging of Skin Cancer Based on Hyperspectral Microscopic Imaging and Machine Learning. *Biosensors* **2022**, *12*, 790. [CrossRef] [PubMed]
28. Qi, M.; Liu, Y.; Li, R.; Liu, L.; Zhang, Z. Classification of Skin Cancer Based on Hyperspectral Microscopic Imaging and Machine Learning. In Proceedings of the SPIE-CLP Conference on Advanced Photonics 2022, Virtual, 28 March 2023; Liu, X., Yuan, X., Zayats, A., Eds.; SPIE: Washington, DC, USA, 2023; p. 16.
29. Huang, H.-Y.; Hsiao, Y.-P.; Mukundan, A.; Tsao, Y.-M.; Chang, W.-Y.; Wang, H.-C. Classification of Skin Cancer Using Novel Hyperspectral Imaging Engineering via YOLOv5. *J. Clin. Med.* **2023**, *12*, 1134. [CrossRef] [PubMed]
30. Fabelo, H.; Melian, V.; Martinez, B.; Beltran, P.; Ortega, S.; Marrero, M.; Callico, G.M.; Sarmiento, R.; Castano, I.; Carretero, G.; et al. Dermatologic Hyperspectral Imaging System for Skin Cancer Diagnosis Assistance. In Proceedings of the 2019 XXXIV Conference on Design of Circuits and Integrated Systems (DCIS), Bilbao, Spain, 20–22 November 2019; IEEE: New York, NY, USA, 2019; pp. 1–6.
31. Petracchi, B.; Gazzoni, M.; Torti, E.; Marenzi, E.; Leporati, F. Machine Learning-Based Classification of Skin Cancer Hyperspectral Images. *Procedia Comput. Sci.* **2023**, *225*, 2856–2865. [CrossRef]
32. La Salvia, M.; Torti, E.; Leon, R.; Fabelo, H.; Ortega, S.; Balea-Fernandez, F.; Martinez-Vega, B.; Castaño, I.; Almeida, P.; Carretero, G.; et al. Neural Networks-Based On-Site Dermatologic Diagnosis through Hyperspectral Epidermal Images. *Sensors* **2022**, mboxemph22, 7139. [CrossRef] [PubMed]
33. Samat, A.; Li, E.; Du, P.; Liu, S.; Xia, J. GPU-Accelerated CatBoost-Forest for Hyperspectral Image Classification Via Parallelized MRMR Ensemble Subspace Feature Selection. *IEEE J. Sel. Top. Appl. Earth Obs. Remote Sens.* **2021**, *14*, 3200–3214. [CrossRef]
34. Camps-Valls, C.; Bruzzone, L. Kernel-Based Methods for Hyperspectral Image Classification. *IEEE Trans. Geosci. Remote Sens.* **2005**, *43*, 1351–1362. [CrossRef]
35. Florimbi, G.; Fabelo, H.; Torti, E.; Ortega, S.; Marrero-Martin, M.; Callico, G.M.; Danese, G.; Leporati, F. Towards Real-Time Computing of Intraoperative Hyperspectral Imaging for Brain Cancer Detection Using Multi-GPU Platforms. *IEEE Access* **2020**, *8*, 8485–8501. [CrossRef]
36. Wu, D.; Sun, D.-W. Advanced Applications of Hyperspectral Imaging Technology for Food Quality and Safety Analysis and Assessment: A Review—Part I: Fundamentals. *Innov. Food Sci. Emerg. Technol.* **2013**, *19*, 1–14. [CrossRef]
37. Adão, T.; Hruška, J.; Pádua, L.; Bessa, J.; Peres, E.; Morais, R.; Sousa, J. Hyperspectral Imaging: A Review on UAV-Based Sensors, Data Processing and Applications for Agriculture and Forestry. *Remote Sens.* **2017**, *9*, 1110. [CrossRef]
38. Sousa, J.J.; Toscano, P.; Matese, A.; Di Gennaro, S.F.; Berton, A.; Gatti, M.; Poni, S.; Pádua, L.; Hruška, J.; Morais, R.; et al. UAV-Based Hyperspectral Monitoring Using Push-Broom and Snapshot Sensors: A Multisite Assessment for Precision Viticulture Applications. *Sensors* **2022**, *22*, 6574. [CrossRef] [PubMed]

39. Abdlaty, R.; Abbass, M.A.; Awadallah, A.M. High Precision Monitoring of Radiofrequency Ablation for Liver Using Hyperspectral Imaging. *Ann. Biomed. Eng.* **2021**, *49*, 2430–2440. [CrossRef] [PubMed]
40. Bassler, M.C.; Stefanakis, M.; Sequeira, I.; Ostertag, E.; Wagner, A.; Bartsch, J.W.; Roeßler, M.; Mandic, R.; Reddmann, E.F.; Lorenz, A.; et al. Comparison of Whiskbroom and Pushbroom Darkfield Elastic Light Scattering Spectroscopic Imaging for Head and Neck Cancer Identification in a Mouse Model. *Anal. Bioanal. Chem.* **2021**, *413*, 7363–7383. [CrossRef]
41. Wahabzada, M.; Besser, M.; Khosravani, M.; Kuska, M.T.; Kersting, K.; Mahlein, A.-K.; Stürmer, E. Monitoring Wound Healing in a 3D Wound Model by Hyperspectral Imaging and Efficient Clustering. *PLoS ONE* **2017**, *12*, e0186425. [CrossRef] [PubMed]
42. He, Q.; Wang, R.K. Analysis of Skin Morphological Features and Real-Time Monitoring Using Snapshot Hyperspectral Imaging. *Biomed. Opt. Express* **2019**, *10*, 5625. [CrossRef] [PubMed]
43. La Salvia, M.; Torti, E.; Gazzoni, M.; Marenzi, E.; Leon, R.; Ortega, S.; Fabelo, H.; Callico, G.M.; Leporati, F. Attention-Based Skin Cancer Classification Through Hyperspectral Imaging. In Proceedings of the 2022 25th Euromicro Conference on Digital System Design (DSD), Maspalomas, Spain, 31 August–2 September 2022; IEEE: New York, NY, USA, 2022; pp. 871–876.
44. Chandra, M.A.; Bedi, S.S. Survey on SVM and Their Application in Image Classification. *Int. J. Inf. Technol.* **2021**, *13*, 1–11. [CrossRef]
45. Brown, M.; Lewis, H.G.; Gunn, S.R. Linear Spectral Mixture Models and Support Vector Machines for Remote Sensing. *IEEE Trans. Geosci. Remote Sens.* **2000**, *38*, 2346–2360. [CrossRef]
46. Mountrakis, G.; Im, J.; Ogole, C. Support Vector Machines in Remote Sensing: A Review. *ISPRS J. Photogramm. Remote Sens.* **2011**, *66*, 247–259. [CrossRef]
47. Breiman, L. Random Forests. *Mach. Learn.* **2001**, *45*, 5–32. [CrossRef]
48. Zhang, H.; Si, S.; Hsieh, C.-J. GPU-Acceleration for Large-Scale Tree Boosting. *arXiv* **2017**, arXiv:1706.08359.
49. Mitchell, R.; Frank, E. Accelerating the XGBoost Algorithm Using GPU Computing. *PeerJ Comput. Sci.* **2017**, *3*, e127. [CrossRef]
50. Chen, T.; Guestrin, C. XGBoost. In Proceedings of the 22nd ACM SIGKDD International Conference on Knowledge Discovery and Data Mining, San Francisco, CA, USA, 13–17 August 2016; ACM: New York, NY, USA, 2016; pp. 785–794.

sensors

MDPI

Article

A Comparative Analysis of UAV Photogrammetric Software Performance for Forest 3D Modeling: A Case Study Using AgiSoft Photoscan, PIX4DMapper, and DJI Terra

Sina Jarahizadeh * and Bahram Salehi

State University of New York, College of Environmental Science and Forestry (SUNY ESF), Department of Environmental Resources Engineering, 1 Forestry Dr., Syracuse, NY 13210, USA; bsalehi@esf.edu

* Correspondence: sjarahizadeh@esf.edu

Abstract: Three-dimensional (3D) modeling of trees has many applications in various areas, such as forest and urban planning, forest health monitoring, and carbon sequestration, to name a few. Unmanned Aerial Vehicle (UAV) photogrammetry has recently emerged as a low cost, rapid, and accurate method for 3D modeling of urban and forest trees replacing the costly traditional methods such as plot measurements and surveying. There are numerous commercial and open-source software programs available, each processing UAV data differently to generate forest 3D modeling and photogrammetric products, including point clouds, Digital Surface Models (DSMs), Canopy Height Models (CHMs), and orthophotos in forest areas. The objective of this study is to compare the three widely-used commercial software packages, namely, AgiSoft Photoscan (Metashape) V 1.7.3, PIX4DMapper (Pix4D) V 4.4.12, and DJI Terra V 3.7.6 for processing UAV data over forest areas from three perspectives: point cloud density and reconstruction quality, computational time, DSM assessment for height accuracy (z) and ability of tree detection on DSM. Three datasets, captured by UAVs on the same day at three different flight altitudes, were used in this study. The first, second, and third datasets were collected at altitudes of 60 m, 100 m, and 120 m, respectively over a forested area in Tully, New York. While the first and third datasets were taken horizontally, the second dataset was taken 20 degrees off-nadir to investigate the impact of oblique images. Results show that Pix4D and AgiSoft generate 2.5 times denser point clouds than DJI Terra. However, reconstruction quality evaluation using the Iterative Closest Point method (ICP) shows DJI Terra has fewer gaps in the point cloud and performed better than AgiSoft and Pix4D in generating a point cloud of trees, power lines and poles despite producing a fewer number of points. In other words, the outperformance in key points detection and an improved matching algorithm are key factors in generating improved final products. The computational time comparison demonstrates that the processing time for AgiSoft and DJI Terra is roughly half that of Pix4D. Furthermore, DSM elevation profiles demonstrate that the estimated height variations between the three software range from 0.5 m to 2.5 m. DJI Terra's estimated heights are generally greater than those of AgiSoft and Pix4D. Furthermore, DJI Terra outperforms AgiSoft and Pix4D for modeling the height contour of trees, buildings, and power lines and poles, followed by AgiSoft and Pix4D. Finally, in terms of the ability of tree detection, DJI Terra outperforms AgiSoft and Pix4D in generating a comprehensive DSM as a result of fewer gaps in the point cloud. Consequently, it stands out as the preferred choice for tree detection applications. The results of this paper can help 3D model users to have confidence in the reliability of the generated 3D models by comprehending the accuracy of the employed software.

Keywords: UAV; photogrammetry; DSM; forest; AgiSoft; PIX4DMapper; DJI Terra

Citation: Jarahizadeh, S.; Salehi, B. A Comparative Analysis of UAV Photogrammetric Software Performance for Forest 3D Modeling: A Case Study Using AgiSoft Photoscan, PIX4DMapper, and DJI Terra. *Sensors* **2024**, *24*, 286. https://doi.org/10.3390/s24010286

Academic Editors: Stelios Krinidis and Christos Nikolaos E. Anagnostopoulos

Received: 18 December 2023
Revised: 29 December 2023
Accepted: 1 January 2024
Published: 3 January 2024

1. Introduction

Three-dimensional (3D) information technologies and the evolution of digital data acquisition have recently caught the attention of researchers [1,2]. In order to eliminate

human errors in the capture of 3D information, researchers are continually working to find an accurate, precise, sustainable solution [3]. The appearance and geometry of an object or scene can be recovered via 3D reconstruction. The most precise and thorough ways to extract the 3D scene and point cloud among the 3D reconstruction techniques now in use are photogrammetry and laser scanning [4]. A laser scanner is an active sensor that transmits pulses to determine distance, generate a 3D point cloud, and estimate coordinates using onboard navigation systems like Global Positioning System (GPS) or Inertial Navigation System (INS). The flight height, platform speed, sensor field of view, and sensor sampling frequency are just a few of the variables that affect laser scanner point density. However, there are certain drawbacks to laser scanning, including challenges when working in indoor environments, operational sensitivity, a requirement for a significant amount of memory storage, longer computation times, and higher costs [5,6]. Photogrammetry and computer vision, in comparison, have been proposed as solutions to existing limitations [2]. Utilizing overlapping photos taken by visual sensors, photogrammetry is a technology that extracts 3D geometrical data and point clouds. Photogrammetry offers several key advantages over laser scanning, including the ability to use video frames as input and the versatility of using digital images captured with various imaging devices, even smartphones. Additionally, it produces 3D point clouds that contain color information that can be densified. Photogrammetry is also known for its automation capabilities, and most importantly, its cost effectiveness [7,8]. On the other hand, Unmanned Aerial Vehicles (UAVs) are increasingly being used for photogrammetric tasks due to their low cost, low flying altitude, real-time data acquisition capabilities, quick, wide-range sensor availability, and capacity to collect geographic data [9–11]. The combination of a low-cost platform, navigation system such as GPS system and IMU system, and high-resolution sensors led to this development [12].

Researchers have introduced a variety of techniques and processes to produce the 3D model from UAV optical data. The significant success of UAV photogrammetry can be largely attributed to the development of Multi-View Stereo (MVS) and Structure From Motion (SfM) algorithms in the field of computer vision, coupled with the advancements in UAV photogrammetric processes. The generation of 3D point clouds, 3D models, and high-quality DSMs has now become straightforward, fast, and user friendly, thanks to the progress in the commercial tools [11,13]. There are over 40 different types of photogrammetric software and tools, both open source and commercial for 3D reconstruction. In order to perform 3D photogrammetric reconstruction, all of these programs generally follow a five-step process: (1) feature detection and matching; (2) triangulation; (3) dense point cloud generation; (4) surface/mesh generation; (5) DSM and orthophoto generation [14].

The advantages of UAV photogrammetry extend across diverse applications and fields including land surface reconstruction [15,16], disaster management [17], and infrastructure applications, such as bridges, roads, railways, and tower inspection [18–20], engineering [21], archaeology [11], and most importantly, agriculture and forest management [22–24]. However, selecting the best and most suitable tools by industry and user experts for a variety of applications has always been difficult, particularly when it comes to forest modeling with its repeated textures and patterns. Accurate, efficient, and up-to-date data on forest characteristics such as tree height, species, and number of trees have been crucial to the success or failure of urban and forest trees 3D modeling. Canopy Height Models (CHMs) are one of the main techniques for evaluating forest attributes derived using the Digital Surface Model (DSM) that can depict the canopy surface, tree height, and density assessment [25,26]. It can be claimed that the accuracy of the DSM directly affects the accuracy of the retrieved forest parameters, and as a result, can determine whether forest 3D modeling is successful or unsuccessful. Therefore, it is crucial to generate DSM as a photogrammetric product over the forested areas using the best technology available.

Few studies have evaluated various photogrammetric tools, even though many have focused on using UAVs to generate 3D models of forests and the potential for doing so. Svenk 2023 used Keystone, SURF, AgiSoft, and MicMac to generate the point cloud and calculate tree parameters for the forest inventory. An evaluation of the Root Mean

Square Error (RMSE) of tree parameters showed that Keystone, SURF, MicMac, and AgiSoft exhibited superior performance in their respective comparison [27]. Terrestrial photos obtained from various visual sensors were employed to compare the 3D models generated by AgiSoft V 1.16, Pix4D V 2.0.89, a combination of Visual SFM V 0.5.22 and SURF V 1.2.0.286, and MicMac V 1.0 on vegetated rock. A point cloud comparison was conducted based on visual evaluation and height profiles. The results indicate that AgiSoft and MicMac exhibit better point cloud accuracy, while Pix4D and the combination of Visual SFM and SURF perform less accurately [28]. Another study compared the DSM produced by AgiSoft, Pix4D, and Leica Photogrammetry Suite (LPS) using ground control points. However, LPS is suitable for airborne (i.e., airplane) photogrammetry and is not effective when it applies to images captured by UAV [29]. A comparison is conducted on height profiles and visual assessments between open-source and commercial photogrammetric software. The results reveal that the software performance depends on applications and texture. Although the ranking of the software depends on the application, Remondino states that AgiSoft generates more reliable and appealing results [30].

It is clear that consumers prefer using the well-known commercial software AgiSoft and Pix4D over other photogrammetric tools for a variety of purposes. Additionally, DJI Terra is a brand new software introduced in 2019, exclusively designed to work with DJI platforms and sensors, making it incomparable to other software [31]. However, given the repeating texture of the forest, a better selection among the existing photogrammetric tools needs to be evaluated considering the application. Also, none of the existing literature has specifically focused on forested areas. In this study, we compare the point clouds and DSM generated over the forest region by AgiSoft, Pix4D, and DJI Terra as well as computational time over the forested areas for forest 3D modeling. The results of this study will assist business and user professionals in identifying constraints and choosing AgiSoft [32], Pix4D [33], or DJI Terra [34] software as the most suitable solution for their project. They will also boost their confidence in their ability to make the right choice instead of investing in expensive projects.

2. Methodology and Data Acquisition

The methodology compares the generated dense point cloud and DSM by AgiSoft V 1.7.3 (AgiSoft LLC, St. Petersburg, Russia) [32], Pix4DMapper V 4.4.12 (Pix4D SA, Lausanne, Switzerland) [33], and DJI Terra V 3.7.6 (DJI, Shenzhen, Guangdong, China) [34] as well as their computational time over forested areas. Figure 1 shows a flowchart of the steps that we conduct in this paper. The main steps are (a) data acquisition, (b) product generation, and (c) product evaluation. In the first step, to compare the program under leaf-on situation, three flights using a 20-megapixel optical sensor with 5472 $\times$ 3648 resolution and 13.2 $\times$ 8.8 mm sensor size were conducted over a section of SUNY ESF Heiberg Forest in Tully, New York about 40 hectares (600 m $\times$ 680 m) in total. This area comprises clearcuts, isolated trees, roads, isolated structures, and electricity lines (Figure 2). The first, second, and third flights were conducted at altitudes of 60 m, 100 m, and 120 m, respectively with about 70 to 80 percent overlaps using Site Scan auto pilot application [35]. The first and third datasets were taken horizontally, while the second dataset was taken 20 degrees off-nadir to investigate the impact of oblique images. Table 1 contains a summary of the flight parameters and dataset. The image position and orientation are also provided from the on-board Global Positioning System (GPS) and Inertial Measurement Unit (IMU).

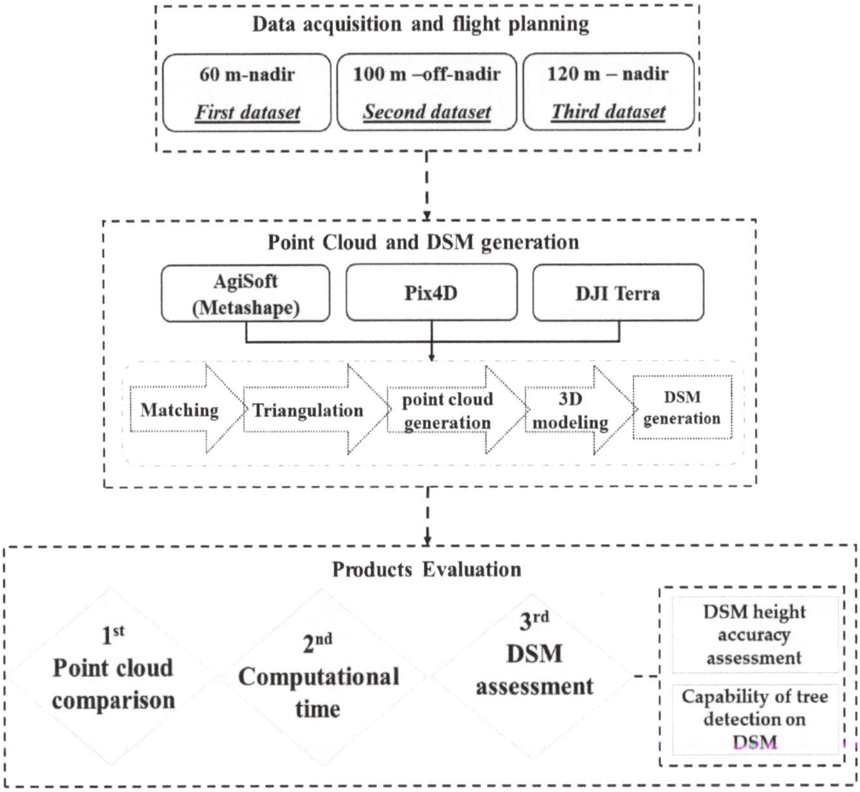

Figure 1. Flowchart of the software comparison strategy in summary.

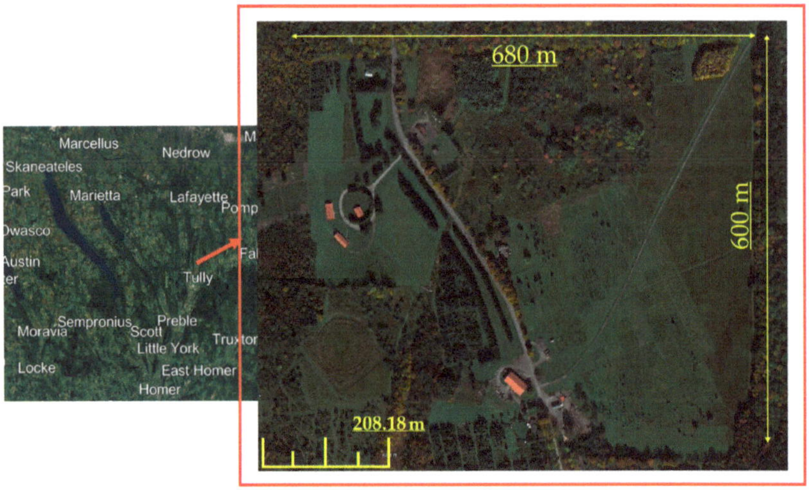

Figure 2. Study area.

Table 1. Datasets and flight parameters.

Platform	Flight Height	Front Overlap	Side Overlap	Gimbal Angle	Resolution	Number of Images	Condition
Dataset 1 (First Flight)	~60 m	70	80	90 degrees	GSD ~1.98 cm	1829	leaf on
Dataset 2 (Second Flight)	~100 m	70	65	70 degrees	GSD ~4.60 cm	768	leaf on
Dataset 3 (Third Flight)	~120 m	70	65	90 degrees	GSD ~3.99 cm	704	leaf on

In the next step, AgiSoft Metashape Professional V 1.7.3 [32], PIX4DMapper V 4.4.12 [33], and DJI Terra V 3.7.6 [34] are used for 3D forest modeling. The common workflow of any photogrammetric software for 3D reconstruction and product generation includes feature recognition, matching, triangulation (pose estimation), sparse point cloud generation, point cloud densification, 3D modeling, and DSM generation. While each of these procedures may have distinct names across various software platforms, they must be executed in their respective sequences. While commercial software employs specific equations, it typically uses common algorithms such as a variant of the Scale-Invariant Feature Transform (SIFT) [36] for feature recognition and matching. Additionally, Collinearity conditions (Equation (1)) or Coplanarity conditions are applied in photogrammetry, while the Essential Matrix or Fundamental Matrix is used in computer vision for pose estimation and point cloud generation [37]. For example, the collinearity condition expresses the basic relationship in which an object point and its image point lie on a straight line passing through the sensor perspective center (Equation (1)) [37]. Equation (1) is as follows, where:

- $\boldsymbol{R}$ is the rotation matrix, k is the scale factor, $\boldsymbol{a}$ is the vector in the object coordinate system, and $\boldsymbol{a}'$ is the corresponding vector in the sensor coordinate system.
- X, Y, Z are the coordinates of the object point and X_C, Y_C, Z_C are the coordinates of the perspective center (sensor center).
- c is the principal distance of the sensor (focal length), x'_0 and y'_0 are the coordinates of the principal point, and x' and y' are the corresponding coordinates.

$$\begin{pmatrix} x' - x'_0 \\ y' - y'_0 \\ -c \end{pmatrix} = k\boldsymbol{R} \begin{pmatrix} X - X_C \\ Y - Y_C \\ Z - Z_C \end{pmatrix} \text{ or } \boldsymbol{a}' = k\boldsymbol{R}\boldsymbol{a} \tag{1}$$

Sparse point clouds, dense point clouds, and DSMs are generated using the recommended parameters. Table 2 contains a list of all used preconfigured software settings for AgiSoft, Pix4D, and DJI Terra. All three datasets have been processed on an Intel i9 core CPU laptop processor unit with NVIDIA GeForce GTX 1650 Ti graphic processing units and 64 gigabytes of random-access memory. Finally, the generated point cloud, DSM, and computational time of the listed software are evaluated both independently and in relation to each other, paying particular attention to forest modeling.

Table 2. Photogrammetric tools processing setting.

	Sparse Point Cloud	Dense Point Cloud	DSM
AgiSoft	High (Full image size)	Medium (down sampled image by factor 2)	High
Pix4D	Full (Full image size)	Multiscale with half image size (down sampled image by factor 2)	Automatic
DJI Terra	High (Full image size)	Height	High

3. Experiments and Results

The software's performance assessments focused on comparing three main criteria: (a) point cloud density and reconstruction quality, (b) computational time, and (c) DSM assessment for height accuracy (z) and ability of tree detection on the DSM.

3.1. Point Cloud Density and Reconstruction Quality

The performance of dense point cloud generation is evaluated independently by assessing the number of generated points, and by comparing the software's generated points. Figure 3 compares the point cloud density per dataset for the three software. In all three datasets, Pix4D and AgiSoft produced point clouds that were roughly 2.5 times denser than those produced by DJI Terra. Moreover, Pix4D generates slightly denser point clouds than AgiSoft. The overall generated 3D point cloud quality over various land cover types such as buildings, hills, and trees have shown that there is no significant difference in spatial errors for point clouds of all software. However, due to the different error sources in matching process and repetitive texture in forested areas, there are some gaps created by Pix4D and AgiSoft that can state that the quality of 3D reconstruction is impacted. The software's generated point cloud can be evaluated for correctness, inaccuracy, and mistake by comparing it to ground truth data. However, distance comparison techniques like the Iterative Closest Point method (ICP) and Multiscale model-to-model Cloud comparison (M3C2) can be used to compare the uniformity, density, and geometry of the point cloud created by various software [38–40]. Using the cloud-to-cloud (C2C) distance toolkit in CloudCompare [41] software, which is based on the Iterative Closest Point method (ICP), we have evaluated the overall quality of the generated 3D point cloud over numerous features, such as trees, power lines, buildings, roads, and grass, relatively. On Dataset 2 (oblique images), all software performed nearly identically in terms of completeness (i.e., successfulness in matching process and consequently generated the points for all the existing objects such as trees and buildings). Comparing the other two datasets (Datasets 1 and 3) shows that the DJI Terra generated fewer gaps on forested regions and power lines than Pix4D and AgiSoft, despite producing a fewer number of overall points. In other words, there are some trees and power lines that Pix4D and AgiSoft did not generate any points for (shown by red circles in Figure 4). This indicates that the increased number of points does not necessarily translate into fewer gaps in the point cloud, as DJI Terra utilizes a better key point recognition and matching algorithm. Additionally, in a study, it has been demonstrated that Pix4D generated significant gaps in vegetation regions than AgiSoft which supports our results [42]

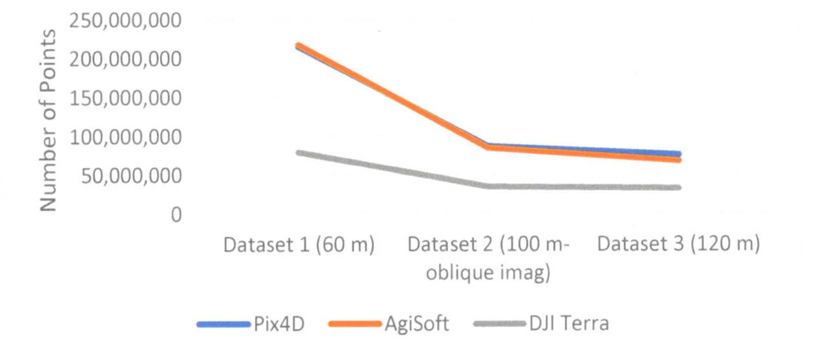

Figure 3. Number of generated points in the dense point cloud.

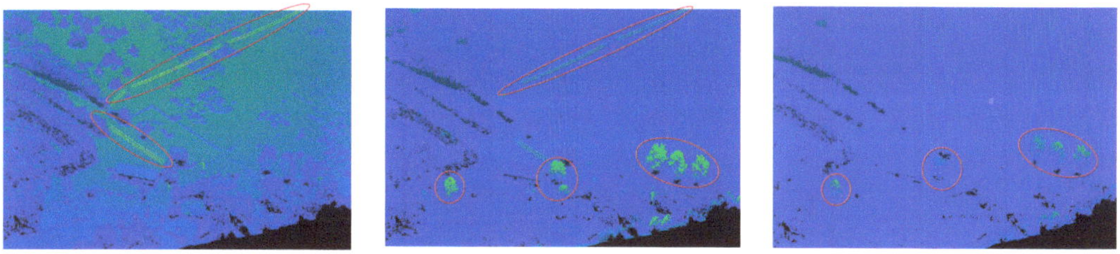

(a) (b) (c)

Figure 4. Computed C2C distance between the generated point cloud by (**a**) AgiSoft and DJI Terra, (**b**) Pix4D and DJI Terra, and (**c**) Pix4D and AgiSoft (red circles show the differences).

3.2. Computational Time

In our evaluation of point cloud density, Pix4D and AgiSoft generated approximately 2.5 times denser point cloud compared to DJI Terra. Consequently, longer computational times for AgiSoft and Pix4D are expected in contrast to DJI Terra. Surprisingly, Pix4D demonstrated an unexpected trend, being roughly three times slower than both AgiSoft and DJI Terra for all datasets (Figure 5). This longer processing time indicates a notable disparity in processing efficiency.

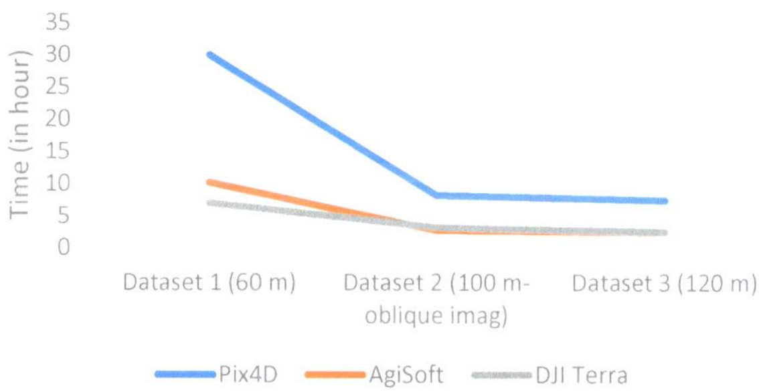

Figure 5. Computational time.

3.3. DSM Assessment

DSM assessment has been carried out both quantitatively and qualitatively for all software. The quantitative evaluation involved comparing the standard deviation (SD) and root mean square error (RMSE). The SD and RMS are calculated using height differences between AgiSoft, Pix4D, and DJI Terra from elevation profiles derived from DSMs of various land cover types including single trees, patches of trees, buildings, and roads. A lower RMSE means a better match between generated elevations by two software. On the other hand, the SD gives a measure of how much the elevations deviate from their mean. A significant difference indicates a systematic error. Subsequently, we assessed the DSM quality for tree detection applications using DSM.

3.3.1. DSM Height Accuracy Assessment Using Elevation Profile

Several elevation profile examples are retrieved for various land covers including buildings (Figure 6), trees (Figure 7), tree patches (Figure 8), and roads (Figure 9) to quantitatively evaluate the generated DSMs. Elevation profiles showed consistent vertical shifts among the generated DSMs for various land cover types and datasets. Specifically, the elevation profile extracted from DJI Terra's DSM consistently is higher than AgiSoft, whereas Pix4D consistently has a lower elevation compared to AgiSoft and DJI Terra. The elevation differences between AgiSoft and DJI Terra are up to 2.5 m for the first dataset,

0.9 m for the second dataset, and 1.5 m for the third dataset. In contrast, the elevation differences between Pix4D and AgiSoft are up to 1 m for the first dataset and 0.5 m for the second and third datasets. It shows that the 3D elevation from Pix4D AgiSoft is distinct from the DJI Terra result while also being similar to each other. The number of generated points may be the root cause of the significant elevation differences between DJI Terra and two other software. Fewer points within a pixel can lead to distinct elevations in the DSM, given that the elevation of each pixel is computed as the weighted total of its internal points. Furthermore, vertical shifts between the generated DSMs may be impacted by the points distribution. The various closed sophisticated algorithms that are applied in commercial software are another potential cause of vertical shifts. In general, when features are found at a higher elevation section of the research area (i.e., on top of a hill), the amount of the vertical shift is reduced since the features are closer to the drone, and thus have a lower flying height than in other areas.

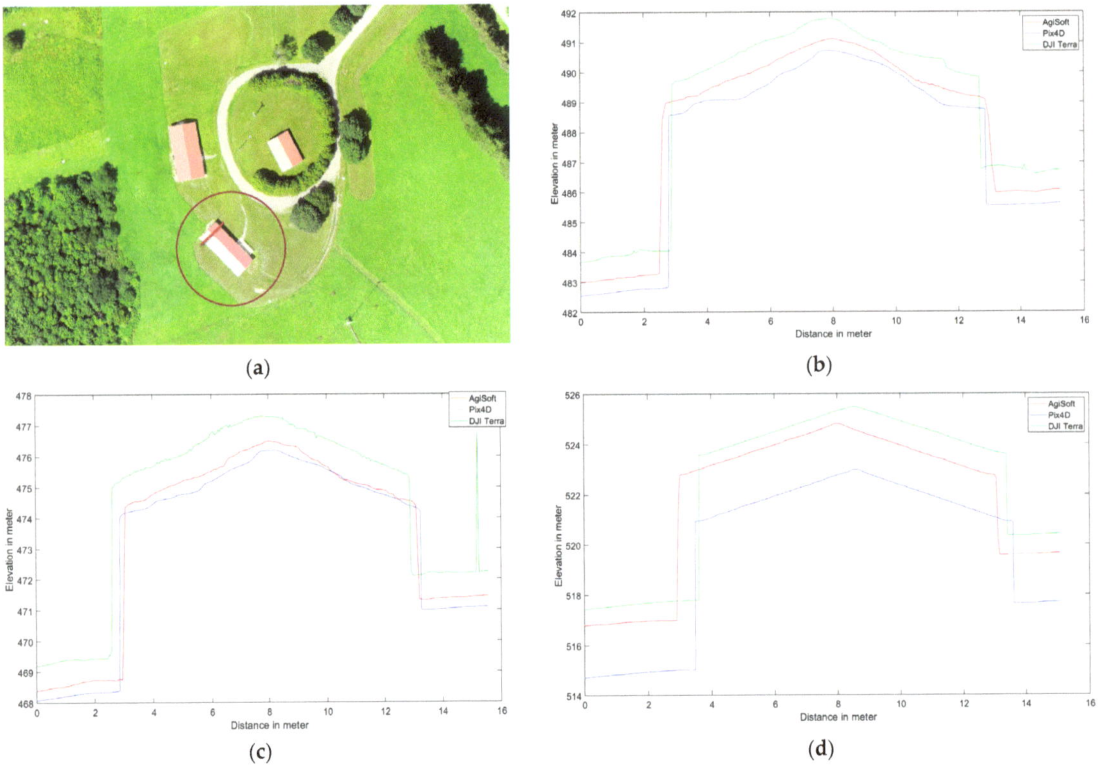

Figure 6. Elevation profile from Pix4D (blue), AgiSoft (red), and DJI Terra (green) on a building, (**a**) profile line (red circle shows the picked feature), (**b**) Dataset 1 (60 m), (**c**) Dataset 2 (100 m oblique images), and (**d**) Dataset 3 (120 m).

The utilization of oblique images rather than vertical ones reduces the vertical shifts across all software. The greater intersection angles in oblique images enhance the accuracy of elevation estimation through improved collinearity equations [43]. The third dataset displays fewer vertical shifts than the first dataset, a reason that may be attributed to a higher flight altitude. Generally, higher flight altitudes often result in lower spatial resolution and consequently reduced detail and repetitive textures, especially in areas with dense forest cover, where repetitive textures can affect the accuracy of matching and elevation data. In the analysis of the first and second datasets, elevation spikes can be seen

on trees in Pix4D and AgiSoft. All applications and datasets also exhibit slight horizontal shifts. Although there are horizontal and vertical shifts, the Pix4D and AgiSoft images are more pleasing and smoother for flat surfaces like roadways than DJI Terra.

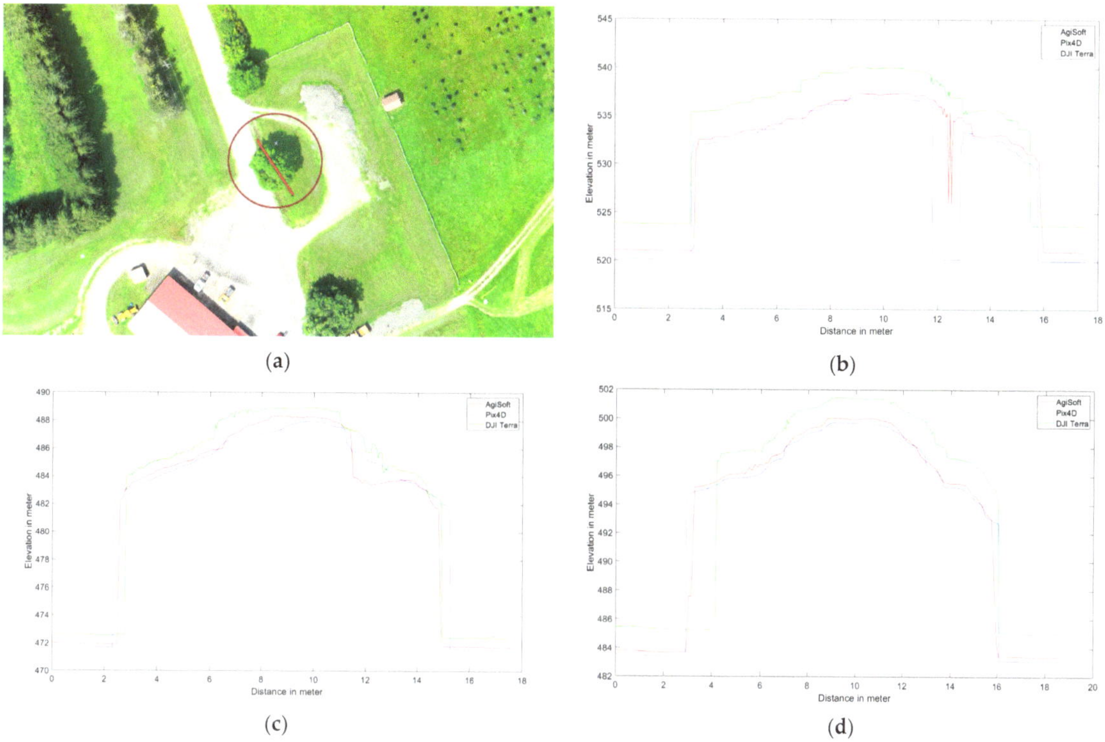

Figure 7. Elevation profile from Pix4D (blue), AgiSoft (red), and DJI Terra (green) on a tree, (**a**) profile line(red circle shows the picked feature), (**b**) Dataset 1 (60 m), (**c**) Dataset 2 (100 m oblique images), and (**d**) Dataset 3 (120 m).

It can be said that the results from DJI Terra are more compelling, especially when applied to natural features such as trees. It is common to see numerous slight height discrepancies in areas covered with vegetation, such as dense trees. However, Pix4D and AgiSoft do not appear to have as many details extracted as DJI Terra which suggests a potential advantage to capture finer details in vegetated areas. The accuracy and adaptability across various datasets are measured by the root mean square error (RMSE) metric and the standard deviation (SD) calculated for height differences between AgiSoft, Pix4D, and DJI Terra. Utilizing the standard deviation (SD) metric defines a range that encompasses the average to identify outliers. It can be concluded that the distribution of errors is normal and there are no systematic errors or outliers in the outputs if the RMSE and standard deviation (SD) values are similar [44,45]. The small discrepancies between RMSE and SD confirm the absence of systematic inaccuracy (bias) among the DSMs produced by all software (Figure 10). Furthermore, it shows how close the 3D profile models from Pix4D, AgiSoft, and DJI Terra are to one another.

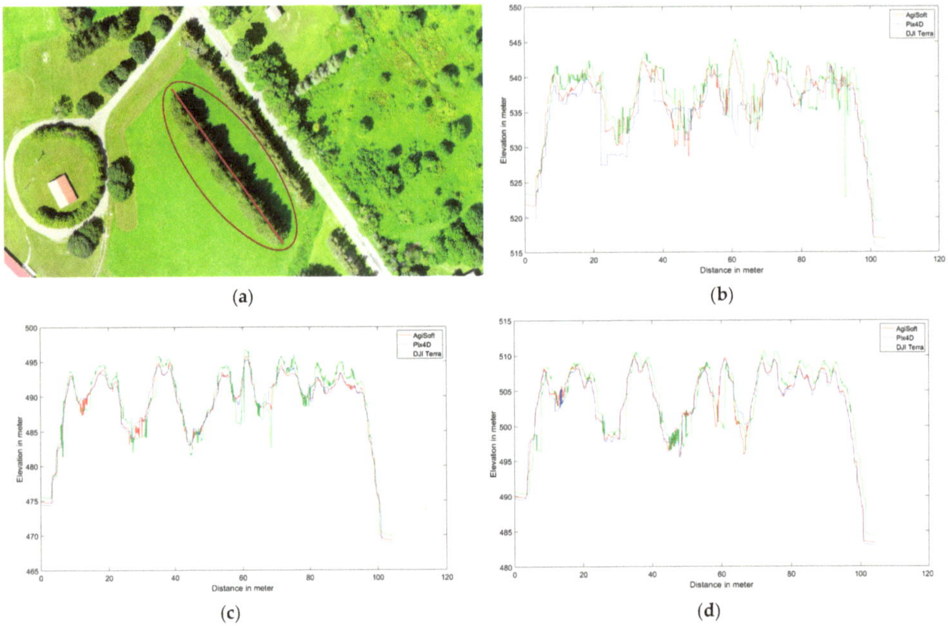

Figure 8. Elevation profile from Pix4D (blue), AgiSoft (red), and DJI Terra (green) on a patch of the trees, (**a**) profile line (red circle shows the picked feature), (**b**) Dataset 1 (60 m), (**c**) Dataset 2 (100 m oblique images), and (**d**) Dataset 3 (120 m).

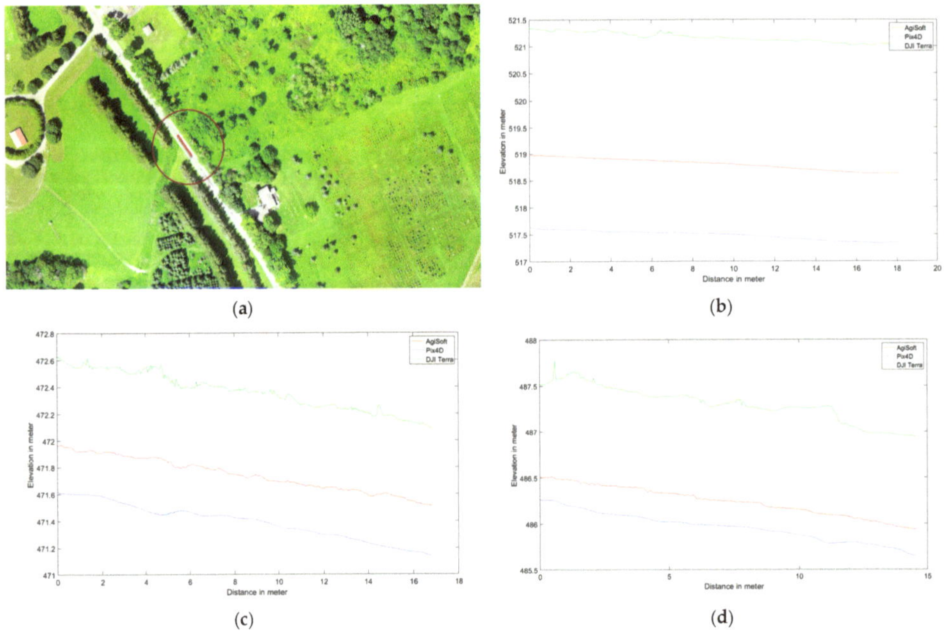

Figure 9. Elevation profile from Pix4D (blue), AgiSoft (red), and DJI Terra (green) on a road, (**a**) profile line (red circle shows the picked feature), (**b**) Dataset 1 (60 m), (**c**) Dataset 2 (100 m oblique images), and (**d**) Dataset 3 (120 m).

Figure 10. Difference between RMSE and SD.

3.3.2. Capability of Tree Detection on DSM

The evaluation of tree detection capabilities in the generated DSMs has been carried out through visual comparisons. The DSMs were generated using Pix4D, AgiSoft, and DJI Terra and were visually assessed for their effectiveness in accurately detecting trees. The results show an obvious elimination of some trees (i.e., missing some trees) in the DSMs generated by Pix4D and AgiSoft which can raise considerations regarding the completeness and accuracy of tree detection in these software outputs. Despite generating around 2.5 times fewer points than Pix4D and AgiSoft, DJI Terra was still able to generate and detect a more detailed DSM, resulting in the identification of several trees that were not present in the DSMs generated by Pix4D and AgiSoft. Examples of missing trees are highlighted with black circles in Figure 11, representing Dataset 1 (60 m), Figure 12 for Dataset 2 (100 m oblique images), and Figure 13 for Dataset 3 (120 m). Furthermore, DJI Terra's DSM is smoother than that generated by Pix4D and AgiSoft. The possible causes include (1) the use of a better outlier rejection approach in the DJI Terra that causes the generation of a better DSM [46], and (2) the improved point distribution achieved by DJI Terra. Furthermore, DJI Terra and AgiSoft demonstrated superior precision in capturing the corners and edges of buildings compared to Pix4D. In general, it can be said that the DJI Terra outperforms Pix4D and AgiSoft in forestry areas by spotting more single trees and identifying the edge of the single trees within tree patches.

(a)

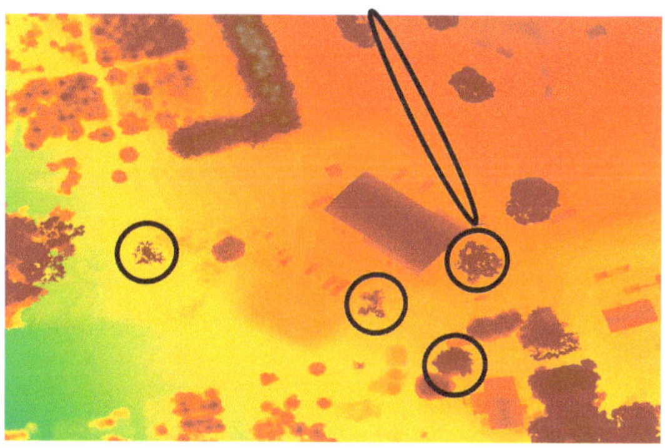
(b)

Figure 11. *Cont.*

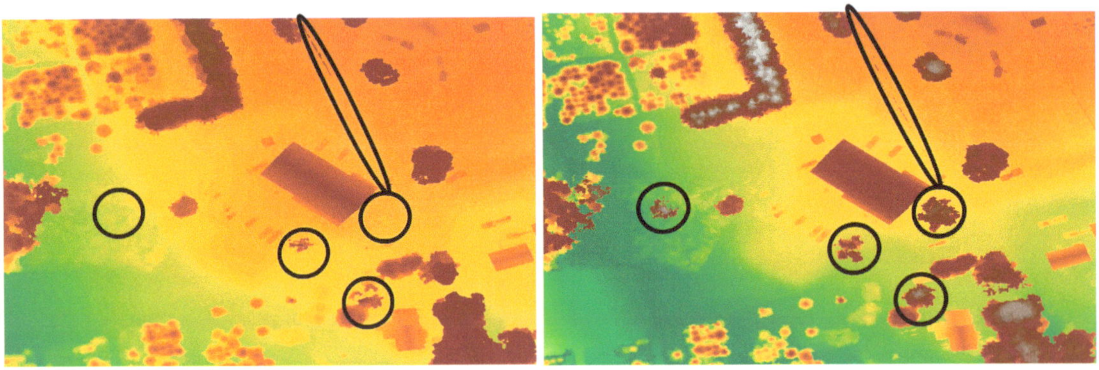

(c) (d)

Figure 11. (**a**) Orthophoto, generated DSM on Dataset 1 (60 m) by (**b**) AgiSoft, (**c**) Pix4D, and (**d**) DJI Terra, black circles indicate the differences.

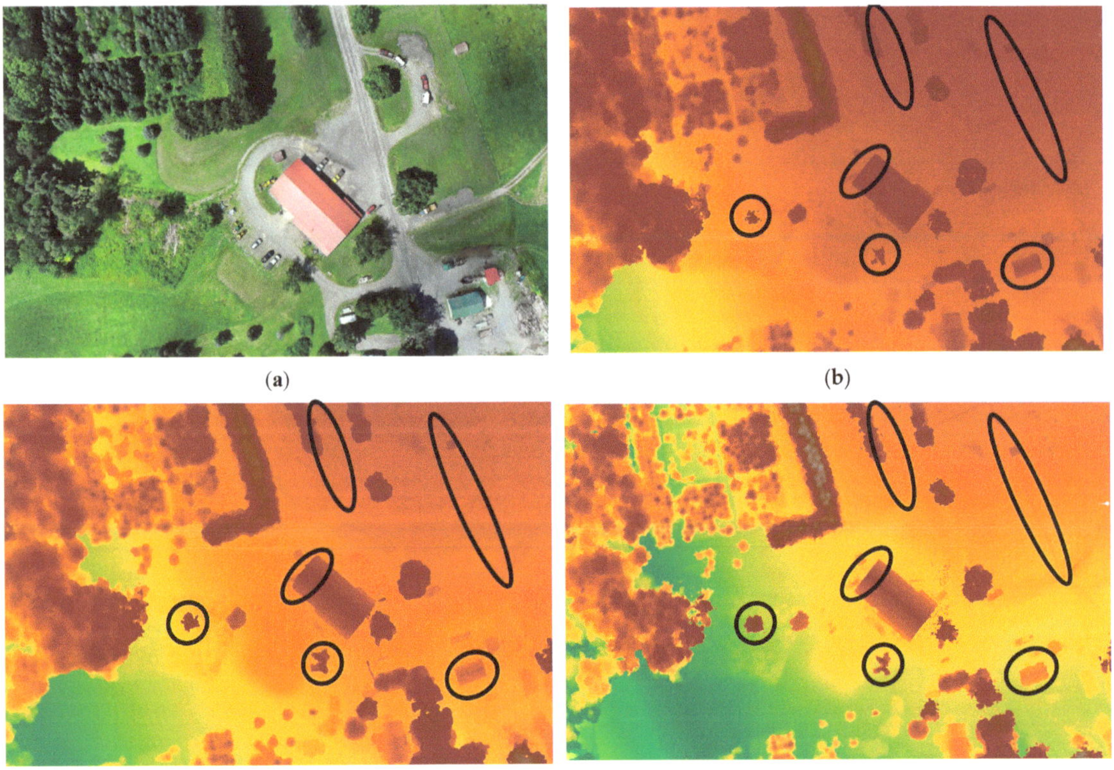

(a) (b) (c) (d)

Figure 12. (**a**) Orthophoto, generated DSM on Dataset 2 (100 m oblique images) by (**b**) AgiSoft, (**c**) Pix4D, and (**d**) DJI Terra, black circles indicate the differences.

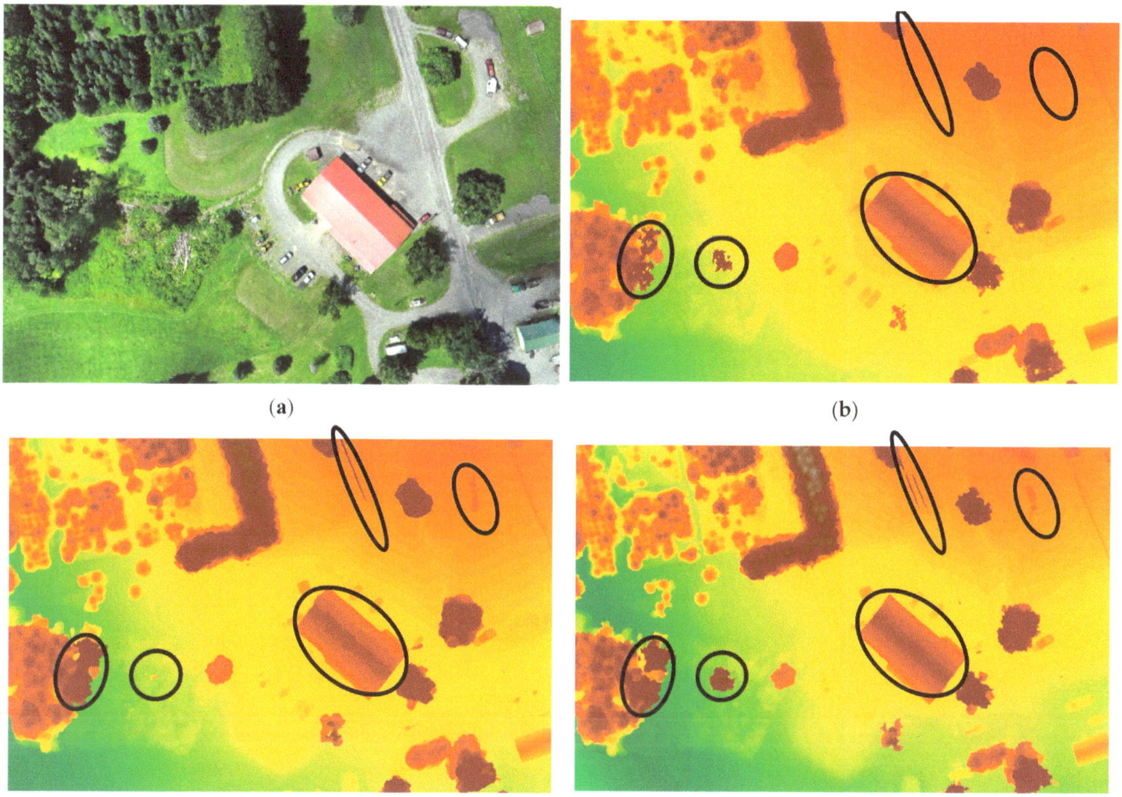

(a) (b) (c) (d)

Figure 13. (**a**) Orthophoto, generated DSM on Dataset 3 (120 m) by (**b**) AgiSoft, (**c**) Pix4D, and (**d**) DJI Terra, black circles indicate the differences.

4. Conclusions

This study was conducted to assist industry and professional users in discovering and choosing the best software among AgiSoft, Pix4D, and DJI Terra for forest 3D modeling purposes as well as to boost their confidence in making the right choice instead of investing in expensive projects. Three flights within altitudes of 60, 100, and 120 m were conducted to evaluate the point cloud density and reconstruction quality, computational time, and DSMs for height accuracy (z) and ability of tree detection both quantitively and qualitatively over the forested area. The results show that Pix4D and AgiSoft generated denser point clouds than DJI Terra. However, DJI Terra provided a better point cloud of trees than the other two software, likely due to utilizing an enhanced matching algorithm. As a result, DJI Terra generated an accurate DSM with fewer gaps than AgiSoft and Pix4D. Despite the vertical shift in height values on generated DSM, DJI Terra performed better in terms of modeling trees and building shapes. However, AgiSoft and Pix4D performed better in generating the road elevation profile than the DJI Terra. In general, Pix4D generated the highest elevation, followed by AgiSoft, and lastly DJI Terra. Finally, the computational time comparison reveals that the processing time of AgiSoft and DJI Terra is roughly half that of Pix4D. Future research can contribute to enhancing our understanding by evaluating the accuracy of each product against referenced ground truth data and comparing them to other commercial software as we only relatively evaluated AgiSoft, Pix4D, and DJI Terra.

Author Contributions: Conceptualization, S.J. and B.S.; methodology, S.J. and B.S.; software, S.J.; validation, S.J. and B.S.; formal analysis, S.J. and B.S.; data curation, S.J.; writing—original draft preparation, S.J.; writing—review and editing, S.J. and B.S.; visualization, S.J.; supervision, B.S.; funding acquisition, B.S. All authors have read and agreed to the published version of the manuscript.

Funding: This project was funded through USDA National Institute of Food and Agriculture (NIFA), McIntire-Stennis Grant.

Data Availability Statement: Data are contained within the article.

Conflicts of Interest: The authors declare no conflicts of interest.

References

1. Manzoor, B.; Othman, I.; Pomares, J.C. Digital Technologies in the Architecture, Engineering and Construction (Aec) Industry—A Bibliometric—Qualitative Literature Review of Research Activities. *Int. J. Environ. Res. Public Health* **2021**, *18*, 6135. [CrossRef] [PubMed]
2. Shao, Z.; Yang, N.; Xiao, X.; Zhang, L.; Peng, Z. A Multi-View Dense Point Cloud Generation Algorithm Based on Low-Altitude Remote Sensing Images. *Remote Sens.* **2016**, *8*, 381. [CrossRef]
3. Mahami, H.; Nasirzadeh, F.; Hosseininaveh Ahmadabadian, A.; Nahavandi, S. Automated Progress Controlling and Monitoring Using Daily Site Images and Building Information Modelling. *Buildings* **2019**, *9*, 70. [CrossRef]
4. Bianco, S.; Ciocca, G.; Marelli, D. Evaluating the Performance of Structure from Motion Pipelines. *J. Imaging* **2018**, *4*, 98. [CrossRef]
5. Lu, R.; Brilakis, I. Digital Twinning of Existing Reinforced Concrete Bridges from Labelled Point Clusters. *Autom. Constr.* **2019**, *105*, 102837. [CrossRef]
6. Woodhead, R.; Stephenson, P.; Morrey, D. Digital Construction: From Point Solutions to IoT Ecosystem. *Autom. Constr.* **2018**, *93*, 35–46. [CrossRef]
7. Kortaberria, G.; Mutilba, U.; Gomez-Acedo, E.; Tellaeche, A.; Minguez, R. Accuracy Evaluation of Dense Matching Techniques for Casting Part Dimensional Verification. *Sensors* **2018**, *18*, 3074. [CrossRef]
8. Zhu, H.; Wu, W.; Chen, J.; Ma, G.; Liu, X.; Zhuang, X. Integration of Three Dimensional Discontinuous Deformation Analysis (DDA) with Binocular Photogrammetry for Stability Analysis of Tunnels in Blocky Rockmass. *Tunn. Undergr. Space Technol.* **2016**, *51*, 30–40. [CrossRef]
9. Ruzgienė, B.; Berteška, T.; Gečyte, S.; Jakubauskienė, E.; Aksamitauskas, V.Č. The Surface Modelling Based on UAV Photogrammetry and Qualitative Estimation. *Measurement* **2015**, *73*, 619–627. [CrossRef]
10. Nikolakopoulos, K.G.; Lampropoulou, P.; Fakiris, E.; Sardelianos, D.; Papatheodorou, G. Synergistic Use of UAV and USV Data and Petrographic Analyses for the Investigation of Beachrock Formations: A Case Study from Syros Island, Aegean Sea, Greece. *Minerals* **2018**, *8*, 534. [CrossRef]
11. Pepe, M.; Alfio, V.S.; Costantino, D. UAV Platforms and the SfM-MVS Approach in the 3D Surveys and Modelling: A Review in the Cultural Heritage Field. *Appl. Sci.* **2022**, *12*, 12886. [CrossRef]
12. Liu, Y.; Gong, W.; Xing, Y.; Hu, X.; Gong, J. Estimation of the Forest Stand Mean Height and Aboveground Biomass in Northeast China Using SAR Sentinel-1B, Multispectral Sentinel-2A, and DEM Imagery. *ISPRS J. Photogramm. Remote Sens.* **2019**, *151*, 277–289. [CrossRef]
13. Girelli, V.A.; Borgatti, L.; Dellapasqua, M.; Mandanici, E.; Spreafico, M.C.; Tini, M.A.; Bitelli, G. Integration of Geomatics Techniques for Digitizing Highly Relevant Geological and Cultural Heritage Sites: The Case of San Leo (Italy). *Int. Arch. Photogramm. Remote Sens. Spat. Inf. Sci.* **2017**, *42*, 281–286. [CrossRef]
14. Qureshi, A.H.; Alaloul, W.S.; Hussain, S.J.; Murtiyoso, A.; Saad, S.; Alzubi, K.M.; Ammad, S.; Baarimah, A.O. Evaluation of Photogrammetry Tools Following Progress Detection of Rebar towards Sustainable Construction Processes. *Sustainability* **2023**, *15*, 21. [CrossRef]
15. Rossi, G.; Tanteri, L.; Tofani, V.; Vannocci, P.; Moretti, S.; Casagli, N. Multitemporal UAV Surveys for Landslide Mapping and Characterization. *Landslides* **2018**, *15*, 1045–1052. [CrossRef]
16. Agüera-Vega, F.; Carvajal-Ramírez, F.; Martínez-Carricondo, P. Assessment of Photogrammetric Mapping Accuracy Based on Variation Ground Control Points Number Using Unmanned Aerial Vehicle. *Measurement* **2017**, *98*, 221–227. [CrossRef]
17. Quaritsch, M.; Kruggl, K.; Wischounig-Strucl, D.; Bhattacharya, S.; Shah, M.; Rinner, B. Networked UAVs as Aerial Sensor Network for Disaster Management Applications. *E I Elektrotech. Inf.* **2010**, *127*, 56–63. [CrossRef]
18. Máthé, K.; Buşoniu, L. Vision and Control for UAVs: A Survey of General Methods and of Inexpensive Platforms for Infrastructure Inspection. *Sensors* **2015**, *15*, 14887–14916. [CrossRef]
19. Bellavia, F.; Colombo, C.; Morelli, L.; Remondino, F. Challenges in Image Matching for Cultural Heritage: An Overview and Perspective. In *Image Analysis and Processing. ICIAP 2022 Workshops*; Springer LNCS: Berlin/Heidelberg, Germany, 2022.
20. Morgenthal, G.; Hallermann, N.; Kersten, J.; Taraben, J.; Debus, P.; Helmrich, M.; Rodehorst, V. Framework for Automated UAS-Based Structural Condition Assessment of Bridges. *Autom. Constr.* **2019**, *97*, 77–95. [CrossRef]
21. Seo, J.; Duque, L.; Wacker, J.P. Field Application of UAS-Based Bridge Inspection. *Transp. Res. Rec.* **2018**, *2672*, 72–81. [CrossRef]

22. Honkavaara, E.; Saari, H.; Kaivosoja, J.; Pölönen, I.; Hakala, T.; Litkey, P.; Mäkynen, J.; Pesonen, L. Processing and Assessment of Spectrometric, Stereoscopic Imagery Collected Using a Lightweight UAV Spectral Camera for Precision Agriculture. *Remote Sens.* **2013**, *5*, 5006–5039. [CrossRef]
23. Feng, Q.; Liu, J.; Gong, J. UAV Remote Sensing for Urban Vegetation Mapping Using Random Forest and Texture Analysis. *Remote Sens.* **2015**, *7*, 1074–1094. [CrossRef]
24. Salehi, B.; Jarahizadeh, S. IMPROVING THE UAV-DERIVED DSM BY INTRODUCING A MODIFIED RANSAC ALGORITHM. *Int. Arch. Photogramm. Remote Sens. Spat. Inf. Sci.* **2022**, *XLIII-B2-2022*, 147–152. [CrossRef]
25. Tu, Y.-H.; Johansen, K.; Phinn, S.; Robson, A. Measuring Canopy Structure and Condition Using Multi-Spectral UAS Imagery in a Horticultural Environment. *Remote Sens.* **2019**, *11*, 269. [CrossRef]
26. Tu, Y.-H.; Phinn, S.; Johansen, K.; Robson, A.; Wu, D. Optimising Drone Flight Planning for Measuring Horticultural Tree Crop Structure. *ISPRS J. Photogramm. Remote Sens.* **2020**, *160*, 83–96. [CrossRef]
27. Svensk, J. Evaluation of Aerial Image Stereo Matching Methods for Forest Variable Estimation 2017. Available online: https://urn.kb.se/resolve?urn=urn:nbn:se:liu:diva-138166 (accessed on 28 December 2023).
28. Niederheiser, R.; Mokroš, M.; Lange, J.; Petschko, H.; Prasicek, G.; Elberink, S.O. Deriving 3D Point Clouds From Terrestrial Photographs-Comparison of Different Sensors and Software. *Int. Arch. Photogramm. Remote Sens. Spat. Inf. Sci.* **2016**, *41*, 685–692. [CrossRef]
29. Bhandari, B.; Oli, U.; Pudasaini, U.; Panta, N. Generation of High Resolution DSM Using UAV Images. In Proceedings of the FIG Working Week, Sofia, Bulgaria, 17–21 May 2015; pp. 17–21.
30. Remondino, F.; Spera, M.G.; Nocerino, E.; Menna, F.; Nex, F. State of the Art in High Density Image Matching. *Photogramm. Rec.* **2014**, *29*, 144–166. [CrossRef]
31. Hao, Z.; Lin, L.; Post, C.J.; Jiang, Y.; Li, M.; Wei, N.; Yu, K.; Liu, J. Assessing Tree Height and Density of a Young Forest Using a Consumer Unmanned Aerial Vehicle (UAV). *New For.* **2021**, *52*, 843–862. [CrossRef]
32. Agisoft LLC. *Agisoft Metashape User Manuals*; Agisoft LLC: St. Petersburg, Russia, 2021; Available online: http://www.agisoft.com/ (accessed on 28 December 2023).
33. Pix4D SA. Pix4Dmapper. Lausanne: Pix4D SA. Available online: https://www.pix4d.com/ (accessed on 28 December 2023).
34. DJI. DJI Terra. Shenzhen: DJI. Available online: https://www.dji.com/ (accessed on 28 December 2023).
35. Esri. Site Scan for ArcGIS. Redlands: Esri. Available online: https://www.esri.com/en-us/arcgis/products/arcgis-sitescan/overview (accessed on 28 December 2023).
36. Lowe, D.G. Object recognition from local scale-invariant features. In Proceedings of the Seventh IEEE International Conference on Computer vision, Corfu, Greece, 20–27 September 1999; IEEE: Piscataway, NJ, USA, 1999; Volume 2, pp. 1150–1157.
37. Elnima, E.E. A Solution for Exterior and Relative Orientation in Photogrammetry, a Genetic Evolution Approach. *J. King Saud. Univ. Eng. Sci.* **2015**, *27*, 108–113. [CrossRef]
38. Lague, D.; Brodu, N.; Leroux, J. Accurate 3D Comparison of Complex Topography with Terrestrial Laser Scanner: Application to the Rangitikei Canyon (NZ). *ISPRS J. Photogramm. Remote Sens.* **2013**, *82*, 10–26. [CrossRef]
39. DiFrancesco, P.-M.; Bonneau, D.; Hutchinson, D.J. The Implications of M3C2 Projection Diameter on 3D Semi-Automated Rockfall Extraction from Sequential Terrestrial Laser Scanning Point Clouds. *Remote Sens.* **2020**, *12*, 1885. [CrossRef]
40. Besl, P.J.; McKay, N.D. Method for Registration of 3-D Shapes. Sensor Fusion IV: Control Paradigms and Data Structures. *Int. Soc. Opt. Photonics* **1992**, *1611*, 586–606.
41. Girardeau-Montaut, D. CloudCompare. *Fr. EDF RD Telecom ParisTech* **2016**, *11*, 5.
42. Georgopoulos, A.; Oikonomou, C.; Adamopoulos, E.; Stathopoulou, E.K. Evaluating Unmanned Aerial Platforms for Cultural Heritage Large Scale Mapping. *Int. Arch. Photogramm. Remote Sens. Spat. Inf. Sci.* **2016**, *XLI-B5*, 355–362. [CrossRef]
43. Sadeq, H.A. Accuracy Assessment Using Different UAV Image Overlaps. *J. Unmanned Veh. Syst.* **2019**, *7*, 175–193. [CrossRef]
44. Chai, T.; Draxler, R.R. Root Mean Square Error (RMSE) or Mean Absolute Error (MAE). *Geosci. Model. Dev. Discuss.* **2014**, *7*, 1525–1534.
45. Lee, D.K.; In, J.; Lee, S. Standard Deviation and Standard Error of the Mean. *Korean J. Anesthesiol.* **2015**, *68*, 220–223. [CrossRef]
46. Salehi, B.; Jarahizadeh, S.; Sarafraz, A. An Improved RANSAC Outlier Rejection Method for UAV-Derived Point Cloud. *Remote Sens.* **2022**, *14*, 4917. [CrossRef]

sensors

MDPI

Article

Analytic Design Technique for 2D FIR Circular Filter Banks and Their Efficient Implementation Using Polyphase Approach

Radu Matei [1,2] and Doru Florin Chiper [1,3,4,*]

1 Faculty of Electronics, Telecommunications and Information Technology, "Gheorghe Asachi" Technical University of Iaşi, 700506 Iaşi, Romania; rmatei@etti.tuiasi.ro
2 Institute of Computer Science, Iaşi Branch of the Romanian Academy, 700481 Iaşi, Romania
3 Technical Sciences Academy of Romania (ASTR), 700050 Iaşi, Romania
4 Academy of Romanian Scientists (AOSR), 030167 Bucureşti, Romania
* Correspondence: chiper@etti.tuiasi.ro

Abstract: This paper proposes an analytical design procedure for 2D FIR circular filter banks and also a novel, computationally efficient implementation of the designed filter bank based on a polyphase structure and a block filtering approach. The component filters of the bank are designed in the frequency domain using a specific frequency transformation applied to a low-pass, band-pass and high-pass 1D prototype with a specified Gaussian shape and imposed specifications (peak frequency, bandwidth). The 1D prototype filter frequency response is derived in a closed form as a trigonometric polynomial with a specified order using Fourier series, and then it is factored. Since the design starts from a 1D prototype with a factored transfer function, the frequency response of the designed 2D filter bank components also results directly in a factored form. The designed filters have an accurate shape, with negligible distortions at a relatively low order. We present the design of two types of circular filter banks: uniform and non-uniform (dyadic). An example of image analysis with the uniform filter bank is also provided, showing that the original image can be accurately reconstructed from the sub-band images. The proposed implementation is presented for a simpler case, namely for a smaller size of the filter kernel and of the input image. Using the polyphase and block filtering approach, a convenient implementation at the system level is obtained for the designed 2D FIR filter, with a relatively low computational complexity.

Keywords: 2D FIR filters; circular filters; analytical design; filter banks; polyphase decomposition; block filters

Citation: Matei, R.; Chiper, D.F. Analytic Design Technique for 2D FIR Circular Filter Banks and Their Efficient Implementation Using Polyphase Approach. *Sensors* **2023**, *23*, 9851. https://doi.org/10.3390/s23249851

Academic Editors: Stelios Krinidis and Christos Nikolaos E. Anagnostopoulos

Received: 16 October 2023
Revised: 10 December 2023
Accepted: 12 December 2023
Published: 15 December 2023

1. Introduction

The technology and architecture of modern image sensors and sensing techniques have evolved dramatically in recent years, driven by the ever-demanding requirements and challenges of this field. For instance, aerial or satellite image sensors for remote sensing must provide clear and low-noise images, with high spatial resolution, either in visible, infrared or microwave domains. In order to provide accurate and relevant information, images acquired by sensors have to be pre-processed using various restoration and enhancement techniques. Various digital filters and filter banks may be used in image analysis and feature extraction tasks, for instance, to decompose the image into several subband components in order to extract relevant details, etc. These are also useful in the automotive field, for rapid feature extraction in real-time computer vision applications, for instance, in driver assistance systems and autonomous driving vehicles.

Along with the unprecedented development of the digital signal processing field, 2D filters have been thoroughly investigated by many researchers, owing to their essential applications in image processing, and various techniques for their design have been elaborated [1]. Analytical design methods rely on 1D prototypes with specified shapes and parameters; applying various frequency transformations, they lead directly to the desired

2D filters. The major advantage of the analytical approach is that a closed-form frequency response is derived, and the 2D filter results are parametric and therefore adjustable.

A large variety of 2D filters, both of FIR and IIR type, with various characteristics and shapes have been developed, with each type of filter having specific applications in the image processing field. One of the best-known methods, widely used in the design of 2D FIR filters with various shapes is the McClellan transform [2,3]. More recent papers approaching computationally efficient 2D FIR filter design techniques based on frequency transformations are [4,5]. The efficient, low-complexity design of 2D FIR filters and the implementation using Farrow structure are described in papers like [6,7]. Other relevant recent papers on efficient 2D filter design are [8–10].

Filters with circular-shaped frequency response have also been widely used owing to their capabilities in image analysis; various design techniques have been proposed for circular filters (CF) in early papers such as [11–13]. Circular filters find applications in texture segmentation and classification [14]. A recent advanced application of circular Gabor filters in SAR interferograms is described in [15].

Two-dimensional filter banks of various types were extensively used in important applications, like texture segmentation and classification or various feature extraction tasks. Such filter banks decompose the frequency spectrum of the image into a number of sub-bands. Two-dimensional filter banks are widely used in fundamental applications such as sub-band coding and compression of images and video sequences. Separable 2D filter banks are obtained by cascading 1D filter banks, and data are processed in each dimension separately. Compared to separable filters, filter banks with nonseparable 2D filters are more flexible and versatile, offering superior performance for imposed specifications. However, their design is substantially more difficult than for separable filter banks [16]. As detailed in the comprehensive review [16], the 2D filter banks currently used are mainly directional, with specific shapes in the frequency plane, such as square (diamond), parallelogram, wedge/fan filters, etc. Multidimensional stable, perfect reconstruction filter banks are also developed in [17].

Directional filter banks (DFBs) with an arbitrary number of sub-bands [18] or arbitrary frequency partitioning [19] have been proposed. A class of multiresolution DFBs is developed in [20]. Multidimensional DFB, multiscale pyramids and the surfacelet transform were introduced in [21]. A very recent application of DFBs was proposed in [22], namely fingerprint image quality assessment. The fingerprint image is decomposed into subbands using the DFB, and similarity between the different subbands is used to calculate the fingerprint image quality. Regarding methods to reduce computational complexity and increase processing speed, the fast block implementation of 2D digital FIR filters was proposed in early papers such as [23]. A high-performance 2D parallel block-filtering system for real-time applications was presented in [24]. The steerable pyramid, a well-known multiscale structure for image decomposition was proposed in the early paper [25]. More recent papers describe specific applications of other two important multiscale architectures, namely the Laplacian pyramid [26] and the wavelet pyramid [27].

Some very recent works propose advanced algorithms implemented on various convolutional neural networks to solve complex image-processing tasks. For instance, in [28], a novel deep-feature model has been proposed for coastal wetland classification using multisource satellite remote sensing data. In [29], multi-scale features from coarse-to-fine receptive field level are extracted, with applications in super-resolution. An advanced algorithm for effective pathology classification from hyperspectral medical images is proposed in [30]. A novel multi-focus image fusion method based on sparse representation and local energy is introduced in [31], which uses the shearlet transform to decompose the source images into low- and high-frequency sub-bands.

The first author of this paper has also proposed various analytical design techniques for 2D filters in previous works [32–35]. Directional IIR filters based on Gaussian and wide-band prototypes were designed in [32]. A useful application of the directional filters in [32] is the detection of straight lines with specified orientation from images; this feature

extraction capability may be useful in the computer vision field. Adjustable, parametric 2D digital IIR filters with elliptical and circular symmetry are proposed in [33]. Two versions of circular IIR filter banks and their applications have been described in [34,35]. An efficient 2D FIR filter implementation based on a polyphase approach and block filtering is proposed in [36].

In this paper, an analytic design procedure is proposed for a particular class of 2D filter banks, namely 2D FIR Gaussian circular filter banks (CFBs). Two versions of CFBs will be designed, namely a uniform CFB and then a non-uniform (dyadic) CFB, each with a specified number of component filters. As a prototype, a 1D low-pass filter with a Gaussian frequency response and specified selectivity is chosen; its frequency response is easily approximated by a trigonometric polynomial, with an imposed precision, using a simple Fourier series expansion. By a simple shifting to a given peak frequency, the band-pass filters of the FB prototype are also derived. Once the prototype FB is obtained, a 1D to 2D frequency mapping derived from the McClellan transform is applied [2,3], which leads directly to the desired circular filters of the CFB. The non-uniform (dyadic) CFB is designed in a similar manner. The filters' characteristics result in an accurate circular shape, with some distortions near the frequency plane margins. Next, as an application example, a grayscale test image is applied to the CFB, obtaining a set of subband images. Summing back all these images, the original input image is reconstructed almost perfectly, which suggests a potential use in an alternative subband coding scheme.

A novel, efficient implementation solution is also proposed for the 2D FIR filters of the designed CFB, which continues the method from previous work [36]. Our implementation uses a polyphase decomposition of a given 2D filtering operation with large kernel size and a block filtering with smaller size matrices.

The paper is organized as follows: Section 2 presents the proposed analytical design procedure, first deriving the uniform and non-uniform prototype FB, then applying the frequency mapping and obtaining the frequency responses of the 2D CFBs. In Section 3, an example of image analysis is given using CFB by decomposing it into subband images. The novel implementation technique based on the polyphase and block filtering approach is described in Section 4. Discussions regarding the computational complexity of the proposed implementation are included in Section 5. Finally, conclusions are drawn in the last section.

2. Analytical Design Technique for 2D Circular FIR Filter Banks

A novel analytical design procedure is proposed for a class of 2D FIR circular filters. This design technique starts from an imposed prototype with specified parameters (peak frequency, bandwidth), to which a 1D to 2D frequency transformation is applied, leading to the desired 2D filters. In order to obtain through frequency transformation, the desired 2D circular filter bank, first a 1D prototype filter bank must be derived. A Gaussian-shaped filter was chosen as prototype, due to its useful property of scalability on the frequency axis.

2.1. Approximation of the Gaussian FIR Filter Prototype Using Fourier Series

The Gaussian filter in the frequency domain has the well-known expression $G(\omega) = \exp(-\sigma^2\omega^2/2)$, where σ is the dispersion parameter; for a simpler form, easier to handle, the substitution $p = \sigma^2/2$, or equivalently $\sigma = \sqrt{2p}$, will be used. Thus the Gaussian low-pass filter function takes the more convenient form $G_{LP}(\omega) = \exp(-p \cdot \omega^2)$, where p will be referred to as selectivity or scaling parameter. Considering a periodic function with period 2π and regarding the LP Gaussian function as a generating pulse, the following expression $H_{LP}(\omega)$ will be easily obtained, which is the Fourier series expansion of the Gaussian $G_{LP}(\omega)$ up to a given order N:

$$G_{LP}(\omega) = \exp\left(-p \cdot \omega^2\right) \cong \frac{1}{2\sqrt{p\pi}} \cdot \left(1 + 2 \cdot \sum_{n=1}^{N} \exp\left(-\frac{n^2}{4p}\right) \cdot \cos n\omega\right) = H_{LP}(\omega) \quad (1)$$

From this Gaussian LP prototype, a band-pass (BP) prototype is easily produced by shifting the Gaussian laterally around the frequencies $\pm\omega_0$:

$$
\begin{aligned}
H_{BP}(\omega) = H_{LP}(\omega-\omega_0) + H_{LP}(\omega+\omega_0) = \exp\left(-p\cdot(\omega-\omega_0)^2\right) + \exp\left(-p\cdot(\omega+\omega_0)^2\right) \cong \\
\frac{1}{\sqrt{p\pi}}\cdot\left(1+2\cdot\sum_{n=1}^{N}\exp\left(-\frac{n^2}{4p}\right)\cdot\cos\left(n\omega_0\right)\cdot\cos n\omega\right)
\end{aligned}
\tag{2}
$$

Directly using Expressions (1) and (2) implemented in a Matlab routine, in the following section, the low-pass, band-pass and high-pass components of the desired FIR filter bank prototype are calculated.

2.2. Design of a Gaussian Uniform FIR Filter Bank Prototype

Next, a uniform filter bank prototype with 11 Gaussian components will be designed, namely one low-pass filter, nine band-pass filters and one high-pass filter. In this uniform FB, the peak frequencies are equally spaced on the frequency axis. A bandwidth is imposed for the nine band-pass components equal to $B = \pi/10 = 0.1\pi$, while the low-pass and high-pass filters will have each half of this bandwidth, namely $B/2 = \pi/20 = 0.05\pi$. The k-th ideal Gaussian BP filter is produced by shifting the LP prototype to the frequency $\omega_{0,k} = k\cdot\omega_0$, and will have the following expression:

$$
G_{BP\,k}(\omega) = G_{LP}(\omega-k\omega_0) + G_{LP}(\omega+k\omega_0) = \exp\left(-p\cdot(\omega-k\omega_0)^2\right) + \exp\left(-p\cdot(\omega+k\omega_0)^2\right)
\tag{3}
$$

At this point, the scaling parameter p for the imposed bandwidth needs to be calculated. In our case, the filter bandwidth is considered defined at 0.5 of the peak value (at 6 dB). Thus, the characteristics of any two adjacent filters will marginally overlap and will intersect at the value 0.5. Referring to the LP filter $G_{LP}(\omega) = \exp(-p\cdot\omega^2)$, the condition $G_{LP}(B/2) = \exp(-p\cdot B^2/4) = 0.5$ is imposed, otherwise written $\exp(p\cdot B^2/4) = 2$, from which the value for the scaling parameter p is obtained as $p = 4\ln 2/B^2$; since for our filter bank a bandwidth $B = \pi/10$ was imposed, the value $p = 400\ln 2/\pi^2 \cong 28.1$ will be produced. The ideal uniform Gaussian filter bank is plotted in Figure 1a.

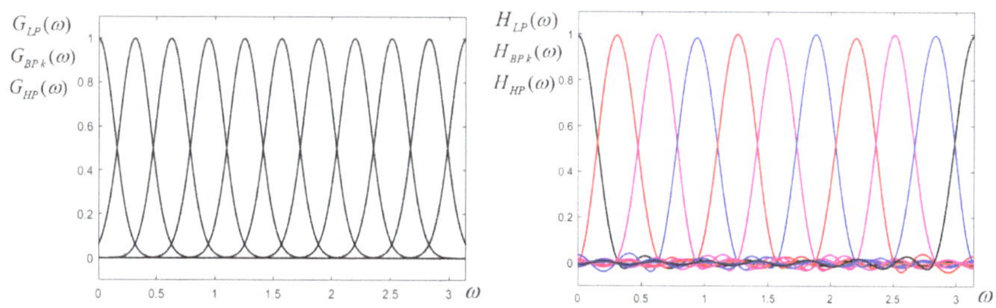

(**a**) (**b**)

Figure 1. (**a**) Ideal Gaussian uniform FB prototype; (**b**) designed Gaussian uniform FB prototype for the 2D CFB.

The filter selectivity is given by the scaling parameter value calculated before, namely $p = 28.1$, with the Fourier series truncated at a number of terms $N = 15$. The larger the number of terms taken into account, the smaller will be the distortions (ripple, etc.), but the filter matrices will be larger in size and will increase the implementation complexity.

Following the above design procedure, once specifying the desired number of filters of the FB and their peak frequencies, using Equations (1) and (2), the frequency responses of all the FB components are calculated. As an example, in our case of a uniform FB with 11 components, the frequency responses of a few filters of the 1D prototype filter bank

are given below, in factored expression. First, the frequency response of an LP prototype expressed as a truncated Fourier series using (1) has the form:

$$\begin{aligned}H_{LP}(\omega) &= 0.055823 + 0.11066 \cdot \cos\,\omega + 0.107743 \cdot \cos\,2\omega + 0.103056 \cdot \cos\,3\omega + 0.096833 \cdot \cos\,4\omega \\ &+0.089382 \cdot \cos\,5\omega + 0.081049 \cdot \cos\,6\omega + 0.072197 \cdot \cos\,7\omega + 0.063177 \cdot \cos\,8\omega + 0.054309 \cdot \cos\,9\omega \\ &+0.045863 \cdot \cos\,10\omega + 0.038047 \cdot \cos\,11\omega + 0.031007 \cdot \cos\,12\omega + 0.024823 \cdot \cos\,13\omega + 0.019523 \cdot \cos\,14\omega \\ &+0.015083 \cdot \cos\,15\omega\end{aligned} \tag{4}$$

which using trigonometric identities can be further expressed as:

$$\begin{aligned}H_{LP}(\omega) = 247.118 \cdot &(\cos\omega+0.99492)(\cos\omega+0.95454)(\cos\omega+0.87546)(\cos\omega+0.76089)(\cos\omega+0.61554) \\ &\cdot(\cos\omega+0.44535)(\cos\omega+0.25729)(\cos\omega+0.05906)(\cos\omega-0.14123)(\cos\omega-0.33537) \\ &\cdot(\cos\omega-0.51542)(\cos\omega-0.67403)(\cos\omega-0.80475)(\cos\omega-0.90811)(\cos\omega-0.93698)\end{aligned} \tag{5}$$

As an example, the frequency responses of the first and last BP filter components of the bank are given below, the intermediate BP filters having similar forms:

$$\begin{aligned}H_{BP\,1}(\omega) = -98.8414 \cdot &(\cos\omega+0.99462)(\cos\omega+0.95194)(\cos\omega+0.86846)(\cos\omega+0.74782)(\cos\omega+0.59531) \\ &\cdot(\cos\omega+0.41762)(\cos\omega+0.22253)(\cos\omega+0.01863)(\cos\omega-0.18508)(\cos\omega-0.37954) \\ &\cdot(\cos\omega-0.55591)(\cos\omega-0.71234)(\cos\omega-0.77319)(\cos\omega-1.00157)\end{aligned} \tag{6}$$

$$\begin{aligned}H_{BP\,9}(\omega) = -98.8414 \cdot &(\cos\omega-0.99462)(\cos\omega-0.95194)(\cos\omega-0.86846)(\cos\omega-0.74782)(\cos\omega-0.59531) \\ &\cdot(\cos\omega-0.41762)(\cos\omega-0.22253)(\cos\omega-0.01863)(\cos\omega+0.18508)(\cos\omega+0.37954) \\ &\cdot(\cos\omega+0.55591)(\cos\omega+0.71234)(\cos\omega+0.77319)(\cos\omega+1.00157)\end{aligned} \tag{7}$$

Finally, the highest component of the FB is the high-pass (HP) filter, which formally has the peak frequency $\omega_0 = \pi$:

$$\begin{aligned}H_{HP}(\omega) = -247.118 \cdot &(\cos\omega+0.93698)(\cos\omega+0.90811)(\cos\omega+0.80475)(\cos\omega+0.67403)(\cos\omega+0.51542) \\ &\cdot(\cos\omega+0.33537)(\cos\omega+0.14123)(\cos\omega-0.05906)(\cos\omega-0.25729)(\cos\omega-0.44535) \\ &\cdot(\cos\omega-0.61554)(\cos\omega-0.76089)(\cos\omega-0.87546)(\cos\omega-0.95455)(\cos\omega-0.99491)\end{aligned} \tag{8}$$

It can be observed that the component filters of the prototype FB whose central frequencies are symmetric with respect to the middle value $\omega = \pi/2$ have symmetric zeros, as is well-known from filter theory. Therefore, the zeros of the HP filter are the zeros of the LP filter with a changed sign; the zeros of the 9th BP filter are the zeros of the first BP filter with a changed sign, etc. Since there is an odd number of filters, the middle filter, namely the 5-th BP filter, with central frequency $\omega_0 = \pi/2$, has no pair, and its transfer function, as expected, has pairs of complementary zeros:

$$\begin{aligned}H_{BP\,5}(\omega) = -319.858 \cdot &(\cos\omega-0.99471)(\cos\omega+0.99471)(\cos\omega-0.95281)(\cos\omega+0.95281)(\cos\omega-0.87107) \\ &\cdot(\cos\omega+0.87107)(\cos\omega-0.75362)(\cos\omega+0.75362)(\cos\omega-0.60689)(\cos\omega+0.60689) \\ &\cdot(\cos\omega-0.43317)(\cos\omega+0.43317)(\cos\omega-0.34221)(\cos\omega+0.34221)\end{aligned} \tag{9}$$

Generally, the k-th band-pass component of the 1D filter bank can be expressed as the following product of first-order factors (where N is the filter order):

$$H_{BP\,k}(\omega) = \xi_k \cdot \prod_{j=1}^{N}(\cos\omega+a_j) \tag{10}$$

The uniform Gaussian filter bank designed above is plotted in Figure 1b and it looks very similar to its ideal counterpart in (a), with a low level of ripple.

2.3. Design of a Gaussian Non-Uniform FIR Filter Bank Prototype

In image analysis, mainly in multirate signal processing, non-uniform filter banks are also currently used. Next, using the method described in Section 2.1 a non-uniform, more specifically a so-called dyadic filter bank will be designed. Such an FB has the property that the bandwidths of the component filters increase proportionally to their peak frequencies, such that generally the ratio between bandwidth and peak frequency remains constant; these filters are also known as constant-Q filter banks.

For our design example, it is considered that the filter bandwidths increase by a factor of 2 from low to high frequencies. An FB with five filters will be designed here: one LP filter, three BP filters, and one HP filter. Specifying the peak frequency of the 3rd BP filter as $\omega_{03} = \pi/2$, the following peak frequencies ω_{0k} and bandwidths B_{0k} are easily found for the five filters, respectively: $\omega_{00} = 0$, $B_{00} = \pi/22$ (LP); $\omega_{01} = \pi/11$, $B_{01} = \pi/11$ (BP1); $\omega_{02} = 5\pi/22$, $B_{02} = 2\pi/11$ (BP2); $\omega_{03} = \pi/2$, $B_{03} = 4\pi/11$ (BP3); $\omega_{04} = \pi$, $B_{04} = 7\pi/22$ (HP). Using the same Formulas (1) and (2) as before, the frequency responses of the component filters are easily found as factored trigonometric polynomials. Unlike the previous case of uniform FB, for this nonuniform FB the higher filters have increasing bandwidths; being less selective, they can be approximated with polynomials of lesser order, therefore their implementation complexity will be significantly lower. The same marginal overlapping between filters at exactly 0.5 was considered. For instance, for the most selective filter (LPF) the parameter p results as $p = \ln 2/(\pi/22)^2 \cong 34$; this filter can still be approximated by truncating the Fourier series at order $N = 15$, as before; the ripple ("ringing") in the stopband will be a little higher, but still acceptable. The following approximations for the frequency responses of the five Gaussian filters were derived:

$$\begin{aligned}H_{LP}(\omega) = 322.53 \cdot (\cos\omega+0.99491)(\cos\omega+0.95448)(\cos\omega+0.87527)(\cos\omega+0.76053)(\cos\omega+0.61495) \\ \cdot(\cos\omega+0.4445)(\cos\omega+0.25615)(\cos\omega+0.05762)(\cos\omega-0.14297)(\cos\omega-0.3374) \\ \cdot(\cos\omega-0.51771)(\cos\omega-0.67653)(\cos\omega-0.80734)(\cos\omega-0.90613)(\cos\omega-0.95148)\end{aligned} \tag{11}$$

$$\begin{aligned}H_{BP1}(\omega) = -261.27 \cdot (\cos\omega+0.99507)(\cos\omega+0.95596)(\cos\omega+0.87933)(\cos\omega+0.76831)(\cos\omega+0.62742) \\ \cdot(\cos\omega+0.46239)(\cos\omega+0.2799)(\cos\omega+0.08752)(\cos\omega-0.10714)(\cos\omega-0.29616) \\ \cdot(\cos\omega-0.47196)(\cos\omega-0.62751)(\cos\omega-0.76157)(\cos\omega-0.81472)(\cos\omega-1.00137)\end{aligned} \tag{12}$$

$$\begin{aligned}H_{BP2}(\omega) = 9.186 \cdot (\cos\omega+0.98714)(\cos\omega+0.88634)(\cos\omega+0.69593)(\cos\omega+0.43711) \\ \cdot(\cos\omega+0.14021)(\cos\omega-1.00605)(\cos\omega-1.05338)((\cos\omega)^2-0.38658\cdot\cos\omega+0.05005)\end{aligned} \tag{13}$$

$$H_{BP3}(\omega) = 0.9531 \cdot ((\cos\omega)^2 + 2.018676\cdot\cos\omega + 1.024286)((\cos\omega)^2 - 2.018676\cdot\cos\omega + 1.024286) \tag{14}$$

$$H_{HP}(\omega) = -0.2117 \cdot (\cos\omega - 1.00109)((\cos\omega)^2 - 2.02723\cdot\cos\omega + 1.68236) \tag{15}$$

The characteristics of this non-uniform FB are plotted in Figure 2.

As mentioned, only the most selective filters (LP and BP1) have visible ripple, while the others have no ripple at all.

As a further remark, in previous papers [32–35], various 2D filters were designed using another efficient procedure, namely the Chebyshev series, which has the advantage of yielding a uniform and efficient approximation for a given function, with equal error along the whole specified range of values. The symbolic calculations are performed in the MAPLE software (version MAPLE 2018), and a change of frequency variable is first required, before effectively deriving the approximation. However, the major drawback of this method is that it is not parametric; it does not have a closed form as in the case of the Fourier series method, therefore is more laborious; for each specified value of selectivity parameter p, the calculation must be carried out in a symbolic calculation software. Therefore, in this paper, the Fourier series approximation was preferred.

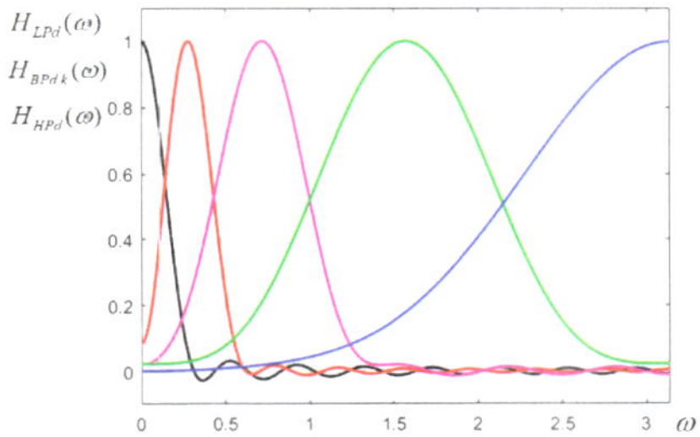

Figure 2. Designed Gaussian nonuniform (dyadic) filter bank prototype for the 2D CFB.

2.4. Gaussian Circular FIR Filter Bank Obtained Using Frequency Transformation

Once specified a convenient 1D prototype with the frequency response $H_p(\omega)$, a 2D circular filter $H(\omega_1,\omega_2)$ is produced by applying to the given prototype the 1D to 2D frequency transformation $\omega \rightarrow \sqrt{\omega_1^2+\omega_2^2}$:

$$H(\omega_1,\omega_2) = H_p\left(\sqrt{\omega_1^2+\omega_2^2}\right) \tag{16}$$

The function $\cos\sqrt{\omega_1^2+\omega_2^2}$ is described by the 3×3 centrally symmetric matrix:

$$C = \begin{bmatrix} 0.125 & 0.25 & 0.125 \\ 0.25 & -0.5 & 0.25 \\ 0.125 & 0.25 & 0.125 \end{bmatrix} \tag{17}$$

and can be approximated by the following expression, which is a simple particular case of the McClellan transform, currently used in 2D FIR filter design [2,3,36]:

$$\cos\sqrt{\omega_1^2+\omega_2^2} \cong C(\omega_1,\omega_2) = -0.5 + 0.5(\cos\omega_1+\cos\omega_2) + 0.5\cos\omega_1\cos\omega_2 \tag{18}$$

The Expression (18) is in fact the discrete space Fourier transform (DSFT) of the matrix **C**. Next, a zero-phase FIR filter $H_P(\omega)$ is considered, whose frequency response is given by the trigonometric polynomial expression [36]:

$$H_P(\omega) = b_0 + 2\sum_{k=1}^{R} b_k \cos k\omega \tag{19}$$

At this point, the trigonometric identities for $\cos k\omega$ ($k = 1\ldots R$) can be used, and thus the following polynomial expression is produced in powers of $\cos\omega$ [36]:

$$H_P(\omega) = c_0 + \sum_{k=1}^{R} c_k(\cos\omega)^k \tag{20}$$

where (19), (20) b_0, b_k, c_0, c_k are polynomial coefficients. Applying frequency mapping (18), the frequency response of the 2D circular filter will become [36]:

$$H(\omega_1,\omega_2) = H_P\left(\sqrt{\omega_1^2+\omega_2^2}\right) = c_0 + \sum_{k=1}^{R} c_k \cdot C^k(\omega_1,\omega_2) \tag{21}$$

where $C(\omega_1, \omega_2) = \cos\sqrt{\omega_1^2 + \omega_2^2}$ as given in (18).

Therefore, by a straightforward substitution of $\cos\omega$ by the circular cosine function $C(\omega_1, \omega_2) = \cos\sqrt{\omega_1^2 + \omega_2^2}$ in the prototype $H_P(\omega)$, the 2D filter frequency response is produced directly. Next, supposing that the frequency response $H_P(\omega)$ is decomposed into first-order and second-order factors in variable $\cos\omega$, and achieving the above substitution in all factors of $H_P(\omega)$, the circular filter frequency response $H(\omega_1, \omega_2)$ is finally derived in factored form:

$$H(\omega_1, \omega_2) = k \cdot \prod_{i=1}^{n} (C + b_i) \cdot \prod_{j=1}^{m} (C^2 + b_{1j} \cdot C + b_{2j}) \tag{22}$$

where C is a concise notation for the two-variable function $C(\omega_1, \omega_2)$ and k is the constant resulting from factorization. Since the specified prototype is expressed as a product of elementary factors, the circular filter frequency response will also become directly factored, which is an essential advantage in actual implementation. Thus, the large kernel $\mathbf{H}$ corresponding to $H(\omega_1, \omega_2)$ can be expressed simply as a discrete convolution of small matrices (of size 3×3 or 5×5):

$$\mathbf{H} = k \cdot (\mathbf{C}_1 * \ldots * \mathbf{C}_i * \ldots * \mathbf{C}_n) * (\mathbf{D}_1 * \ldots * \mathbf{D}_j * \ldots * \mathbf{D}_m) \tag{23}$$

The matrix expression (23) is related to the factored frequency response (22). Using the 3×3 matrix $\mathbf{C}$ in (17) and considering also (22), each of the matrices $\mathbf{C}_i$ of size 3×3 in (23) is derived by adding coefficient b_i, which appears in the first-order factors in (22), to the center element in matrix $\mathbf{C}$. Thus, the matrix $\mathbf{D}_j$ (5×5) becomes:

$$\mathbf{D}_j = \mathbf{C} * \mathbf{C} + b_{1j} \cdot \mathbf{C}_1 + b_{2j} \cdot \mathbf{C}_0 \tag{24}$$

where $\mathbf{C}_0$ is a null matrix of size 5×5 with central element of value one; $\mathbf{C}_1 (5 \times 5)$ is produced by the boarding matrix $\mathbf{C}$ (size 3×3) with zeros; here the symbol * denotes convolution.

Thus, the frequency response of each CFB component is directly derived by substitution. Correspondingly, the overall kernel matrix H of the filter will be given by an expression similar to (23), but in our particular case with only first-order factors, as in (10), it becomes

$$\mathbf{H} = \xi_k \cdot (\mathbf{C}_1 * \ldots * \mathbf{C}_i * \ldots * \mathbf{C}_n) \tag{25}$$

The filters of the designed 1D prototype filter bank are of order 15; it follows that the corresponding 2D circular filters derived through the above transformation have kernel matrices relatively large, of size 31×31. Such a large matrix will be implemented efficiently using a polyphase approach described in Section 4.

As a remark, all the component filters of the designed FB are non-separable, except the LP filter. Indeed, it is easy to see that the circular LP Gaussian filter is separable as a product of two Gaussian LP filters on the two frequency axes:

$$\exp\left(-p \cdot (\omega_1^2 + \omega_2^2)\right) = \exp\left(-p \cdot \omega_1^2\right) \cdot \exp\left(-p \cdot \omega_2^2\right) \tag{26}$$

A very important advantage of the proposed FB is that the filters' transfer functions are real-valued (zero-phase), therefore they will not introduce any phase distortions; this will be visible in the simulation results given in the following section.

The 1D prototypes, frequency characteristics and corresponding contour plots for all 11 filters of the circular filter bank are displayed in Figures 3 and 4. It is easily observed that up to the 6th band-pass filter, the characteristics are visually almost perfectly circular. For the higher band-pass filters, the characteristics have a more pronounced deviation from circularity, tending to the shape of a rounded square. The filter with the highest frequency ($\omega_0 = \pi$) has almost a square shape. This effect of distortion from circularity is well known when applying the frequency mapping (18), the simplest form of the McClellan transform, and could be corrected only by using a more accurate approximation of the circular cosine;

however, this would imply a higher complexity of the filters (larger kernel matrices) and a more difficult implementation.

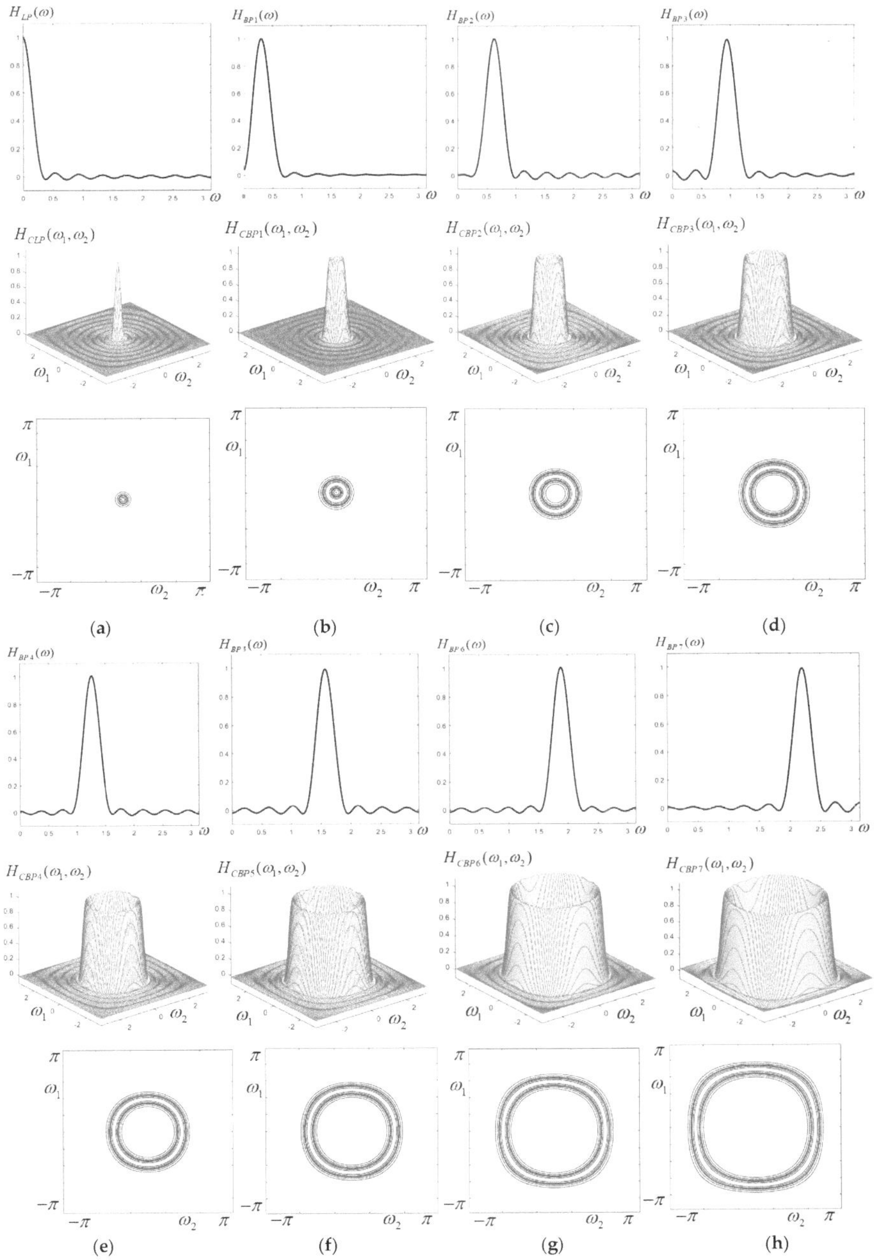

Figure 3. 1D prototypes, characteristics and contour plots for the first eight components of the circular filter bank; (**a**) low-pass filter; (**b–h**) band-pass filters BP1–BP7.

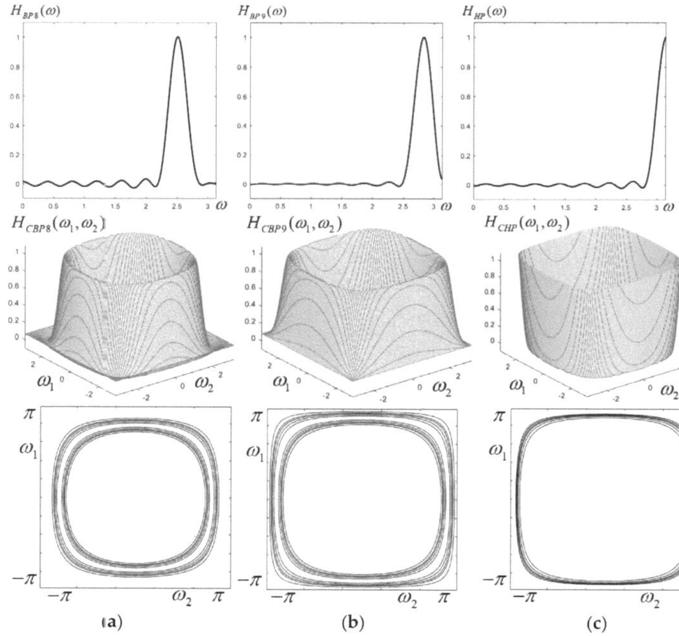

Figure 4. 1D prototypes, characteristics and contour plots for the last three components of the circular filter bank; (**a**,**b**) band-pass filters BP8, BP9; (**c**) high-pass filter.

The characteristics and contour plots of the five component filters of the non-uniform (dyadic) CFB derived from the 1D prototype filters designed in Section 2.3, with frequency responses given by (11)–(15) are displayed in Figure 5, and it can be observed that they have a good circular symmetry.

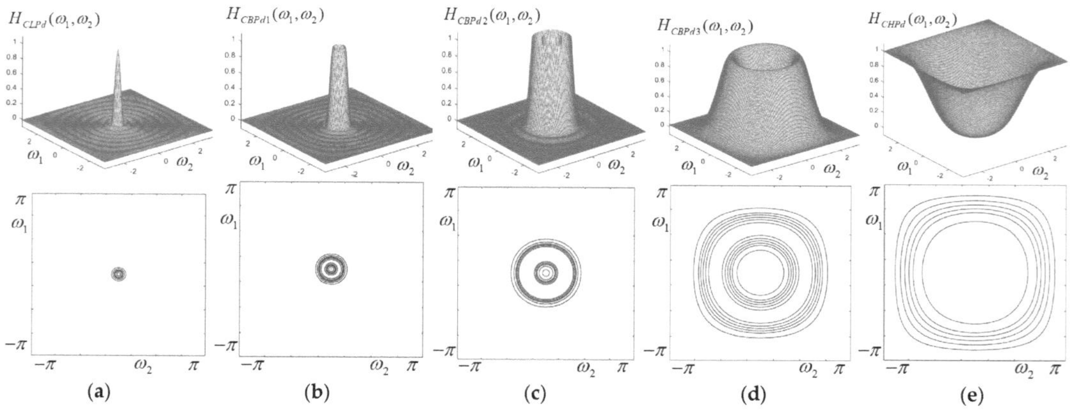

Figure 5. Characteristics and contour plots of the components of the dyadic CFB; (**a**) LP filter; (**b**–**d**) BP filters; (**e**) LP filter.

3. Image Analysis Using the Designed Circular Filter Banks

In this section, examples of image analysis using the uniform and dyadic CFBs designed before are presented. First, the grayscale test image in Figure 6a is considered, of size 399 × 399 pixels, representing a group of trees without foliage; this image was chosen as it has a lot of fine details, represented by the tree ramifications into thinner and thinner twigs. This image is filtered by applying all the 11 components of the designed uniform CFB (one LP filter, nine BP filters, one HP filter); it can be considered that our test image is decomposed into sub-bands using the analysis CFB designed before.

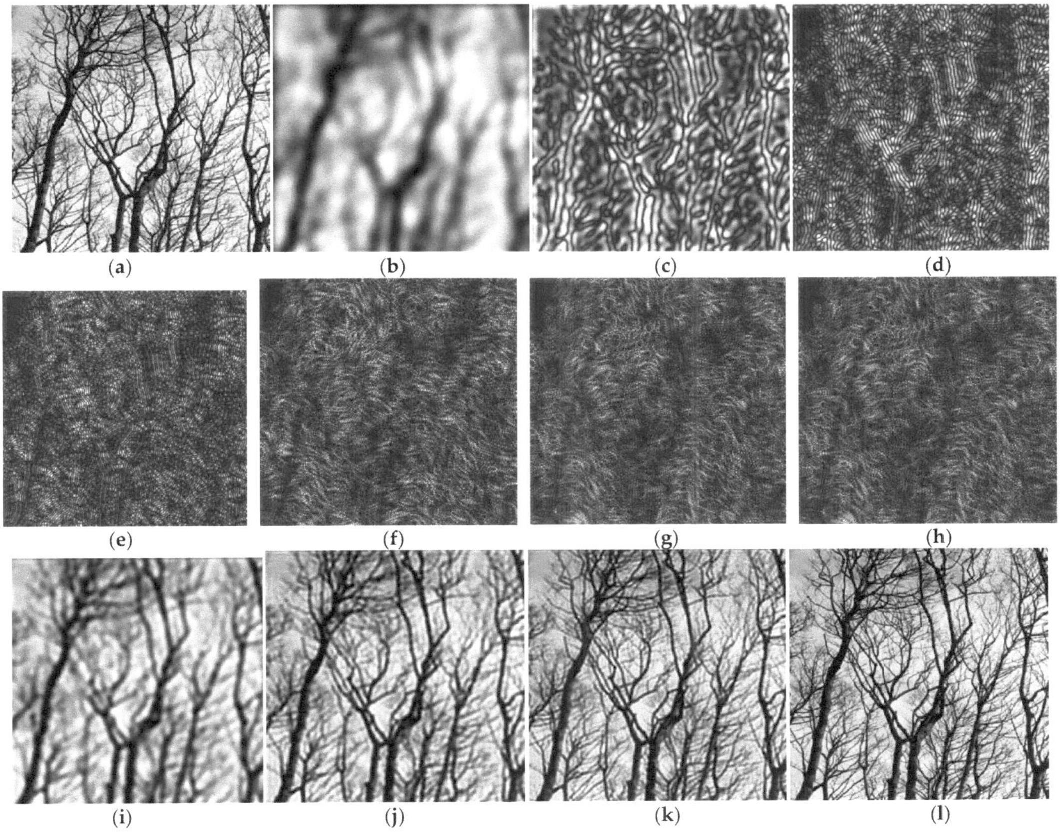

Figure 6. Image analysis using the uniform circular FB: (**a**) original "Trees" image; (**b**) LP filtered; (**c**–**g**) BP filtered with BPF1, BPF2, BPF3, BPF4, BPF5, respectively; (**h**) HP filtered; (**i**–**k**) recovered image by summing the first two, three and four components; (**l**) recovered image by summing all 11 components (sub-band images).

The original image is displayed in Figure 6a. The image obtained at the output of the narrow LP filter is (b), and it can be observed that it is very blurred, the fine details (thin twigs) are no longer visible. The images obtained from the first five BP filters are shown in (c–g), respectively, and contain details corresponding to the selected bandwidth. The image (h) is produced at the output of the HP filter and contains the highest frequencies, corresponding to the finest details.

The original image was converted into "double" format and its pixel values were rescaled to the range [0, 1] for MATLAB processing. The image produced at the output of LPF has the overall mean pixel value 0.529; for all the other 10 images (produced at the outputs of BP filters and HP filter), the mean pixel value is very close to zero, as expected, since these filters eliminate the zero-frequency component corresponding to mean value. In Figure 6i–l, it is shown how the original image is reconstructed by adding the component images into which it was decomposed. Thus, image (i) is produced by adding the first two components (LP and BP1); image (j) is produced as a sum of the first three components (LP,

BP1, BP2); image (k) is produced by adding the component BP3. Finally, by summing all the 11 components, the image (l) is produced, which visually is very similar to the original image, showing all the fine details very clearly.

These simulations prove that the designed CFBs (uniform and non-uniform) could be practically used as analysis filter banks for decomposing a given image into sub-band images. However, the rigorous mathematical conditions required will have to be further investigated in future work.

The energy of each component sub-band image can also be evaluated using the well-known formula $E_k = \sqrt{\sum_{i=1}^{M}\sum_{j=1}^{N} p_{ij}}$, where the image is of size $M \times N$ and p_{ij} is the current pixel value; the expression of the relative energy can also be given as a percentage:

$$E_{R\,k} = \frac{100}{M \cdot N} \cdot \sqrt{\sum_{i=1}^{M}\sum_{j=1}^{N} p_{ij}}\ (\%) \quad (27)$$

Calculating the energies of the 11 filtered images resulting at the output of the designed CFB, the values given in Table 1 are easily found; summing these values, it can be verified that they add up to approximately 1 in normal values, or 100% in percentages. It can be observed that almost 56% of the image energy is contained in the low-pass component (in the frequency domain around zero, with radius 0.1π), while almost 85% is contained in the first four components (within a 0.4π radius), at the output of LP filter and first three BP filters. The relative energies of the sub-band images decrease almost uniformly; as an exception, ER9 > ER8, and ER11 > ER8, ER9, ER10. The highest frequencies in the image give less than 2% of the total image energy. These relative energy values are summarized in Table 1 and represented graphically in the chart from Figure 7.

Table 1. Relative sub-band energies (in %) for the 11 images resulting at the output of the uniform CFB.

ER1	55.81594	ER7	2.51503
ER2	14.03036	ER8	1.63033
ER3	8.12698	ER9	1.68105
ER4	6.36161	ER10	1.17531
ER5	4.81307	ER11	1.84367
ER6	3.64377		

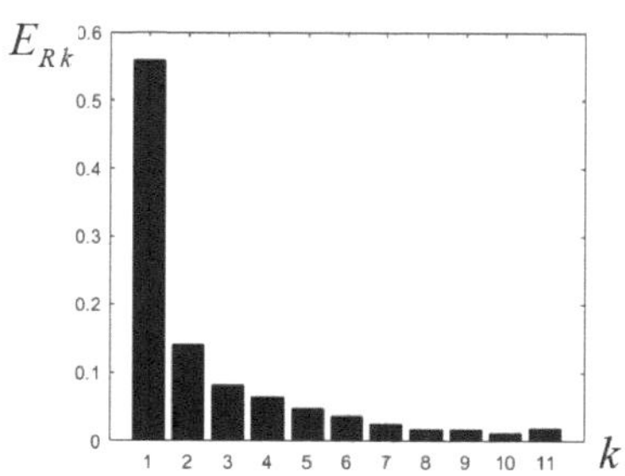

Figure 7. Relative energies calculated for the 11 images resulting at the output of the uniform circular filter bank.

As a remark, the designed circular filter banks are rotation invariant; the image spectrum is separated into concentric, ring-shaped regions, with frequencies increasing while image energy is generally decreasing, from the center to the margins of the frequency plane. Due to rotational invariance, the decomposition coefficient and energy in each subband remain more or less constant. This property is very useful in specific feature extraction and classification tasks in image processing.

A similar experiment was performed using the dyadic circular filter bank with five components shown in Figure 5, applied on the same grayscale test image, for comparison. The filtered images obtained at the output of the LP filter, three BP filters and HP filter are displayed in Figure 8a–e. As in the previous example, the original image is then reconstructed by adding the first two and three sub-band images, then all the five sub-band images, as shown in Figure 8f–h. Summing up all the sub-band images leads to an image very similar to the original one.

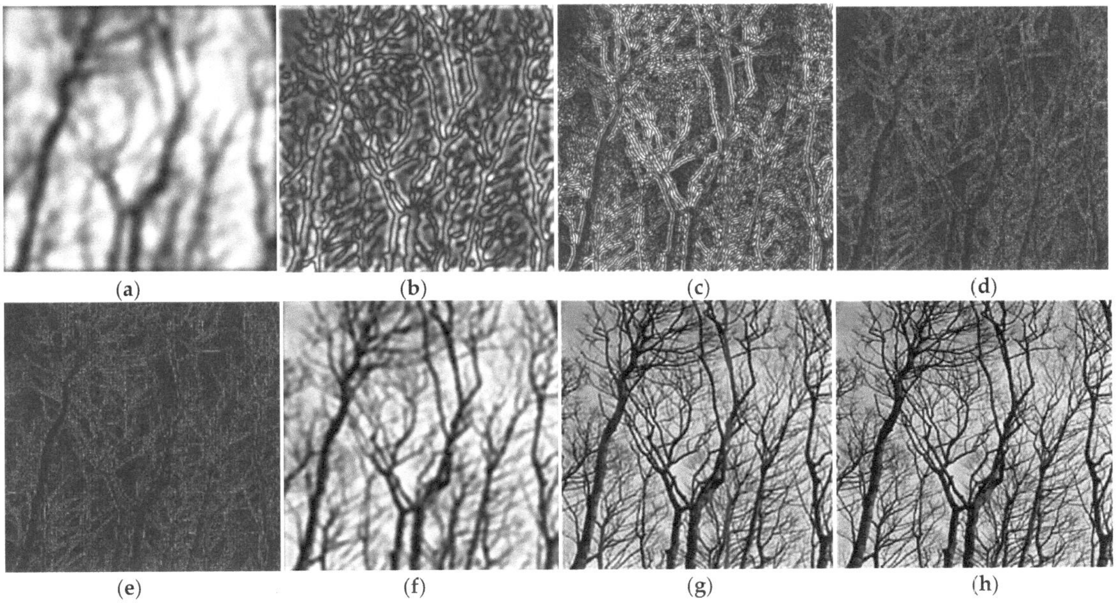

Figure 8. Image analysis using the dyadic circular FB: (**a**) LP filtered; (**b–d**) BP filtered with BPF1, BPF2, BPF3, respectively; (**e**) HP filtered; (**f,g**) recovered image by summing the first two and three components, respectively; (**h**) recovered image by summing all five component.

An additional experiment was performed using the same dyadic CFB, applied on another grayscale test image ("Fields"), of size 699 × 699, showing an aerial view of a rural landscape with fields and a river (Figure 9a). The filtered images obtained at the output of the LP filter, three BP filters and HP filter, respectively, are displayed in Figure 9b–f. As in the previous examples, the original image is then reconstructed by adding the first two sub-band images (Figure 9g), then all the five sub-band images, as shown in Figure 9h. Summing up all the five sub-band images yields an image very similar to the original one. Table 2 displays the relative sub-band energies, calculated for both test images, namely "Trees" and "Fields". Again, most of the image energy is contained in the lowest sub-band (corresponding to the LPF); however, the energy distribution clearly depends on the particular image, as was expected, an can be considered a numerical indicator characterizing the sub-band decomposition of a given image.

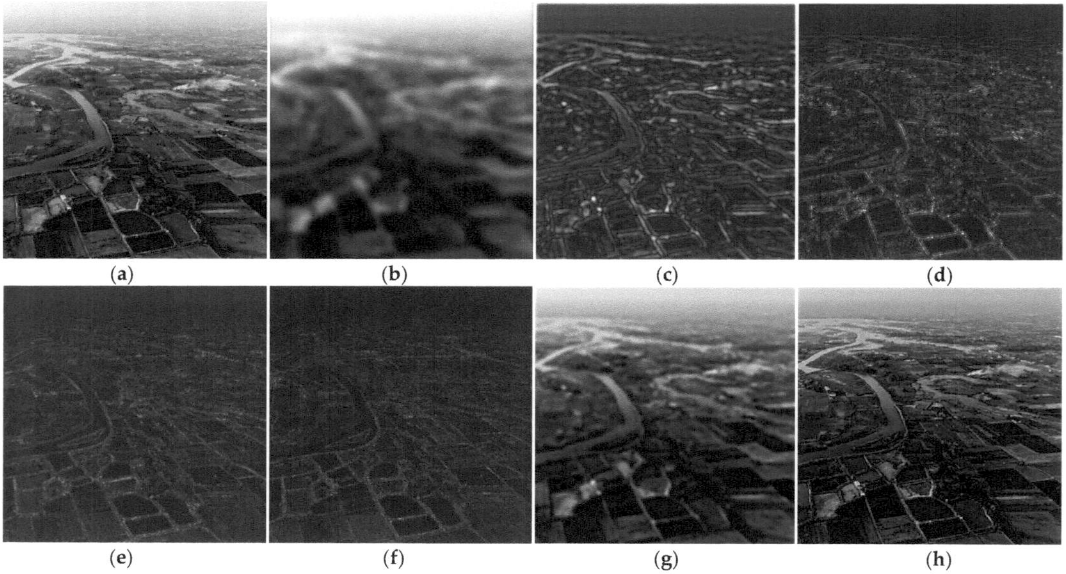

Figure 9. Image analysis using the dyadic circular FB: (**a**) "Fields" test image; (**b**) LP filtered; (**c**–**e**) BP filtered with BPF1, BPF2, BPF3, respectively; (**f**) HP filtered; (**g**) recovered image by summing the first two components; (**h**) recovered image by summing all five components.

Table 2. Relative sub-band energies (in %) for the five images resulting at the output of the dyadic CFB.

Image	ER1	ER2	ER3	ER4	ER5
"Trees"	57.925	13.965	18.430	6.268	3.411
"Fields"	68.003	11.767	11.618	5.433	3.179

Regarding the noise suppression issue, the authors did not intend to investigate it in this paper. Of course, noise removal is a very important task in image enhancement and restoration. Considering the nature of the noise (Gaussian, salt-and-pepper, speckle noise, etc.), a specific type of filter should be chosen to remove it optimally. Anyway, noise removal should be achieved before any further image analysis. For the proposed CFB, since the image spectrum is partitioned into ring-shaped regions corresponding to sub-band images, if the original image was affected by some type of noise, it would be distributed more or less evenly in the sub-band images, mainly in higher frequency bands. Therefore, it would have to be eliminated separately from each sub-band component image, which may be a more difficult task. This issue remains to be studied in future work on this topic.

4. Polyphase Implementation of the Designed 2D Circular FIR Filters

In the following, a low-complexity implementation is proposed for the 2D FIR circular filter bank previously designed, relying on a polyphase structure of a 2D filtering task with a convolution kernel of relatively large size (31×31). In order to achieve convolution with such a large kernel, a block processing technique [24,25] and a polyphase decomposition approach will be employed.

As a first step, using sub-expression sharing techniques, a 2D filtering algorithm with a 4×4 kernel was elaborated, which is detailed as follows. The kernel of the filter resulting from the design and the input image are decimated by factors 3 and 5, respectively; the polyphase filtering approach is subsequently applied. Using this technique, three output component images are derived, namely $\mathbf{Y}_0$, $\mathbf{Y}_1$, $\mathbf{Y}_2$, given by Equations (28)–(30):

$$\mathbf{Y}_0 = \begin{bmatrix} \mathbf{A}_0 & \mathbf{A}_0 & \mathbf{A}_0 & \mathbf{A}_0 \\ \mathbf{O}_{4\times10} & \mathbf{A}_0 & \mathbf{O}_{4\times10} & \mathbf{O}_{4\times10} \\ \mathbf{O}_{4\times10} & \mathbf{O}_{4\times10} & \mathbf{A}_0 & \mathbf{O}_{4\times10} \\ \mathbf{O}_{4\times10} & \mathbf{O}_{4\times10} & \mathbf{O}_{4\times10} & \mathbf{A}_0 \end{bmatrix} \times diag\left(\begin{bmatrix} \mathbf{O}_{10\times4} & \mathbf{O}_{10\times4} & \mathbf{O}_{10\times4} & \mathbf{A}_1 \\ \mathbf{O}_{10\times4} & \mathbf{O}_{10\times4} & \mathbf{A}_1 & \mathbf{A}_1 \\ \mathbf{O}_{10\times4} & \mathbf{A}_1 & \mathbf{O}_{10\times4} & \mathbf{A}_1 \\ \mathbf{A}_1 & \mathbf{O}_{10\times4} & \mathbf{O}_{10\times4} & \mathbf{A}_1 \end{bmatrix} \mathbf{H}^T\right) \times$$
$$\times \begin{bmatrix} \mathbf{A}_2 & -\mathbf{A}_2 & -\mathbf{A}_2 & -\mathbf{A}_2 & \mathbf{O}_{10\times7} & \mathbf{O}_{10\times7} & \mathbf{O}_{10\times7} \\ \mathbf{O}_{10\times7} & \mathbf{A}_2 & \mathbf{O}_{10\times7} & \mathbf{O}_{10\times7} & \mathbf{O}_{10\times7} & \mathbf{O}_{10\times7} & \mathbf{O}_{10\times7} \\ \mathbf{O}_{10\times7} & \mathbf{O}_{10\times7} & \mathbf{A}_2 & \mathbf{O}_{10\times7} & \mathbf{O}_{10\times7} & \mathbf{O}_{10\times7} & \mathbf{O}_{10\times7} \\ \mathbf{O}_{10\times7} & \mathbf{O}_{10\times7} & \mathbf{O}_{10\times7} & \mathbf{A}_2 & \mathbf{O}_{10\times7} & \mathbf{O}_{10\times7} & \mathbf{O}_{10\times7} \end{bmatrix} \times \mathbf{X}_{2D} \quad (28)$$

$$\mathbf{Y}_1 = \begin{bmatrix} \mathbf{O}_{4\times10} & \mathbf{O}_{4\times10} & \mathbf{O}_{4\times10} \\ \mathbf{A}_0 & \mathbf{A}_0 & \mathbf{A}_0 \\ \mathbf{O}_{4\times10} & \mathbf{A}_0 & \mathbf{O}_{4\times10} \\ \mathbf{O}_{4\times10} & \mathbf{O}_{4\times10} & \mathbf{A}_0 \end{bmatrix} \times diag\left(\begin{bmatrix} \mathbf{O}_{10\times4} & \mathbf{O}_{10\times4} & \mathbf{A}_1 & \mathbf{O}_{10\times4} \\ \mathbf{O}_{10\times4} & \mathbf{A}_1 & \mathbf{A}_1 & \mathbf{O}_{10\times4} \\ P & \mathbf{O}_{10\times4} & \mathbf{A}_1 & \mathbf{O}_{10\times4} \end{bmatrix} \mathbf{H}^T\right) \times$$
$$\times \begin{bmatrix} \mathbf{O}_{10\times7} & -\mathbf{A}_2 & \mathbf{A}_2 & -\mathbf{A}_2 & -\mathbf{A}_2 & \mathbf{O}_{10\times7} & \mathbf{O}_{10\times7} \\ \mathbf{O}_{10\times7} & \mathbf{O}_{10\times7} & \mathbf{O}_{10\times7} & \mathbf{A}_2 & \mathbf{O}_{10\times7} & \mathbf{O}_{10\times7} & \mathbf{O}_{10\times7} \\ \mathbf{O}_{10\times7} & \mathbf{O}_{10\times7} & \mathbf{O}_{10\times7} & \mathbf{O}_{10\times7} & \mathbf{A}_2 & \mathbf{O}_{10\times7} & \mathbf{O}_{10\times7} \end{bmatrix} \times \mathbf{X}_{2D} \quad (29)$$

$$\mathbf{Y}_2 = \begin{bmatrix} \mathbf{O}_{4\times10} & \mathbf{O}_{4\times10} & \mathbf{O}_{4\times10} \\ \mathbf{O}_{4\times10} & \mathbf{O}_{4\times10} & \mathbf{O}_{4\times10} \\ \mathbf{A}_0 & \mathbf{A}_0 & \mathbf{O}_{4\times10} \\ \mathbf{O}_{4\times10} & \mathbf{A}_0 & \mathbf{A}_0 \end{bmatrix} \times diag\left(\begin{bmatrix} \mathbf{O}_{10\times4} & \mathbf{A}_1 & \mathbf{O}_{10\times4} & \mathbf{O}_{10\times4} \\ \mathbf{A}_1 & \mathbf{A}_1 & \mathbf{O}_{10\times4} & \mathbf{O}_{10\times4} \\ \mathbf{A}_1 & \mathbf{O}_{10\times4} & \mathbf{O}_{10\times4} & \mathbf{O}_{10\times4} \end{bmatrix} \mathbf{H}^T\right) \times$$
$$\times \begin{bmatrix} \mathbf{O}_{10\times7} & \mathbf{O}_{10\times7} & -\mathbf{A}_2 & -\mathbf{A}_2 & \mathbf{A}_2 & -\mathbf{A}_2 & \mathbf{O}_{10\times7} \\ \mathbf{O}_{10\times7} & \mathbf{O}_{10\times7} & \mathbf{O}_{10\times7} & \mathbf{O}_{10\times7} & \mathbf{O}_{10\times7} & \mathbf{A}_2 & \mathbf{O}_{10\times7} \\ \mathbf{O}_{10\times7} & \mathbf{O}_{10\times7} & \mathbf{O}_{10\times7} & -\mathbf{A}_2 & -\mathbf{A}_2 & -\mathbf{A}_2 & \mathbf{A}_2 \end{bmatrix} \times \mathbf{X}_{2D}^T \quad (30)$$

in which the block matrices have the form given below:

$$\mathbf{A}_0 = \begin{bmatrix} 1 & 1 & 1 & 1 & 0 & 0 & 0 & 0 & 0 & 0 \\ 0 & 1 & 0 & 0 & 1 & 1 & 1 & 0 & 0 & 0 \\ 0 & 0 & 1 & 0 & 0 & 1 & 0 & 1 & 1 & 0 \\ 0 & 0 & 0 & 1 & 0 & 0 & 1 & 0 & 1 & 1 \end{bmatrix}; \ \mathbf{A}_1 = \begin{bmatrix} 0 & 0 & 0 & 1 \\ 0 & 0 & 1 & 1 \\ 0 & 1 & 0 & 1 \\ 1 & 0 & 0 & 1 \\ 0 & 0 & 1 & 0 \\ 0 & 1 & 1 & 0 \\ 1 & 0 & 1 & 0 \\ 0 & 1 & 0 & 0 \\ 1 & 1 & 0 & 0 \\ 1 & 0 & 0 & 0 \end{bmatrix}; \ \mathbf{A}_2 = \begin{bmatrix} 1 & -1 & -1 & -1 & 0 & 0 & 0 \\ 0 & 1 & 0 & 0 & 0 & 0 & 0 \\ 0 & 0 & 1 & 0 & 0 & 0 & 0 \\ 0 & 0 & 0 & 1 & 0 & 0 & 0 \\ 0 & -1 & 1 & -1 & -1 & 0 & 0 \\ 0 & 0 & 0 & 1 & 0 & 0 & 0 \\ 0 & 0 & 0 & 0 & 1 & 0 & 0 \\ 0 & 0 & -1 & -1 & 1 & -1 & 0 \\ 0 & 0 & 0 & 0 & 0 & 1 & 0 \\ 0 & 0 & 0 & -1 & -1 & -1 & 1 \end{bmatrix} \quad (31)$$

and $\mathbf{O}_{4\times10}$, $\mathbf{O}_{10\times4}$, $\mathbf{O}_{10\times7}$ are zero matrices of size 4×10, 10×4, and 10×7, respectively. Adding the partial results $\mathbf{Y}_0$, $\mathbf{Y}_1$ and $\mathbf{Y}_2$ given by (28), (29) and (30), the following output vector $\mathbf{Y}$ containing 16 samples of the filtered image is obtained:

$$\begin{aligned} \mathbf{Y} &= \mathbf{Y}_0 + \mathbf{Y}_1 + \mathbf{Y}_2 \\ &= \begin{bmatrix} Y_{00} & Y_{01} & Y_{02} & Y_{03} & Y_{10} & Y_{11} & Y_{12} & Y_{13} & Y_{20} & Y_{21} & Y_{22} & Y_{23} & Y_{30} & Y_{31} & Y_{32} & Y_{33} \end{bmatrix}^T \end{aligned} \quad (32)$$

The vector $\mathbf{H}$ occurring in Equations (28)–(30) is given below:

$$\mathbf{H} = [h_{00} \ \ h_{01} \ \ h_{02} \ \ h_{03} \ \ h_{10} \ \ h_{11} \ \ h_{12} \ \ h_{13} \ \ h_{20} \ \ h_{21} \ \ h_{22} \ \ h_{23} \ \ h_{30} \ \ h_{31} \ \ h_{32} \ \ h_{33}]^T \quad (33)$$

while the input vector X_{2D} is displayed as follows:

$$X_{2D} = [x_{00} \quad x_{01} \quad \ldots \quad x_{06} \quad x_{10} \quad x_{11} \quad \ldots \quad x_{16} \quad \ldots \quad x_{60} \quad x_{61} \quad x_{62} \quad x_{63} \quad x_{64} \quad x_{65} \quad x_{66}]^T \tag{34}$$

The main reason for proposing this 2D FIR filtering algorithm was the reduced number of arithmetic operations involved. It is well known that in a direct 2D convolution there is a high degree of redundancy in operations. In a direct 2D convolution there are overlapping blocks of input data; by eliminating these redundant calculations, a significant reduction in the arithmetic complexity will be obtained. The filtering algorithm presented above was produced using a block filtering technique.

At this point, the 2D filtering algorithm discussed above will be extended from an elementary kernel of size 4 × 4 to the case of a 31 × 31 kernel. In order to achieve this and to obtain a parallel implementation, a block processing technique will be used, relying on a polyphase structure. To derive this 2D polyphase structure, a decimation of the kernel matrix with factor 4 will be performed. Before decimation, the kernel was enlarged to have a dimension multiple of 4, in our case 32 × 32, by bordering it with a row and a column of zeros. Decimation by a factor of 5 was also applied to the input image and thus a 25 × 25 input image was produced.

Using a block polyphase decomposition and the previous fast algorithm, the following efficient algorithm was obtained for the computation of the designed 2D FIR filter. The vectors H_{00}^T, H_{01}^T, H_{02}^T, $H_{03}^T H_{10}^T$, H_{11}^T, H_{12}^T, $H_{13}^T H_{20}^T H_{21}^T$, H_{22}^T, H_{23}^T, $H_{30}^T H_{31}^T$, H_{32}^T, H_{33}^T for a kernel matrix of size 12 × 12 and an input matrix of size 21 × 21 have the general form H_{ij} given below (where $i = 0, 1, 2, 3$ and $j = 0, 1, 2, 3$):

$$H_{ij} = \begin{bmatrix} h_{0+i,0+j} & h_{0+i,4+j} & h_{0+i,8+j} & h_{4+i,0+j} & h_{4+i,4+j} & h_{4+i,8+j} & h_{8+i,0+j} & h_{8+i,4+j} & h_{8+i,8+j} \end{bmatrix} \tag{35}$$

For example, the vectors H_{00}, H_{12}, H_{33} generated by the Formula (35) will be:

$$\begin{array}{ll} H_{00} = \begin{bmatrix} h_{0,0} & h_{0,4} & h_{0,8} & h_{4,0} & h_{4,4} & h_{4,8} & h_{8,0} & h_{8,4} & h_{8,8} \end{bmatrix} & (i = 0,\ j = 0) \\ H_{12} = \begin{bmatrix} h_{1,2} & h_{1,6} & h_{1,10} & h_{5,2} & h_{5,5} & h_{5,8} & h_{9,2} & h_{9,5} & h_{9,8} \end{bmatrix} & (i = 1,\ j = 2) \\ H_{33} = \begin{bmatrix} h_{3,3} & h_{3,7} & h_{3,11} & h_{7,3} & h_{7,7} & h_{7,11} & h_{11,3} & h_{11,7} & h_{11,11} \end{bmatrix} & (i = 3,\ j = 3) \end{array} \tag{36}$$

In order to explain the proposed method in an easier way, our demonstration was restricted to a less complex particular situation where the kernel matrix is of size 12 × 12 and the input image is 21 × 21, but the results can be readily extended for the kernel of the circular FIR filter designed above of size 31 × 31, previously extended to size 32 × 32 (by padding with zeros), to be able to achieve the decimation by a factor of 4.

The simpler algorithm described above for a 2D filter with a 3 × 3 kernel and 5 × 5 input matrix, can be extended by performing a decimation by factor 4 for the kernel matrix and a decimation by factor 5 for the input matrix. Thus, performing a decimation with factor 4, instead of the kernel of size 12 × 12, 16 matrices of size 3 × 3 are derived. For instance, in the case of H_{01}^T, applying decimation by 4, the following block matrix of size 3 × 3 will produce:

$$H'_{01} = \begin{bmatrix} h_{01} & h_{05} & h_{09} \\ h_{41} & h_{45} & h_{49} \\ h_{81} & h_{85} & h_{89} \end{bmatrix} \tag{37}$$

Next, by concatenating the rows of matrix H'_{01}, the matrix H_{01}^T is derived from Equation (35). The vector X_{2D} is also substituted with vector $\mathbf{X}_{2D}$ given by (34).

The vectors $X_{00}, X_{01}, X_{02}, X_{03}, \ldots, X_{66}$, composing the matrix $\mathbf{X}_{2D}$ and related to the input image, are defined through the following general formula:

$$X_{ij} = \begin{bmatrix} x_{14+i,14+j} & x_{14+i,7+j} & x_{14+i,0+j} & x_{7+i,14+j} & x_{7+i,7+j} & x_{7+i,0+j} & x_{0+i,14+j} & x_{0+i,7+j} & x_{0+i,0+j} \end{bmatrix} \tag{38}$$

For example, the vectors X_{03}, X_{31}, X_{66} generated by the Formula (38) will be:

$$\begin{aligned} X_{03} &= \begin{bmatrix} x_{14,17} & x_{14,10} & x_{14,3} & x_{7,17} & x_{7,10} & x_{7,3} & x_{0,17} & x_{0,10} & x_{0,3} \end{bmatrix} & (i=0,\ j=3) \\ X_{31} &= \begin{bmatrix} x_{17,15} & x_{17,8} & x_{17,1} & x_{10,15} & x_{10,8} & x_{10,1} & x_{3,15} & x_{3,8} & x_{3,1} \end{bmatrix} & (i=3,\ j=1) \\ X_{66} &= \begin{bmatrix} x_{20,20} & x_{20,13} & x_{20,6} & x_{13,20} & x_{13,13} & x_{13,6} & x_{6,20} & x_{6,13} & x_{6,6} \end{bmatrix} & (i=6,\ j=6) \end{aligned} \tag{39}$$

The vectors $X_{00}, \ldots, X_{66}$ were produced as described below. To explain the idea, our demonstration is restricted to the situation in which the input image is a matrix of size 21×21 and a decimation by factor 5 is performed. In doing so, instead of the input matrix of dimension 21×21, a number of 77 matrices of size 3×3 are obtained. For instance, in the case of X_{01}, applying decimation by the factor 7, the following 3×3 block matrix is derived:

$$X'_{01} = \begin{bmatrix} x_{0,1} & x_{0,8} & x_{0,15} \\ x_{7,1} & x_{7,8} & x_{7,15} \\ x_{14,1} & x_{14,8} & x_{14,15} \end{bmatrix} \tag{40}$$

At this point, the rows of matrix X'_{01} are concatenated, then the resulting vector is reversed and thus the vector $X_{1,0}$ is derived from the general Equation (38):

$$X_{1,0} = \begin{bmatrix} x_{15,14} & x_{15,7} & x_{15,0} & x_{8,14} & x_{8,7} & x_{8,0} & x_{1,14} & x_{1,7} & x_{1,0} \end{bmatrix} \tag{41}$$

Even if our discussion was restricted to the particular case where the input matrix is 21×21 to be easier for the reader to follow our discussion, it is easy to extend it for a more general case. Thus, a 2D FIR filtering operation with a 4×4 kernel and a 7×7 input matrix was decomposed into 100 1D inner products (FIR filtering operations) using the following equations:

$$\begin{aligned} \mathbf{Y}_0 = & \begin{bmatrix} \mathbf{B}_0 & \mathbf{B}_0 & \mathbf{B}_0 & \mathbf{B}_0 \\ \mathbf{O}_{4\times 90} & \mathbf{B}_0 & \mathbf{O}_{4\times 90} & \mathbf{O}_{4\times 90} \\ \mathbf{O}_{4\times 90} & \mathbf{O}_{4\times 90} & \mathbf{B}_0 & \mathbf{O}_{4\times 90} \\ \mathbf{O}_{4\times 90} & \mathbf{O}_{4\times 90} & \mathbf{O}_{4\times 90} & \mathbf{B}_0 \end{bmatrix} \times diag\left(\begin{bmatrix} \mathbf{O}_{90\times 36} & \mathbf{O}_{90\times 36} & \mathbf{O}_{90\times 36} & \mathbf{B}_1 \\ \mathbf{O}_{90\times 36} & \mathbf{O}_{90\times 36} & \mathbf{B}_1 & \mathbf{A}_1 \\ \mathbf{O}_{90\times 36} & \mathbf{B}_1 & \mathbf{O}_{90\times 36} & \mathbf{A}_1 \\ \mathbf{B}_1 & \mathbf{O}_{90\times 36} & \mathbf{O}_{90\times 36} & \mathbf{A}_1 \end{bmatrix} \mathbf{H}_2^T\right) \times \\ & \times \begin{bmatrix} \mathbf{B}_2 & -\mathbf{B}_2 & -\mathbf{B}_2 & -\mathbf{B}_2 & \mathbf{O}_{90\times 63} & \mathbf{O}_{90\times 63} & \mathbf{O}_{90\times 63} \\ \mathbf{O}_{90\times 63} & \mathbf{B}_2 & \mathbf{O}_{90\times 63} & \mathbf{O}_{90\times 63} & \mathbf{O}_{90\times 63} & \mathbf{O}_{90\times 63} & \mathbf{O}_{90\times 63} \\ \mathbf{O}_{90\times 63} & \mathbf{O}_{90\times 63} & \mathbf{B}_2 & \mathbf{O}_{90\times 63} & \mathbf{O}_{90\times 63} & \mathbf{O}_{90\times 63} & \mathbf{O}_{90\times 63} \\ \mathbf{O}_{90\times 63} & \mathbf{O}_{90\times 63} & \mathbf{O}_{90\times 63} & \mathbf{B}_2 & \mathbf{O}_{90\times 63} & \mathbf{O}_{90\times 63} & \mathbf{O}_{90\times 63} \end{bmatrix} \times \mathbf{X}_2^T \end{aligned} \tag{42}$$

$$\begin{aligned} \mathbf{Y}_1 = & \begin{bmatrix} \mathbf{O}_{4\times 90} & \mathbf{O}_{4\times 90} & \mathbf{O}_{4\times 90} \\ \mathbf{B}_0 & \mathbf{B}_0 & \mathbf{B}_0 \\ \mathbf{O}_{4\times 90} & \mathbf{B}_0 & \mathbf{O}_{4\times 90} \\ \mathbf{O}_{4\times 90} & \mathbf{O}_{4\times 90} & \mathbf{B}_0 \end{bmatrix} \times diag\left(\begin{bmatrix} \mathbf{O}_{90\times 36} & \mathbf{O}_{90\times 36} & \mathbf{B}_1 & \mathbf{O}_{90\times 36} \\ \mathbf{O}_{90\times 36} & \mathbf{B}_1 & \mathbf{B}_1 & \mathbf{O}_{90\times 36} \\ \mathbf{B}_1 & \mathbf{O}_{90\times 36} & \mathbf{B}_1 & \mathbf{O}_{90\times 36} \end{bmatrix} \mathbf{H}_2^T\right) \times \\ & \times \begin{bmatrix} \mathbf{O}_{90\times 63} & -\mathbf{B}_2 & \mathbf{B}_2 & -\mathbf{B}_2 & -\mathbf{B}_2 & \mathbf{O}_{90\times 63} & \mathbf{O}_{90\times 63} \\ \mathbf{O}_{90\times 63} & \mathbf{O}_{90\times 63} & \mathbf{O}_{90\times 63} & \mathbf{B}_2 & \mathbf{O}_{90\times 63} & \mathbf{O}_{90\times 63} & \mathbf{O}_{90\times 63} \\ \mathbf{O}_{90\times 63} & \mathbf{O}_{90\times 63} & \mathbf{O}_{90\times 63} & \mathbf{O}_{90\times 63} & \mathbf{B}_2 & \mathbf{O}_{90\times 63} & \mathbf{O}_{90\times 63} \end{bmatrix} \times \mathbf{X}_2^T \end{aligned} \tag{43}$$

$$\begin{aligned} \mathbf{Y}_2 = & \begin{bmatrix} \mathbf{O}_{4\times 90} & \mathbf{O}_{4\times 90} & \mathbf{O}_{4\times 90} \\ \mathbf{O}_{4\times 90} & \mathbf{O}_{4\times 90} & \mathbf{O}_{4\times 90} \\ \mathbf{B}_0 & \mathbf{B}_0 & \mathbf{O}_{4\times 90} \\ \mathbf{O}_{4\times 90} & \mathbf{B}_0 & \mathbf{B}_0 \end{bmatrix} \times diag\left(\begin{bmatrix} \mathbf{O}_{90\times 36} & \mathbf{B}_1 & \mathbf{O}_{90\times 36} & \mathbf{O}_{90\times 36} \\ \mathbf{B}_1 & \mathbf{B}_1 & \mathbf{O}_{90\times 36} & \mathbf{O}_{90\times 36} \\ \mathbf{B}_1 & \mathbf{O}_{90\times 36} & \mathbf{O}_{90\times 36} & \mathbf{O}_{90\times 36} \end{bmatrix} \mathbf{H}_2^T\right) \times \\ & \times \begin{bmatrix} \mathbf{O}_{90\times 63} & \mathbf{O}_{90\times 63} & -\mathbf{B}_2 & -\mathbf{B}_2 & \mathbf{B}_2 & -\mathbf{B}_2 & \mathbf{O}_{90\times 63} \\ \mathbf{O}_{90\times 63} & \mathbf{O}_{90\times 63} & \mathbf{O}_{90\times 63} & \mathbf{O}_{90\times 63} & \mathbf{O}_{90\times 63} & \mathbf{B}_2 & \mathbf{O}_{90\times 63} \\ \mathbf{O}_{90\times 63} & \mathbf{O}_{90\times 63} & \mathbf{O}_{90\times 63} & -\mathbf{B}_2 & -\mathbf{B}_2 & -\mathbf{B}_2 & \mathbf{B}_2 \end{bmatrix} \times \mathbf{X}_2^T \end{aligned} \tag{44}$$

Finally, the following output vector is produced:

$$\mathbf{Y} = \mathbf{Y}_0 + \mathbf{Y}_1 + \mathbf{Y}_2 \\ = \begin{bmatrix} Y_{00} & Y_{01} & Y_{02} & Y_{03} & Y_{10} & Y_{11} & Y_{12} & Y_{13} & Y_{20} & Y_{21} & Y_{22} & Y_{23} & Y_{30} & Y_{31} & Y_{32} & Y_{33} \end{bmatrix}^T \quad (45)$$

In Equations (42)–(44), the block matrices are, respectively: $\mathbf{B}_0 = \mathbf{A}_0 \otimes U_9$, where the vector U_9 is $U_9 = \begin{bmatrix} 1 & 1 & 1 & 1 & 1 & 1 & 1 & 1 & 1 \end{bmatrix}$ and $\mathbf{B}_1 = \mathbf{A}_1 \otimes \mathbf{I}_9$, where $\mathbf{I}_9$ is the 9×9 identity matrix (with ones on the main diagonal and zeros elsewhere); we also have $\mathbf{B}_2 = \mathbf{A}_2 \otimes \mathbf{I}_9$. The matrices $\mathbf{O}_{4\times 90}$, $\mathbf{O}_{90\times 36}$ and $\mathbf{O}_{90\times 63}$ are zero matrices of size 4×90, 90×36 and 90×63, respectively.

In order to obtain the above equations, we considered a polyphase decomposition of a 1D filter that can compute four samples in parallel using a decimation factor of 4 as:

$$\begin{bmatrix} y_{4n} \\ y_{4n+1} \\ y_{4n+2} \\ y_{4n+3} \end{bmatrix} = \begin{bmatrix} H_3 & H_2 & H_1 & H_0 & 0 & 0 & 0 \\ 0 & H_3 & H_2 & H_1 & H_0 & 0 & 0 \\ 0 & 0 & H_3 & H_2 & H_1 & H_0 & 0 \\ 0 & 0 & 0 & H_3 & H_2 & H_1 & H_0 \end{bmatrix} \times \begin{bmatrix} X_{4n-3} \\ X_{4n-2} \\ X_{4n-1} \\ X_{4n} \\ X_{4n+1} \\ X_{4n+2} \\ X_{4n+3} \end{bmatrix} \quad (46)$$

By extending the Equation (46) to 2D and using sub-expression sharing, we obtained Equations (42)–(44). Although at first sight, the matrix equations describing the proposed polyphase implementation may seem very complex, mainly due to their block structure, they actually lead to a very efficient and economic filtering structure, with a high degree of parallelism and therefore with a low computational complexity in terms of number of arithmetic operations. All these equations were verified in Matlab (version R2017a).

5. Discussion

The proposed design method for 2D circular filters is entirely analytical, without using any global numerical optimization techniques. Analytical design methods lead to closed-form and parametric filters, with adjustable, tunable frequency responses. To the best of the authors' knowledge, the analytical design of FIR circular filter banks has not been systematically approached previously by other researchers. As a reference to existing works, analytical techniques for designing 2D filters of IIR type with circular frequency response, including CFBs, have been previously proposed by the first author [33–35].

The Gaussian filter was chosen as a prototype for the CFB due to its advantages. It is a smooth function that can be easily approximated by a trigonometric polynomial and can be scaled on the frequency axis to adjust its selectivity. For very selective filters, the Gaussian shape is probably the best choice. Its frequency response is zero-phase; since frequency components will not be phase-shifted, image distortions will not occur. The resulting filters have accurate shapes, with negligible distortions. Moreover, they can be approximated efficiently, leading to low-order filters.

The circular filter bank (CFB) designed in our paper can be compared with other types of filter banks, from a qualitative point of view. The comparative discussion will mainly refer to works [25–27], as well-known multiscale pyramidal decomposition methods. Our proposed filter banks, like the steerable pyramid [25], have rotation invariance, while the Laplacian pyramid [26] and wavelet pyramid [27] are not rotationally invariant. Another important aspect regards frequency plane partitioning. While the steerable, wavelet and Laplacian pyramids all split the image spectrum into fixed sub-band regions, the proposed circular FB is flexible, in the sense that the bandwidths of the sub-band regions can be chosen wider or narrower, with adjustable selectivity, depending on application. This is due to the scalability of Gaussian-shaped filters along the frequency axis, which allows us to obtain filters with imposed selectivity starting from the same prototype. In Section 3, a uniform CFB with 11 components was generated, partitioning the image spectrum into concentric ring-shaped sub-bands. Moreover, in some applications, the non-uniform (dyadic) filter bank is also useful from the multi-resolution point of view. Since the energy of an image spectrum is mainly contained in the low-frequency region and decreases towards higher frequencies, the dyadic-type CFB allows for a more uniform energy distribution on frequency bands. Also, the proposed polyphase implementation structure for the filter bank has a lower arithmetic complexity than other implementations found in the literature.

A rigorous comparison in terms of performance with other circular filters found in literature is quite difficult to make. Design approaches like circular filters in [11–13] are very different from the one proposed here and lead to filters with other characteristics and purposes, so they are quite difficult to compare exactly with our proposed method.

To summarize, the novelty of the proposed CFB consists of an analytic design method (yielding parametric, closed-form expressions of frequency response), frequency scalability, flexible partitioning of spectrum sub-bands, low order due to efficient approximation and low arithmetic complexity due to polyphase implementation.

The proposed novel implementation technique significantly reduces the number of arithmetic operations required. A short comparison can be made between the direct convolution operation and the proposed filtering method in terms of computational complexity. The 2D filtering of an image of size $M \times N$ pixels, with an FIR filter with kernel size $m \times n$ implies a 2D convolution between a $m \times n$ matrix and a $M \times N$ matrix. This means that the filter kernel slides on the horizontal and vertical axes along the image, so for each pixel $m \times n$ multiplications are required; therefore the whole 2D filtering would have approximately a complexity of $O(MNmn)$. It is easy to calculate that the total number of additions are $(N + n^2)(M + m^2)$.

In the simpler case used to exemplify our implementation, the filter kernel has size 12×12, while the image is 21×21; thus for usual convolution, there will be 63,504 multiplications with 27,225 additions. In our approach, only 100 inner products are used, with 3×3 multiplications and 3×3 additions for each, that is $100 \times 33 = 900$ multiplications and 900 additions, plus 12×90 additions in the pre-processing stage and 7×10 additions in the post-processing stage. As an additional example, for a larger value of the filter kernel where the filter kernel has the size 20×20 while the image is 35×35, 100 inner products are used, with 5×5 multiplications and 5×5 additions for each inner product, that is $100 \times 5 \times 5 = 2500$ multiplications and 2500 additions, plus 20×90 additions in pre-processing stage and 70 additions in the post-processing stage.

6. Conclusions

The proposed design method for 2D circular filter banks is entirely analytical, without using any global numerical optimization. The advantages of the proposed method compared to other works are: it yields a factored 2D frequency response; the designed CFB is parametric, with adjustable characteristics; the CFB components are solved easily for any choice of number of filters and selectivity; the proposed implementation using polyphase and block filtering leads to a low complexity filter structure. This approach solves the problem of designing an adjustable and efficient circular FB with imposed specifications. The obtained results prove that the proposed rotationally invariant filter banks can be used in decomposing a given image into its subband components.

Taking into account the simulation results on test images and the fact that the original image can be reconstructed very accurately at least from a visual, subjective point of view from its component images, the authors intend in future work to study and investigate whether such CFBs (either uniform or non-uniform) could be used in sub-band coding schemes. While practically and intuitively this would seem possible, the required mathematical conditions for perfect reconstruction will have to be investigated rigorously. Regarding the implementation part, the authors will also study how to choose the decimation factors for the input image and the filter kernel, in order to obtain a very efficient, optimal design, and to minimize the number of arithmetic operations.

Author Contributions: Conceptualization, D.F.C. and R.M.; methodology, D.F.C. and R.M.; software, D.F.C. and R.M.; validation, D.F.C. and R.M.; formal analysis, D.F.C. and R.M.; investigation, D.F.C. and R.M.; resources, D.F.C. and R.M.; writing, original draft preparation; writing, review and editing, D.F.C. and R.M.; project administration, D.F.C.; funding acquisition, D.F.C. All authors have read and agreed to the published version of the manuscript.

Funding: This work was partially supported by a grant of the Romanian Ministry of Education and Research, CNCS—UEFISCDI, project number PCE 172 (PN-III-P4-ID-PCE2020-0713), within PNCDI III.

Institutional Review Board Statement: Not applicable.

Informed Consent Statement: Not applicable.

Data Availability Statement: Data are contained within the article.

Conflicts of Interest: The authors declare no conflict of interest.

References

1. Lu, W.; Antoniou, A. *Two-Dimensional Digital Filters*; CRC Press: Boca Raton, FL, USA, 1992.
2. Shyu, J.; Pei, S.; Huang, Y. Design of variable two-dimensional FIR digital filters by McClellan transformation. *IEEE Trans. Circ. Syst. I* **2009**, *56*, 574–582. [CrossRef]
3. Wang, Y.; Yue, J.; Su, Y.; Liu, H. Design of two-dimensional zero-phase FIR digital filter by McClellan transformation and interval global optimization. *IEEE Trans. Circ. Syst. II Express Briefs* **2013**, *60*, 167–171. [CrossRef]
4. Manuel, M.; Elias, E. Design of sharp 2D multiplier-less circularly symmetric FIR filter using harmony search algorithm and frequency transformation. *J. Signal Inf. Proc.* **2012**, *3*, 344–351. [CrossRef]
5. Kim, K.J.; Kim, J.H.; Nam, S.W. Design of computationally efficient 2D FIR filters using sampling-kernel-based interpolation and frequency transformation. *Electron. Lett.* **2015**, *51*, 1326–1328. [CrossRef]
6. Pun, C.K.S.; Chan, S.C.; Ho, K.L. Efficient 1D and Circular Symmetric 2D FIR Filters with Variable Cutoff Frequencies Using the Farrow Structure and Multiplier-Block. In Proceedings of the IEEE International Symposium Circuits Systems ISCAS 2001, Sydney, Australia, 6–9 May 2001; Volume 2, pp. 561–564. [CrossRef]
7. Bindima, T.; Elias, E. Design and implementation of low complexity 2-D variable digital FIR filters using single-parameter-tunable 2-D Farrow structure. *IEEE Trans. Circuits Syst. I Regul. Papers* **2018**, *65*, 618–627. [CrossRef]
8. Stavrou, V.N.; Tsoulos, I.G.; Mastorakis, N.E. Transformations for FIR and IIR filters' design. *Symmetry* **2021**, *13*, 533. [CrossRef]
9. Apostolov, P.S.; Yurukov, B.P.; Stefanov, A.K. An easy and efficient method for synthesizing two-dimensional finite impulse response filters with improved selectivity. *IEEE Signal Proc. Mag.* **2017**, *34*, 180–183. [CrossRef]
10. Capizzi, G.; Sciuto, G.L. A novel 2-D FIR filter design methodology based on a Gaussian-based approximation. *IEEE Signal Process. Lett.* **2019**, *26*, 362–366. [CrossRef]
11. Guillemot, C.; Ansari, R. Two-dimensional filters with wideband circularly symmetric frequency response. *IEEE Trans. Circuits Syst. II* **1994**, *41*, 703–707. [CrossRef]
12. Bindima, T.; Manuel, M.; Elias, E. An efficient transformation for two dimensional circularly symmetric wideband FIR filters. In Proceedings of the IEEE Region 10 Conference TENCON, Singapore, 22–25 November 2016; pp. 2838–2841. [CrossRef]
13. Hung, T.Q.; Tuan, H.D.; Nguyen, T.Q. Design of diamond and circular filters by semi-definite programming. In Proceedings of the IEEE International Symposium on Circuits and Systems, New Orleans, LA, USA, 27–30 May 2007; pp. 2966–2969. [CrossRef]
14. Randen, T.; Husoy, J.H. Filtering for texture classification: A comparative study. *IEEE Trans. Pattern Anal. Mach. Intell.* **1999**, *21*, 291–310. [CrossRef]
15. Porzycka-Strzelczyk, S.; Rotter, P.; Strzelczyk, J. Automatic detection of subsidence troughs in SAR interferograms based on circular Gabor filters. *IEEE Geosci. Remote Sens. Lett.* **2018**, *15*, 873–876. [CrossRef]
16. Lin, Y.P.; Vaidyanathan, P.P. Theory and design of two-dimensional filter banks: A review. *Multidim Syst. Signal Process.* **1996**, *7*, 263–330. [CrossRef]
17. Basu, S. Multidimensional causal, stable, perfect reconstruction filter banks. *IEEE Trans. Circuits Syst. I Fundam. Theory Appl.* **2002**, *49*, 832–842. [CrossRef]
18. Shi, G.; Liang, L.; Xie, X. Design of directional filter banks with arbitrary number of subbands. *IEEE Trans. Signal Process.* **2009**, *57*, 4936–4941. [CrossRef]
19. Liang, L.; Shi, G.; Xie, X. Nonuniform directional filter banks with arbitrary frequency partitioning. *IEEE Trans. Image Process.* **2010**, *20*, 283–288. [CrossRef] [PubMed]
20. Nguyen, T.T.; Oraintara, S. A class of multiresolution directional filter banks. *IEEE Trans. Signal Process.* **2007**, *55*, 949–961. [CrossRef]
21. Lu, Y.M.; Do, M.N. Multidimensional directional filter banks and surfacelets. *IEEE Trans. Image Process.* **2007**, *16*, 918–931. [CrossRef] [PubMed]

22. Hendre, M.; Patil, S.; Abhyankar, A. Directional filter bank-based fingerprint image quality. *Pattern Anal. Appl.* **2022**, *25*, 379–393. [CrossRef]
23. Mertzios, B.G. Fast block implementation of two-dimensional FIR digital filters by systolic arrays. *Int. J. Electron.* **1992**, *73*, 1233–1246. [CrossRef]
24. Aziz, M.; Boussakta, S.; McLernon, D.C. High performance 2D parallel block-filtering system for real-time imaging applications using the sharc ADSP21060. *Real-Time Imaging* **2003**, *9*, 151–161. [CrossRef]
25. Simoncelli, E.P.; Freeman, W.T. The steerable pyramid: A flexible architecture for multi-scale derivative computation. In Proceedings of the International Conference on Image Processin, Washington, DC, USA, 23–26 October 1995; Volume 3, pp. 444–447. [CrossRef]
26. Zhou, J.; Zhang, D.; Zou, P.; Zhang, W.; Zhang, W. Retinex-based Laplacian pyramid method for image defogging. *IEEE Access* **2019**, *7*, 122459–122472. [CrossRef]
27. Xu, Y.; Yang, X.; Ling, H.; Ji, H. A new texture descriptor using multifractal analysis in multi-orientation wavelet pyramid. In Proceedings of the IEEE Conference on Computer Vision and Pattern Recognition, San Francisco, CA, USA, 13–18 June 2010; pp. 161–168. [CrossRef]
28. Gao, Y.; Li, W.; Zhang, M. Hyperspectral and multispectral classification for coastal wetland using depthwise feature interaction network. *IEEE Trans. Geosci. Remote Sens.* **2022**, *60*, 5512615. [CrossRef]
29. Li, M.; Zhao, Y. Multi-scale feature selection network for lightweight image super-resolution. *Neural Netw.* **2023**, *169*, 352–364. [CrossRef] [PubMed]
30. Zhang, X.; Li, W. Hyperspectral pathology image classification using dimension-driven multi-path attention residual network. *Expert Syst. Appl.* **2023**, *230*, 120615. [CrossRef]
31. Li, L.; Lv, M.; Jia, Z.; Ma, H. Sparse representation-based multi-focus image fusion method via local energy in shearlet domain. *Sensors* **2023**, *23*, 2888. [CrossRef] [PubMed]
32. Matei, R. A class of directional zero-phase 2D filters designed using analytical approach. *IEEE Trans. Circuits Syst. I Regular Papers* **2022**, *69*, 1629–1640. [CrossRef]
33. Matei, R. Design and applications of adjustable 2D digital filters with elliptical and circular symmetry. *Analog. Integr. Circuits Signal Process.* **2023**, *114*, 345–358. [CrossRef]
34. Matei, R. Analytic Design of Uniform Circular Filter Banks. In Proceedings of the 24th IEEE Signal Processing: Algorithms, Architectures, Arrangements, and Applications (SPA), Poznań, Poland, 23–25 September 2020; pp. 58–62. [CrossRef]
35. Matei, R. Efficient Design Procedure for Circular Filter Banks. In Proceedings of the IEEE 63rd International Midwest Symposium on Circuits and Systems (MWSCAS), East Lansing, MI, USA, 9–11 August 2021; pp. 259–262. [CrossRef]
36. Chiper, D.F.; Matei, R. Polyphase Implementation for Gaussian 2D FIR Filters with Circular Symmetry. In Proceedings of the 15th International Conference on Electronics, Computers and Artificial Intelligence ECAI 2023, Bucharest, Romania, 29–30 June 2023. [CrossRef]

sensors

MDPI

Article

A Novel Fuzzy-Based Remote Sensing Image Segmentation Method

Barbara Cardone [1], Ferdinando Di Martino [1,2,*] and Vittorio Miraglia [1]

1 Department of Architecture, University of Naples Federico II, Via Toledo 402, 80134 Naples, Italy; b.cardone@unina.it (B.C.); vittorio.miraglia@unina.it (V.M.)

2 Center for Interdepartmental Research "Alberto Calza Bini", University of Naples Federico II, Via Toledo 402, 80134 Naples, Italy

* Correspondence: fdimarti@unina.it; Tel.: +39-081-2538904

Abstract: Image segmentation is a well-known image processing task that consists of partitioning an image into homogeneous areas. It is applied to remotely sensed imagery for many problems such as land use classification and landscape changes. Recently, several hybrid remote sensing image segmentation techniques have been proposed that include metaheuristic approaches in order to increase the segmentation accuracy; however, the critical point of these approaches is the high computational complexity, which affects time and memory consumption. In order to overcome this criticality, we propose a fuzzy-based image segmentation framework implemented in a GIS-based platform for remotely sensed images; furthermore, the proposed model allows us to evaluate the reliability of the segmentation. The Fast Generalized Fuzzy c-means algorithm is implemented to segment images in order to detect local spatial relations between pixels and the Triple Center Relation validity index is used to find the optimal number of clusters. The framework elaborates the composite index to be analyzed starting by multiband remotely sensed images. For each cluster, a segmented image is obtained in which the pixel value represents, transformed into gray levels, the graph belonging to the cluster. A final thematic map is built in which the pixels are classified based on the assignment to the cluster to which they belong with the highest membership degree. In addition, the reliability of the classification is estimated by associating each class with the average of the membership degrees of the pixels assigned to it. The method was tested in the study area consisting of the south-western districts of the city of Naples (Italy) for the segmentation of composite indices maps determined by multiband remote sensing images. The segmentation results are consistent with the segmentations of the study area by morphological and urban characteristics, carried out by domain experts. The high computational speed of the proposed image segmentation method allows it to be applied to massive high-resolution remote sensing images.

Keywords: remote sensing; RSIS; fuzzy clustering; image segmentation; FGFCM; TCR

Citation: Cardone, B.; Di Martino, F.; Miraglia, V. A Novel Fuzzy-Based Remote Sensing Image Segmentation Method. *Sensors* **2023**, *23*, 9641. https://doi.org/10.3390/s23249641

Academic Editors: Christos Nikolaos E. Anagnostopoulos and Stelios Krinidis

Received: 25 October 2023
Revised: 24 November 2023
Accepted: 4 December 2023
Published: 5 December 2023

1. Introduction

Remotely sensed images are used increasingly in many problems related to the analysis and control of the territory, such as the analysis of climate risks on urban and natural fabrics and the control of the territory for the purposes of prevention from natural disasters or those generated by anthropic, soil protection, and environmental pollution control [1–3]. One of the critical points relating to the processing of remote sensed data is the fact that it is a massive amount of data and is continuously updated over time. This entails the need to use methods and techniques for processing remote sensed images which, on the one hand, optimize CPU times and memory allocation, and on the other provide accurate and reliable results. In particular, one of the most used image processing methods in remote sensing image analysis is image segmentation, which has the objective of partitioning the image into non-overlapping patterns having different characteristics. It allows you to detect and extract areas of the study area with specific characteristics (for example, soil types) [4].

Different Remote Sensing Image Segmentation (RSIS) methods have been proposed in the literature; among them, the most used are pixel-based RSIS algorithms, which use threshold and clustering analysis techniques to segment the image based on the pixel values. Threshold [5–7] is an RSIS technique in which optimal thresholds are obtained by dividing the image histogram into two or more parts; the Otsu method [8] is the most widely used RSIS threshold method. In clustering RSIS algorithms, a clustering method is applied to classify the pixels so that pixels assigned to the same cluster (segment) have characteristics that are as similar as possible and as dissimilar as possible to pixels assigned to other segments. K-means [9], Fuzzy C-means (FCM) [10], and their variants are the more used clustering algorithms applied in RSIS methods [11–14]. They are computationally very fast but are very sensitive to the presence of outliers and noises in the data and do not consider local spatial relations between nearest pixels; in addition, a validity index needs to be used to set the number of clusters.

Region-based RSIS methods are iterative methods in which adjacent regions of the image are merged to form larger regions. The main region-based RSIS methods are region grooving and region splitting and merging segmentation methods [15]. In RSIS region grooving algorithms, seed elements consisting of small regions of the image are initially selected; subsequently, each of these regions are enlarged by applying growth rules that merge adjacent pixels that have specific common characteristics. On the contrary, the region splitting and merging segmentation methods split heterogeneous regions into smaller regions; these methods do not require manual selection of seeds but are computationally more complex than region merging methods.

The best-known region grooving RSIS algorithm is JSEG [16]. JSEG is applied and uses the color and texture characteristics of the image to define the growth rules. JSEG is computationally fast but suffers from the problem of image over-segmentation. To overcome this problem, in [17] a hybrid JSEG algorithm based on wavelet transform, called WJSEG, was proposed; the results of tests performed on high-resolution SPOT 5 pan-sharpened multispectral images and IKONOS panchromatic images showed that JSEG provides more accurate segmentation results with respect to JSEG, reducing the over-segmentation problem. However, it is sensitive to noise and is computationally slow.

To improve the accuracy of the segmentation results, recently meta-heuristic RSIS methods were proposed. An ANN-based RSIS method using an enhanced boosted convolutional neural network was proposed by [18]. In [19], a lightweight deep learning noise robust image segmentation method is proposed to detect and measure dam crack widths. These methods are robust to noise and produce very accurate results but require numerous training data and the training process is computationally very expensive.

In [20], a hybrid thresholding image segmentation method based on an adaptive fractional-order particle swarm optimization algorithm was developed; the results of testing on samples of aerial images show that this method improves the segmentation performances of the Otsu thresholding RSIS algorithm. However, it is very slow in processing massive satellite images. An FCM-based RSIS algorithm in which features are extracted in the remote sensed image and used as samples by machine learning classifiers is proposed in [21]; this method considers some characteristics of the remoted sensed image as entropy, intensity, and edge features, but it neglects the local relations between pixels. Some authors developed variations of FCM for image segmentation which overcome some critical points, such as the neglect of the spatial relations between pixels and the lack of robustness with respect to the presence of noise and outliers. In [22,23], variations of FCM to increase the robustness are proposed. They improve the robustness to the noise of FCM; however, they do not consider the spatial relations between neighboring pixels.

An extension of FCM, called Fast Generalized Fuzzy *c*-means (FGFCM), was proposed in [24] to incorporate local spatial value information in the image. FGFCM was applied in [25] to segment images compressed by using the bidimensional Fuzzy Transform [26]. The authors show that this model provides a good trade-off between the segmentation accuracy and time and memory consumption. In [27], this image segmentation model was applied to segment medical massive bedsores images in order to monitor the status and evolution of bedsores in elderly people unable to access hospital facilities during the COVID-19 pandemic period. A variation of FGFCM is proposed in [28], in which the PSO algorithm is used to find the centers of the initial clusters to avoid FGFCM getting stuck at the local minimum. To improve the robustness with respect to noise in the image, a variation of FCM called Modified Robust Fuzzy C-Means (MRFCM) is tested in [29] to segment brain magnetic resonance (MR) images. MRFCM is more robust to various types of noise than other FCM-based image segmentation algorithms; however, it has a high computational complexity and is unsuitable for the processing of massive and multi-modal images. In [30], a variation of FGFCM called Generalized FCM aiming to be independent from the parameters used in FGFGM is proposed. The authors show that this algorithm provides results comparable with FGFGM; however, the CPU times required are too high to make it suitable for RSIS applications.

Therefore, recently proposed RSIS methods improve the accuracy and robustness to noise compared to canonical RSIS methods but are computationally expensive. In summary, recently proposed RSIS methods improve the accuracy and robustness of the FGFCM algorithm; in contrast, it is very fast but is less robust to noise. Furthermore, it depends on the selection of the number of clusters, which is fixed a priori. The main goal of this research is to test a new RSIS cluster-based method that provides a trade-off between the accuracy of the results and the computational speed and which is robust to the presence of noise in the images.

Moreover, since in many problems the segmentation process must be carried out on raster datasets that represent specific indices built starting from multiband source satellite images, the segmentation process must be executed for any type of raster dataset, which represents a particular index, regardless of the domain of values assumed by this index. In fact, generally, remotely sensed images are used in GIS-based applications in order to construct a composite index as a function of the image in a set of bands. For example, if we intend to analyze the spatial distribution of the Normalized Difference Vegetation Index (NDVI), which provides information on the health and density of vegetation covering a study area, using Landsat satellite images in the Red (R) and Near InfraRed (NIR) bands, it is possible to calculate the NDVI index using the formula NDVI = (NIR − R)/(NIR + R). The result is a raster dataset, i.e., a dataset in image format containing information belonging to any domain, in which the values of the cells range between the interval $[-1, 1]$. Of course, an image dataset is a type of raster dataset. Therefore, an RSIS method must be able to analyze any type of raster dataset, which represents a particular index.

For this purpose, a new GIS-based framework applied to satellite image segmentation based on FGFCM is proposed; a preprocessing phase is performed to create the raster dataset representing a composite index by using multiband remotely sensed images. This raster dataset is, then, transformed in an image dataset and the triple center relation (for short, TCR) clustering validity measure [31] is used to assess the optimal number of fuzzy clusters C. Subsequently, FGFCM is executed on the image representing the synthetic index to obtain C gray images where in the *jth* pixel of the *ith* image is stored the membership degree of the pixel to the *ith* cluster. Then, a final classified raster dataset is constructed in which the value of a pixel is given by the label of the cluster to which it belongs with the highest membership degree.

The proposed framework allows us to overcome the limitations of the RSIS methods proposed in the literature. In particular:

- It provides a method to segment any type of raster dataset representing a specific synthetic index so that its use is not restricted only to source remotely sensed images;
- The use of the FGFCM segmentation algorithm facilitates considering the relations between neighboring pixels, spatial constraints, and local spatial information in the image;
- The triple center relation validity index [31] determines the optimal number of clusters even in the presence of noisy images and cluster centers that are spatially close to each

other. This feature is fundamental in cluster-based RSIS as remotely sensed images can be affected by various types of noise.

In summary, the proposed RSIS framework, unlike the RSIS models proposed in the recent literature, maintains the high computational speed of the FGFCM algorithm; furthermore, it is more robust than FGFCM with respect to the presence of noise in the image, providing more accurate results. Finally, it can be applied to any type of raster dataset constructed from the source multiband satellite image.

The rest of this paper is organized as follows: in Section 2 the FGFCM clustering image segmentation method and the TCR validity index are briefly described. Section 3 introduces the proposed framework and describes in detail its functional components. Section 4 presents and discusses the results of our tests performed on remotely sensed images. The conclusions are presented in Section 5.

2. Preliminaries

In this section, the RSIS FGFGM algorithm is synthetized and the TCR validity index used in our framework in a preprocessing phase to set the optimal number of fuzzy clusters is briefly described.

2.1. The FGFCM Image Segmentation Algorithm

The FGFCM algorithm is proposed in [24] to incorporate local spatial and grey level information together.

Let $\mathbf{X} = \{x_1,\ldots,x_N\} \subset R^n$ be a dataset of N elements where each element is a point in the space R^n of the n features. If the dataset is a gray image having N pixel, n = 1 and the element x_j is given by the gray value of the *jth* pixel.

Let $\mathbf{V} = \{v_1,\ldots,v_C\} \subset R^n$ the C cluster centers to be detected.

To consider local information, in [24] the following transformation to the jth element is performed:

$$\xi_j = \frac{\sum_{k\in N_w} S_{jk} x_k}{\sum_{k\in N_w} S_{jk}} \tag{1}$$

where N_w is a window around the jth pixel and the weight S_{jk} is given by

$$S_{jk} = \begin{cases} S_{s\_jk} \cdot S_{g\_jk} & \text{if } k \neq j \\ 0 & \text{if } k = j \end{cases} \tag{2}$$

in which the term S_{s_jk} measures the influence of the kth pixel in the set of the neighbors to the jth pixel and the term S_{g_jk} measures the grey similarity.

The term S_{s_jk} is given by

$$S_{s\_jk} = \exp\left(\frac{-\max\left(\left|p_j - p_k\right|, \left|q_j - q_k\right|\right)}{\lambda_s}\right) \tag{3}$$

where (p_j, q_j) and (p_k, q_k) are the coordinate, respectively, of the jth and the kth pixel and λ_s sets the spread of the exponential function.

The term S_{g_jk} is given by

$$S_{g\_jk} = \exp\left(\frac{-\|x_j - x_k\|^2}{\lambda_g \cdot \sigma_{g\_j}^2}\right) \tag{4}$$

where λ_g sets the spread of the function S_{g_jk}. The parameter σ_{g_j} is a function of the density of the local region surrounding the jth pixel; the higher this density, the higher its value. It is defined as

$$\sigma_{g\_j} = \sqrt{\frac{\sum_{k\in N_w} \|x_j - x_k\|^2}{N}} \tag{5}$$

The objective function to minimize is

$$J(X,U,V) = \sum_{i=1}^{C}\sum_{r=1}^{q} \gamma_r u_{ir}^m (\xi_r - v_i)^2 \quad (6)$$

where q < N is the number of distinct grey level values in the transformed image, γ_r is the number of pixels in the transformed image having grey level r, and ξ_r is the value of the lth grey level in the transformed image.

Applying the Lagrange multiplier method to find the minimum of (6), the solutions for U and V are obtained:

$$u_{ir} = \frac{(\xi_r - v_i)^{-\frac{2}{m-1}}}{\sum_{k=1}^{C}(\xi_r - v_k)^{-\frac{2}{m-1}}} \quad (7)$$

and

$$v_i = \frac{\sum_{r=1}^{q} \gamma_r u_{ir}^m \xi_r}{\sum_{r=1}^{q} \gamma_r u_{ir}^m} \quad (8)$$

where u_{ir} is the membership degree of the pixels having value ξ_r to the *ith* cluster and v_i is the center of the *ith* cluster.

In output, FGFCM provides C images with N pixels, where the i-th image represents, transformed in the interval [0, 255], the degree of belonging of the pixel to the ith cluster.

Below is shown in pseudocode the FGFCM algorithm (Algorithm 1).

Algorithm 1: FGFCM

```
Input: Original image with N pixels I
Number of clusters C
Fuzzifier m
End iteration threshold ε
Output: The C segmented images
Initialize randomly the center of the clusters c_i i = 1,...,C
For j = 1,...,N
Transform the value of the jth pixel by (1)
q:= number of distinct grey level values in the transformed image
Repeat
    For i = 1,...,C
        For r = 1,...,q
            Compute u_ir by (7)
        Next r
        Compute v_i by (8)
    Next i
Until |U^(t) − U^(t−1)| > ε |U^(t) − U^(t−1)| > ε
For i = 1,...,C
  Create the ith segmented image
Next i
Return the C segmented images
```

2.2. The TCR Validity Index

The TCR index is a fuzzy clustering validity measure related to the well-known Dunn index [32] used to detect compact well-separated clusters. The TCR is applied to assess the compactness of clusters and the separability among clusters.

Let $\mathbf{X} = \{x_1,\ldots,x_N\} \subset R^n$ be a dataset of N elements where each element is a point in the space R^n of the n features.

Let $\mathbf{V} = \{v_1,\ldots,v_C\} \subset R^n$ the C cluster centers.

The mean and the variance of the cluster centers are defined as

$$\hat{v} = \frac{1}{C}\sum_{i=1}^{C} v_i \tag{9}$$

and

$$\sigma_v^2 = \frac{1}{C-1}\sum_{i=1}^{C} \|v_i - \hat{v}\|^2 \tag{10}$$

The compactness of the cluster is measured by the following index:

$$\mathrm{Com}(C) = \sum_{i=1}^{C} \frac{\sum_{j=1}^{N} u_{ij}^m \|x_j - v_i\|^2}{\sum_{j=1}^{N} \max\limits_{i=1,..,C} u_j^m} \tag{11}$$

The separability among clusters is measured by the following indices:

$$\mathrm{Sep}(C) = S_1(C)\cdot S_2(C)\cdot S_3(C) \tag{12}$$

where

$$S_1(C) = N\cdot\sigma_v^2 \tag{13}$$

$$S_2(C) = \frac{1}{C}\sum_{i=1}^{C}\sum_{\substack{k=1\\k\neq i}}^{C} \|v_i - v_k\|^2 \tag{14}$$

$$S_3(C) = \min_{i=1,\dots,C} \sum_{\substack{k=1\\k\neq i}}^{C} \|v_i - v_k\|^2 \tag{15}$$

The three indices measure, respectively, the sample variance, the mean distance among cluster centers, and the minimum distance among cluster centers. Their combination obtains accurate measurements of intra-cluster separability, even in cases where the cluster centers are closely distributed. The lower the value of SEP(C), the higher the intra-cluster separability.

The final TCR index is given by the ratio between the compactness and the separability indices:

$$\mathrm{TCR}(C) = \frac{\mathrm{Com}(C)}{\mathrm{Sep}(C)} \tag{16}$$

The optimal number of clusters is selected by minimizing the TCR index. In Algorithm 2 the algorithm using TCR to find the optimal number of clustering is shown in pseudocode, where any FCM-based algorithm can be used.

The results of tests performed in [19] show that TCR give better performances with respect to other fuzzy clustering validity indices in the presence of noised datasets.

Algorithm 2: TCRValidityIndex

```
Input: Dataset with N elements D
       Fuzzifier m
       End iteration threshold ε
Output: Optimal number of clusters
Set C_MAX //maximum value for the number of clusters
C_OPT:= 1 //initialization of the best number of clusters
TCROLD:= 0 //initialization of the TCR
For c = 1,. . .,C_MAX
    Execute FCM-based algorithm (D, c, m, ε)
    Compute Com(c) by (11)
    Compute Sep(c) by (12)
    TCR = Com/Sep //TCR index obtained for c clusters
    If c = 1 Then
        TCR_OLD = TCR
    Else
      If TCR < TCR_OLD Then
        C_OPT:= c
        TCR_OLD = TCR
      End if
    End if
Next c
Return C_OPT
```

3. The Proposed Framework

The proposed RSIS framework includes:

- A preprocessing phase in which, starting from the multiband remotely sensed image source, the raster dataset of a composite index is constructed and the TCR validity measure to find the optimal number of clusters is used;
- The image segmentation phase in which the FGFCM algorithm is executed to the index image and the final classified image is created.
- Figure 1 schematizes the architecture of the framework.

The source dataset is given by a set of remotely sensed images acquired in one or more bands. The *index construction* component is the GIS-based process in which raster functions and map algebra operators are used to compute the composite index raster dataset.

The *transformation in pixel values* component transforms the index domain in a digital image domain. For example, the NDVI raster dataset is transformed in an image dataset converting the range [−1, 1] in the range [0, 255]. The result of the process is an image in which the pixel values are made up of the transformed values of the index to be analyzed (*Index image*).

The framework is highly flexible so as to allow segmentation of the source image into a band as well. In this case, the Index Image consists of the source image in the specified band.

The final functional component (*Find the optimal number of clusters*) aims to determine the optimal number of fuzzy clusters using the TCR validity index. This component executes iteratively FGFCM, setting a different number of clusters each time and measuring the corresponding TCR value. The number of clusters C chosen is the one that minimizes the TCR index.

An example of execution of the preprocessing phase in which the raster dataset of the NDVI index is created is schematized in Figure 2.

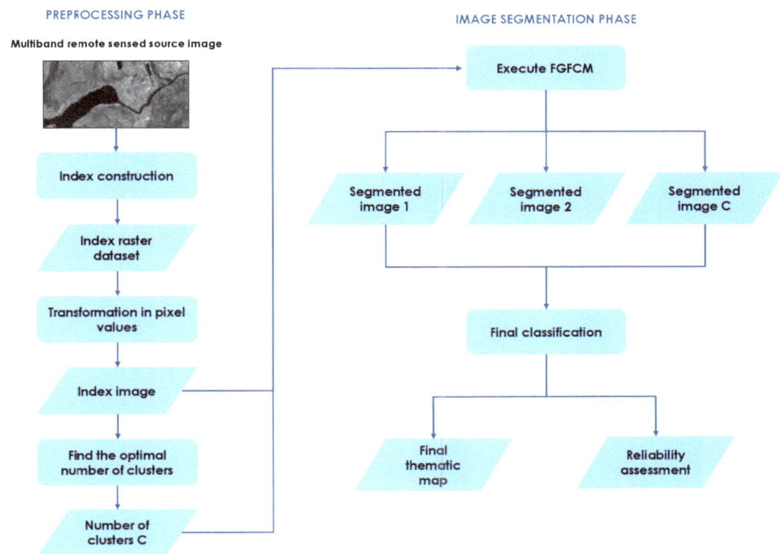

Figure 1. Schema of the proposed framework.

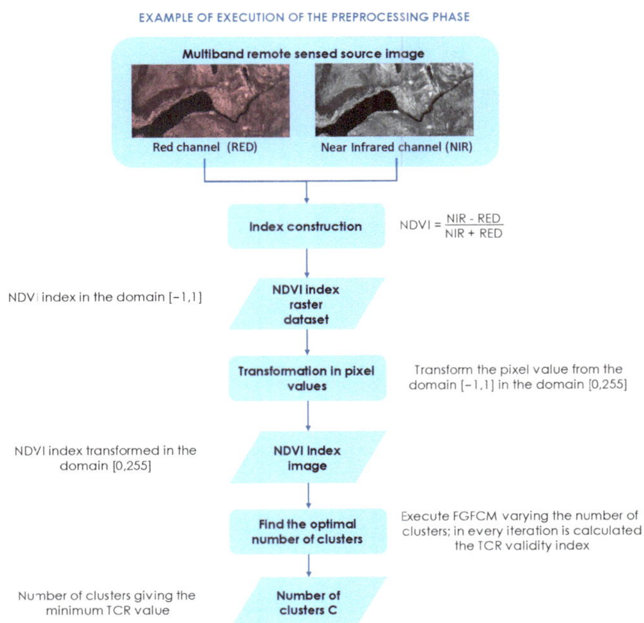

Figure 2. Example of execution of the preprocessing phase.

In the image segmentation phase FGFCM is executed on the index image, setting the number of fuzzy clusters to C. Outputs of the component *Execute FGFCM* are the set of C segmented images where the value of a pixel in the ith segmented image are converted in the digital image domain from the membership degree of the pixel to the ith cluster.

The *Final classification* component assigns to each pixel the label of the cluster to which it belongs with the highest degree of membership. The component provides a raster dataset in which the pixel values are given by the classes they belong to. A thematic map is appropriately constructed creating a one-to-one association between a cluster and a thematic class and assigning a semantic label to the thematic class. (*Final thematic map*). In addition, for each thematic class, the reliability of the assignment of pixels to the corresponding cluster is evaluated as the average of the membership degrees to the cluster of all the pixels assigned to it; the final assessed reliabilities are assigned to all the thematic classes and stored (*Reliability assessment*). The reliability measures for each cluster allows us to evaluate the reliability of the assignment of image pixels to the cluster; in fact, it is calculated as the average value of the membership degrees to the cluster of the pixels assigned to it. The higher this value, the greater the certainty that the pixels assigned to the cluster belong to it; therefore, the greater the accuracy of the detected segments.

Formally, if N_i is the number of pixels assigned to the ith cluster, the reliability of the assignment of these pixels to this cluster is given by

$$\mathrm{Rel}_i = \frac{1}{N_i}\sum_{j=1}^{N_i} u_{ij} \tag{17}$$

Below is shown in pseudocode our RSIS method (Algorithm 3). FGFCM is the FCM-based algorithm used executing the TCRValidityIndex algorithm.

Algorithm 3: The proposed RSIS method

Input: Original multiband image with N pixels
Output: Final classification thematic map and reliability assessment

```
Set m, ε
---------------- Preprocessing phase ------------------------------------
Construct the composite index raster dataset CI
Transform the composite index raster dataset in an image dataset II
C:= TCRValidityIndex(II, m, ε)
---------------- Image segmentation phase ----------------------------
Execute FGFCM(II, C, m, ε)
For j = 1,...,N
      u_MAX:= u_1j
      RCj:= l_1     //label of the first cluster
      For i = 2,...,C
        If u_ij > u_MAX Then
          u_MAX:= u_ij
          RCj:= lb_i      //label of the ith cluster
        End if
      Next i
Next j
For i = 1,...,C
      Rel_i:= 0
      Num_i:= 0
      For j = 1,...,N
        If RCj = lb_i Then
          Rel_i = Rel_i + RCj
          Num_i = Num_i + 1
        End if
       Next j
          Rel_i = Rel_i/Num_i
Next i
Return thematic map RC[N] and cluster assignment reliability Rel[C]
```

The framework was implemented in the ESRI ArcGIS desktop suite by using the Python ArcPy library.

In next section, we show the results of a set of tests of our framework applied on a study area given by the southwestern districts of Naples, Italy.

4. Test Results

The framework was tested on a study area given by the three districts of the southwestern area of the metropolitan city of Naples, Italy: Bagnoli, Fuorigrotta, and Posillipo.

Figure 3 shows the study area that includes the three districts. The area has been identified in order to test the accuracy of the image segmentation process of raster data representing composite indexes extracted by satellite images.

Figure 3. Framing of the study area: southwestern districts of Naples, Italy.

4.1. Morphological Analysis

To improve our understanding of the data from satellite images, we have been provided a morphological description of the whole study area, thanks to an experienced planner.

Posillipo has a very mountainous landscape; the Coroglio ridge, which runs the entire length of the district, is the morphological feature that indicates the district's division from the other two districts. All of Fuorigrotta is straight, with the exception of the eastern border region. The Agnano basin is a largely level volcanic area that is part of the Bagnoli district in the Campi Flegrei volcanic area. The southern area of the district is completely flat; almost all of the area is covered by an old industrial plant, now decommissioned for about 30 years, belonging to the old steel Italsider company.

To better understand the morphological constitution of the territory, in Figure 4 is shown the study area map of the Digital Terrain Model (DTM); a topographical model of the Earth's surface that contains data, in a digital format, of the elevation of the bare ground devoid of any natural or anthropic element present on the surface. For the study area, the DTM domain has an interval between 0 and 600 m that measures the surface height above sea level.

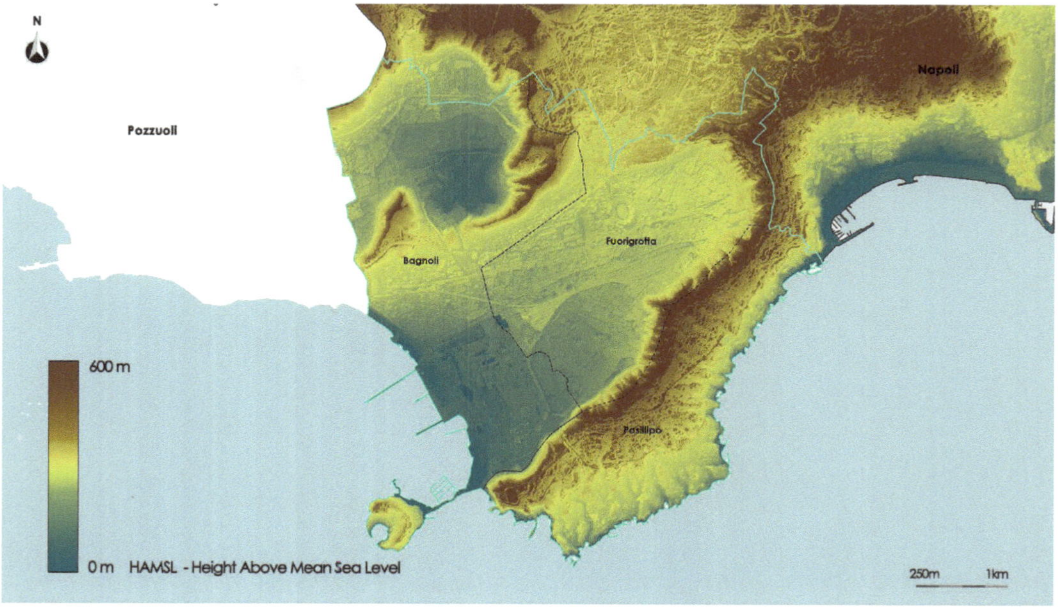

Figure 4. Digital Terrain Model of the study area.

The results obtained by running the proposed RSIS method on raster datasets of composite indices processed starting from satellite images are shown below. For brevity's sake, we show the results obtained for three composite indices: Albedo, NDVI, and Sky View Factor.

The Albedo index identifies the fraction of light on a horizontal surface that is reflected in all directions; it constitutes the reflective power. It is aimed at identifying the reflection characteristic of the solar radiation affecting the materials on the ground. It takes values in the range [0, 1]. The maximum albedo is 1 when all the incident radiation is reflected; this occurs in the case of perfectly white soils. The minimum albedo is 0 when no fraction of the radiation is reflected; this value is obtained in the presence of perfectly black soils. The Albedo index was calculated as the weighted average of the ratios between the visible and near infrared (0.315–2.8 μm) incident and reflected energy, using the visible and infrared emission and absorption spectral bands obtained with the RapidEye satellite, with resolution of 7×9 m.

Figure 5 shows the distribution of the Albedo on the study area.

The NDVI—Normalized Difference Vegetation Index—measures how vigorous the vegetation is. Its purpose is to document the presence of vegetation on the surface of the earth as well as its development over time. The ratio between the difference and a sum of the reflected radiation in the near infrared (NIR), in which the light is reflected by the leaves, and in the red (RED), in which the chlorophyll absorbs light, is used to compute the NDVI. The domain values are in the range of -1 and 1. When vegetation is present, values between 0.2 and 1 are assumed. The range of values between -1 and 0 can be attributed to uncultivated environments like streams and urban areas. The data are processed by the satellite Sentinel2 with a resolution of 7×7 m.

Figure 6 shows the distribution of the NDVI on the study area.

The Sky View Factor (SVF) index indicates the fraction of sky visible from a point on the surface. The index shall be calculated taking into account any obstacle that prevents the full visibility of the sky. The domain is between 0 and 1. With the approximation of the values to 0, there is a smaller portion of the visible sky and an increasingly complete obstruction of visibility; with the application of the values to 1, it will increase the portion of the sky detectable until a complete visibility of 360°. This shows that the higher the SVF value, the greater the heat loss in the atmosphere. The values were processed by the satellite Landsat 8 with a resolution of 1.7 × 1.7 m.

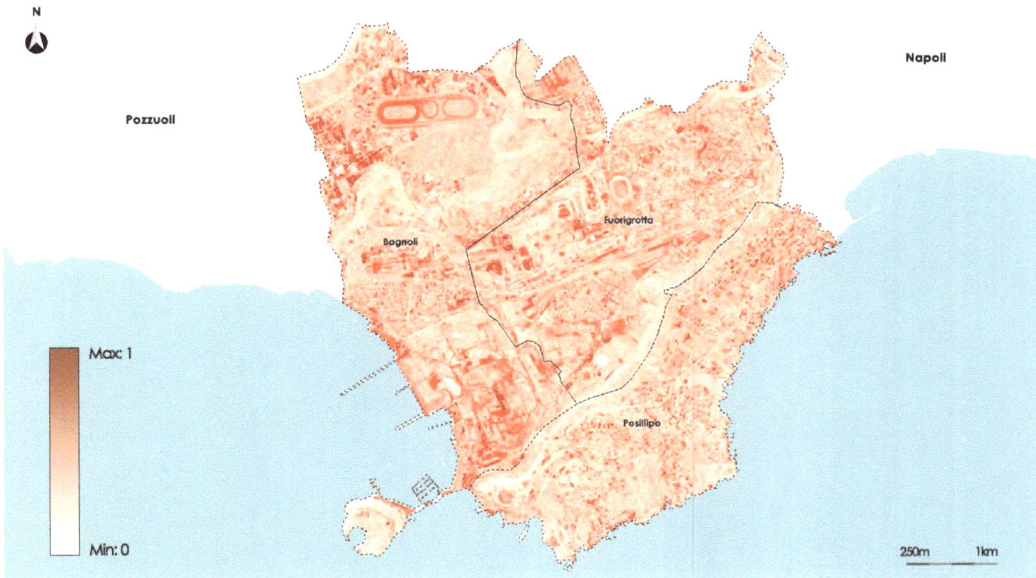

Figure 5. Map of Albedo satellite data.

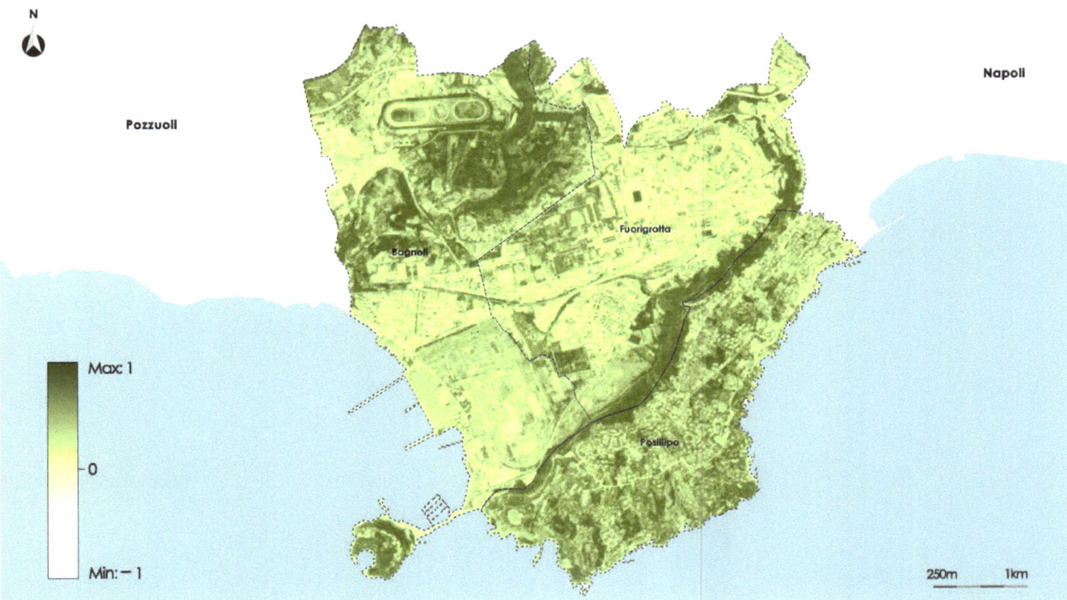

Figure 6. Map of NDVI satellite data.

Figure 7 shows the distribution of the SVF on the study area.

Figure 7. Map of Sky View Factor satellite data.

Following the segmentation process, thematic maps for each index were created: Albedo, NDVI, and SVF.

The optimal number of clusters determined in the preprocessing phase for the Albedo index is five. After executing the segmentation process, a thematic map of Albedo given by five thematic classes called, respectively, Low, Medium-Low, Medium, Medium-high, and High is created. Figure 8 shows the thematic map of the Albedo.

The segmentation algorithm was able to clearly distinguish areas with different values, managing to faithfully perimeter the areas as identified by the input raster. The inability to discern minute differences in values between several locations is the only drawback. According to the morphological analysis, it is clear that the areas with a lower value of Albedo are distributed mainly to the south along the ridge of Posillipo and north along the side of Mount Spina that delimits the basin of Agnano (locality of the district of Bagnoli).

The highest values are mainly concentrated within the complex of the Mostra d'Oltremare in the district of Fuorigrotta and in the disused industrial areas of the former Italsider and in the automotive sector of via Pisciarelli, respectively, to the south and north-east of Bagnoli.

The reliabilities assessed for each class are given in Table 1.

Table 1. Reliabilities of the classes of Albedo.

Class	Mean Reliability	Standard Deviation
Low	0.74	0.11
Medium-low	0.58	0.07
Medium	0.77	0.08
Medium-high	0.75	0.07
High	0.67	0.08

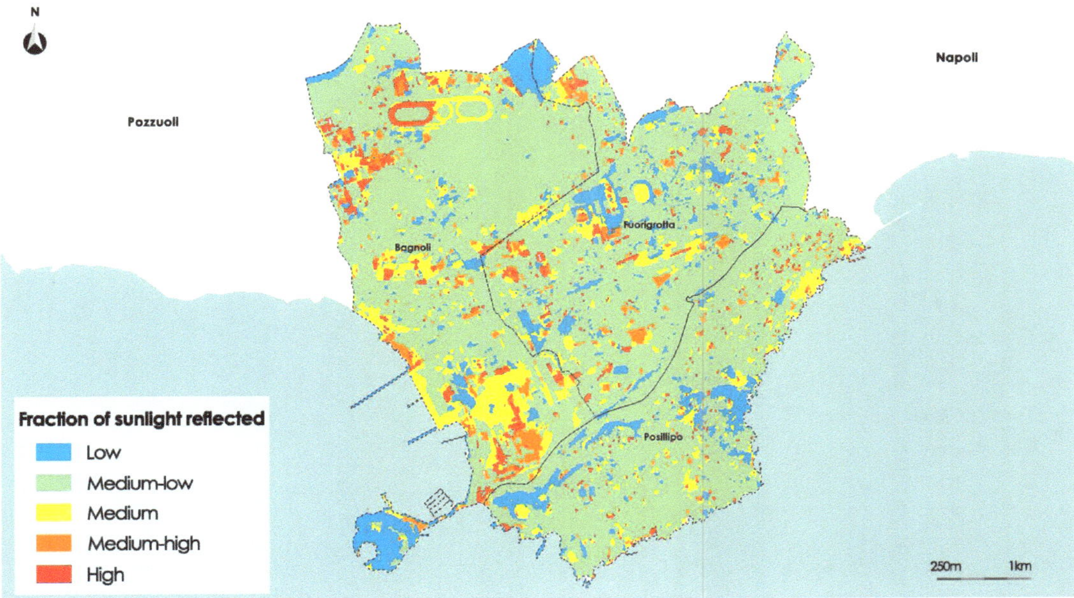

Figure 8. Map of Albedo after the segmentation process.

The average reliability is higher than 0.65 for all thematic classes, except the *Medium-low* thematic class, whose average reliability is equal to 0.58; furthermore, this thematic class presents the highest standard deviation of reliability. This is presumably due to the fact that this class includes large areas with different shapes and types of soil.

Now the results obtained for the NDVI index are shown. The optimal number of clusters determined in the preprocessing phase for the NDVI index is five. After executing, then in the segmentation process a thematic map of NDVI given by five thematic classes called, respectively: *Absent*, *Low*, *Scanty*, *Good*, and *High* is created. Figure 9 shows the thematic map of NDVI.

The technique for segmentation conformed to the same input raster's boundary while accurately identifying areas with varying NDVI values. As per the planner's expectations, the areas with the highest value correspond to the long ridge that splits the district of Posillipo from that of Fuorigrotta and to the basin of Agnano close to the border between the district of Bagnoli and Fuorigrotta. Both surfaces are mainly covered by wooded areas. Due to its high level of urbanization, the majority of the land is categorized as *Scanty*; both built or and natural surfaces belong into this class. However, the disused industrial area in Bagnoli is an example of how badly vegetated this class is.

The reliabilities assessed for each class are given in Table 2.

Table 2. Reliabilities of the classes of NDVI.

Class	Mean Reliability	Standard Deviation
Absent	0.78	0.04
Low	0.71	0.08
Scanty	0.55	0.13
Good	0.68	0.09
High	0.72	0.08

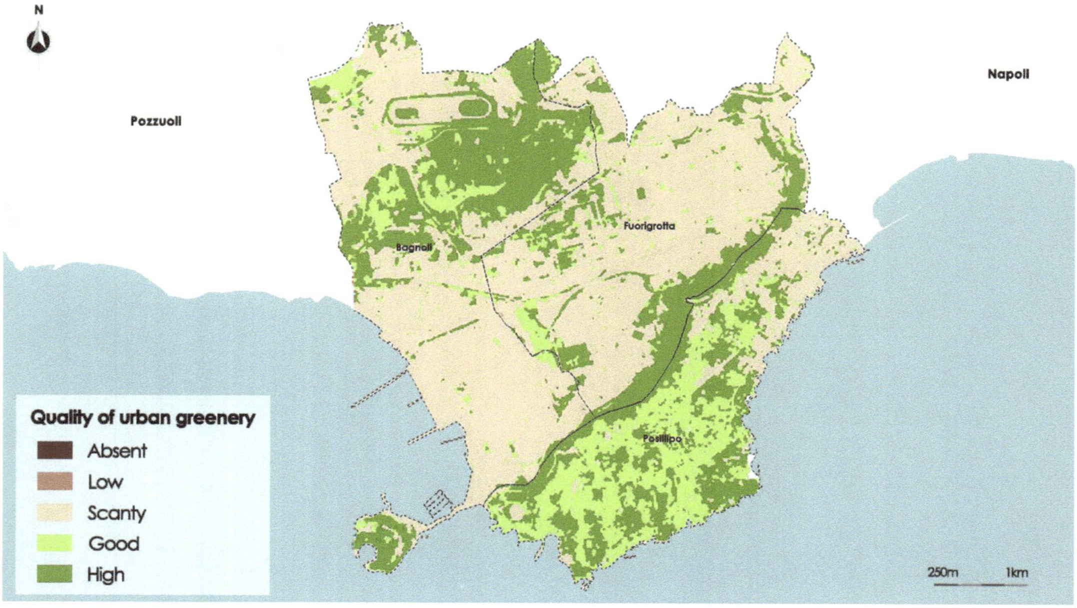

Figure 9. Map of NDVI after the segmentation process.

The average reliability is higher than 0.70 for all thematic classes except the *Scanty* thematic class, whose average reliability is equal to 0.55; furthermore, this thematic class presents the highest standard deviation of reliability (0.13). In fact, very large zones of the study area belong to this class, with a sparse presence of living vegetation, both in the built fabric and in impervious open spaces and in uncultivated or abandoned areas.

Below the results obtained for the SVF index are shown. The optimal number of clusters determined in the preprocessing phase for the SVF index is three. After executing, then in the segmentation process a thematic map of Albedo given by three thematic classes called, respectively, *Low*, *Medium*, and *High* is created. Figure 10 shows the thematic map of the Sky View Factor.

Even more accurately than in the prior instances, the segmentation algorithm has captured the perimeter in this instance as well. The input file's higher resolution than the other two raster images could be the cause of this. In line with the morphological analysis, the areas with higher values of SVF are those with a flat character, such as the disused industrial area in Bagnoli to the south and the flat inside the basin of Agnano to the north. Both areas have a high degree of visible sky fraction. As expected, the areas with the lowest level of visibility are those with a high density of built surfaces due to the dense mesh of buildings that hinders the fraction of sky visible from the road.

Table 3 shows the reliabilities assessed for three SVF thematic classes.

Table 3. Reliabilities of the classes of SVF.

Class	Mean Reliability	Standard Deviation
Low	0.73	0.06
Medium	0.71	0.06
High	0.70	0.07

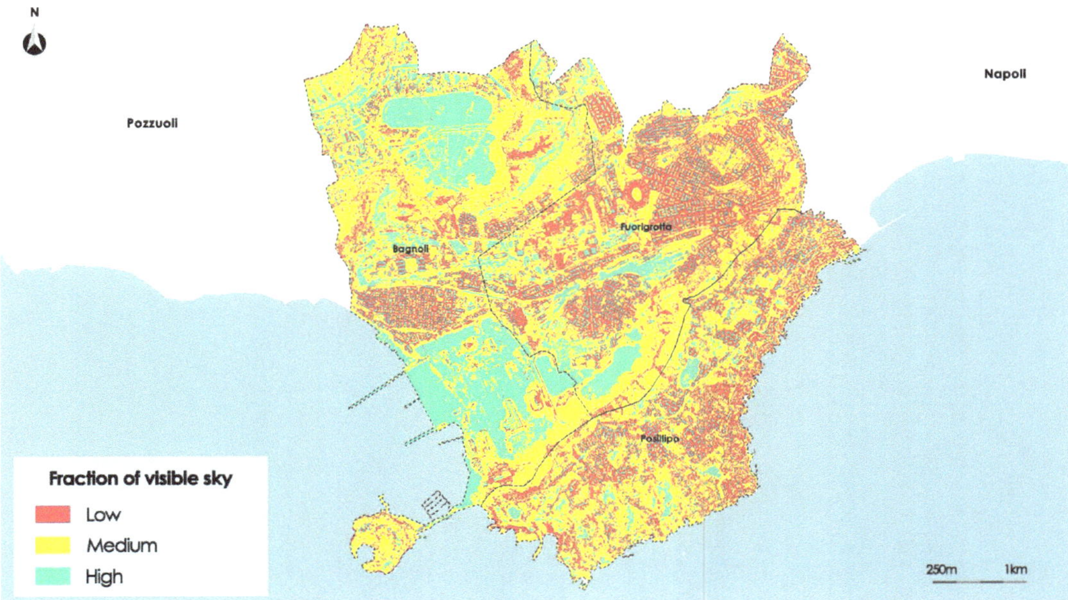

Figure 10. Map of Sky View Factor after the segmentation process.

The mean reliability and the standard deviation of the three thematic classes is very similar; in particular, the mean reliability is higher than 0.7 for all thematic classes. This result highlights that areas with *Low*, *Medium*, and *High* sky view factors are very distinct from each other.

In order to analyze the performance of the proposed method, it was compared with the well-known Otsu thresholding segmentation method, analyzing a specific region of the study area selected by the domain experts. The comparison was performed by measuring the Hamming Distance [33] between the segmentation results obtained executing the Otsu thresholding algorithm and the proposed method. The Hamming distance between two binary segmentations R and S in a region evaluates the similarity between the two segmentations in that region. It is defined as

$$HD(R, S) = 1 - \frac{|R_B \cap S_F| + |R_F \cap S_B|}{|R|} \tag{18}$$

where $|R|$ is the number of pixels in the region, $|R_B \cap S_F|$ is the number of pixels of the region classified in the background in the segmentation R and in the foreground in the segmentation S, and $|R_F \cap S_B|$ is the number of pixels of the region classified in the background in the segmentation S and in the foreground in the segmentation R.

HD ranges between 0 and 1. The more HD approaches 1, the more similar the two segmentations are in the region of the analyzed image.

The two methods are executed to a selected region in the images of the three composite indexes of Albedo, NDVI, and Sky View Factor. To obtain the background and the foreground areas using our FGFCM-based segmentation method, the thematic classes in the resultant segmented image were aggregated to form only the two thematic classes called, respectively, *Foreground* and *Background*.

Figure 11 show the segmentations obtained for the three synthetic indices analyzed: Albedo, NDVI, and Sky View Factor.

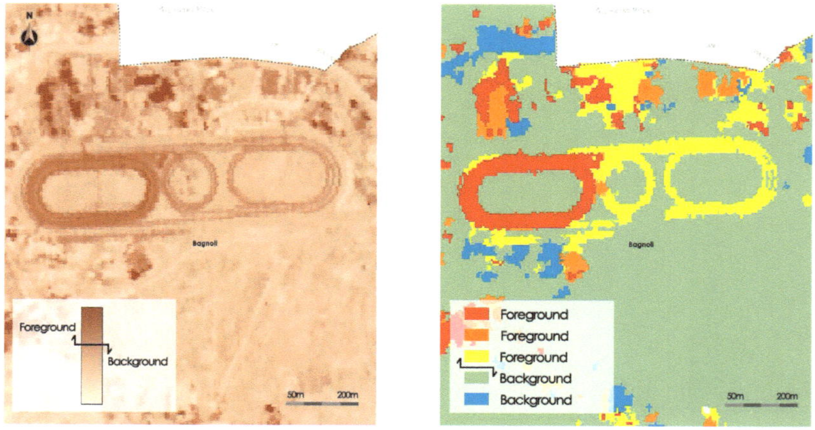

(**a**)

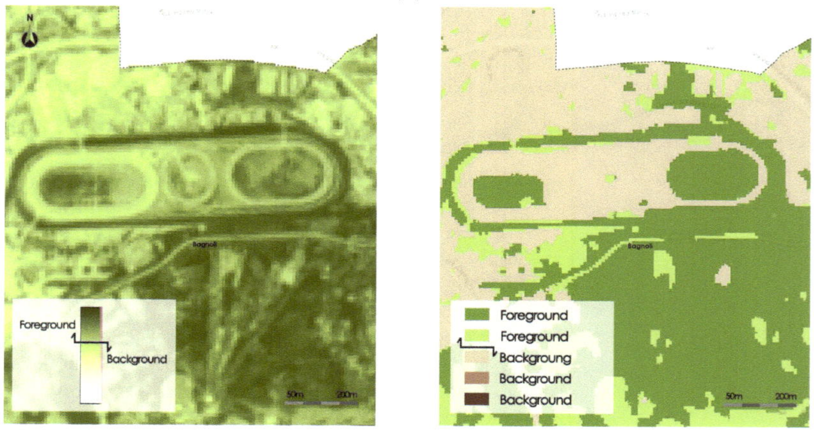

(**b**)

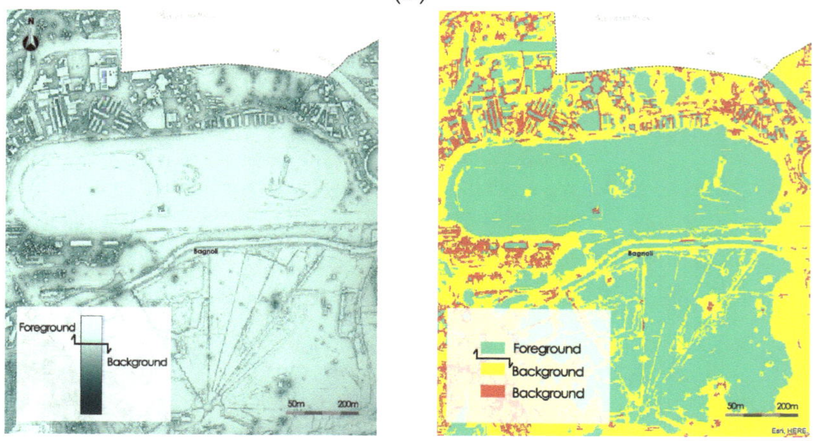

(**c**)

Figure 11. Segmented regions compared for the three synthetic indices: (**a**) Albedo, (**b**) NDVI, (**c**) Sky View factors.

Table 4 shows the results of the comparison. The table shows the Hamming distance similarity measure and the execution times of the two methods necessary for the segmentation of regions in the three raster datasets.

Table 4. Hamming distance and CPU time of the Otsu thresholding and the proposed method.

Synthetic Index	HD	Otsu CPU Time (s)	Our Method CPU Time (s)
Albedo	0.91	2.01	1.38
NDVI	0.93	2.14	1.42
Sky View Factor	0.95	1.97	1.40

The HD measure is higher than 0.9 in all three cases. Furthermore, the execution times obtained with the proposed method are in all cases lower than those obtained by running the Otsu algorithm.

4.2. Discussion of the Results

The results of the classification agree with assessments provided by topic-matter specialists who assessed how closely the areas described in the thematic map conformed to their morphological and urban features. This implies that the method proposed by us can be used to improve the analysis of urban systems thanks to its short computational time.

In fact, our algorithm can guarantee excellent results even with high-resolution satellite images without having to wait as long as other models do. From a classification point of view, our model allows the determination of the optimal number of clusters thanks to the use of the TCR validity index. This is guaranteed even in high-noise conditions.

By analyzing the low standard deviation values of each class found in each of the satellite rasters analyzed, it is possible to demonstrate that our model has a good degree of reliability in the determination of thematic classes and a low level of uncertainty.

Furthermore, the results of comparisons with the Otsu thresholding algorithm show that the proposed RSIS method provides good accuracy and better execution times.

5. Conclusions

A new RSIS method based on the Fast Generalized Fuzzy C-means algorithm is proposed. In a preprocessing phase, a raster dataset representing the distribution of the composite index on the study area is obtained by processing remotely sensed image datasets and the TCR validity index is used to determine the optimal number of clusters. Then, FGFCM is executed to obtain the segmented images; the segmentation result is given by a thematic map of the composite index in which each thematic class is related to a specific fuzzy cluster. A pixel is assigned to the thematic class corresponding to the cluster to which it belongs with the greatest membership degree. Finally, the mean reliability of every thematic class is assessed as the average membership degrees of the pixels belonging to the class.

Our framework was tested on a set of remotely sensed images to construct a segmented thematic map of composite indices in the study area given by the southwestern districts of Naples, Italy. The final thematic maps of the analyzed composite indices are in line with the assessments made by domain experts who evaluated the adherence of the areas classified in the thematic map with their morphological and urban characteristics.

The use of the FGFCM algorithm, which has a high computational speed, allows the proposed method to be applied also to high-resolution remotely sensed images; furthermore, the use of the TCR validity index can determine the optimal number of clusters even in the presence of noisy images. A further benefit is the assessment of the reliability of the final thematic classes, which allows the effectiveness of the classification to be assessed.

Our model, thanks to its ability to process remote sensing images at high resolutions in short computational times, can be a useful supporting tool for urban morphological analysis for the assessment of physical vulnerability compared to multi-risks caused by extreme events such as heatwaves or pluvial flooding.

In the future, we intend to carry out further comparative tests on different types of territories and urban settlements in order to determine the accuracy and efficiency of the proposed method as the type of study area and the resolution and the quality of the source remotely sensed images vary.

Author Contributions: Conceptualization, B.C., F.D.M. and V.M.; methodology, B.C., F.D.M. and V.M.; software, B.C., F.D.M. and V.M.; validation, B.C., F.D.M. and V.M.; formal analysis, B.C., F.D.M. and V.M.; investigation, B.C., F.D.M. and V.M.; resources, B.C., F.D.M. and V.M.; data curation, B.C., F.D.M. and V.M.; writing—original draft preparation, B.C., F.D.M. and V.M.; writing—review and editing, B.C., F.D.M. and V.M.; visualization, B.C., F.D.M. and V.M.; supervision, B.C., F.D.M. and V.M. All authors have read and agreed to the published version of the manuscript.

Funding: This research received no external funding.

Institutional Review Board Statement: Not applicable.

Informed Consent Statement: Not applicable.

Data Availability Statement: Data are contained within the article.

Conflicts of Interest: The authors declare no conflict of interest.

References

1. Qiao, H.; Wan, X.; Wan, Y.; Li, S.; Zhang, W. A Novel Change Detection Method for Natural Disaster Detection and Segmentation from Video Sequence. *Sensors* **2020**, *20*, 5076. [CrossRef] [PubMed]
2. Marcos, D.; Volpi, M.; Kellenberger, B.; Devis, T. Land cover mapping at very high resolution with rotation equivariant CNNs: Towards small yet accurate models. *ISPRS J. Photogramm. Remote Sens.* **2018**, *145*, 96–107. [CrossRef]
3. Ramadas, M.; Abraham, A. Segmentation on remote sensing imagery for atmospheric air pollution using divergent differential evolution algorithm. *Neural Comput. Appl.* **2023**, *35*, 3977–3990. [CrossRef] [PubMed]
4. Kotaridis, J.; Lazaridou, M. Remote sensing image segmentation advances: A meta-analysis. *ISPRS J. Photogramm. Remote Sens.* **2021**, *173*, 309–322. [CrossRef]
5. Gonzalez, R.; Woods, R. E: Thresholding. In *Digital Image Processing*, 3rd ed.; Prentice Hall: Upper Saddle River, NJ, USA, 2007; p. 954. ISBN 978-0131687288.
6. Pare, S.; Kumar, A.; Singh, G.K.; Bajaj, V. Image Segmentation Using Multilevel Thresholding: A Research Review. *Iran. J. Sci. Technol. Trans. Electr. Eng.* **2020**, *44*, 1–29. [CrossRef]
7. Wang, Y.; Lv, H.; Deng, R.; Zhuang, S. A Comprehensive Survey of Optical Remote Sensing Image Segmentation Methods. *Can. J. Remote Sens.* **2020**, *46*, 501–531. [CrossRef]
8. Otsu, N. A threshold selection method from gray-level histogram. *IEEE Trans. Syst. Man Cybern.* **1979**, *9*, 62–66. [CrossRef]
9. Macqueen, J. Some methods for classification and analysis of multivariate observations. In Proceedings of the Berkeley Symposium on Mathematical Statistics & Probability, Berkeley, CA, USA, 21 June–18 July 1965; University of California Press: Oakland, CA, USA; Volume 5.1, pp. 281–297.
10. Bezdek, J.C.; Ehrlich, R.; Full, W. FCM: The fuzzy c-means clustering algorithm. *Comput. Geosci.* **1974**, *10*, 191–203. [CrossRef]
11. Wang, Y.; Li, D.; Wang, Y. Realization of remote sensing image segmentation based on K-means clustering, SAMSED 2018. In *IOP Conference Series: Materials Science and Engineering*; IOP Publishing: Bristol, UK, 2019; Volume 490, p. 072008. [CrossRef]
12. Hamada, M.; Kanat, Y.; Adejor, A.E. Multi-Spectral Image Segmentation Based on the K-means Clustering. *Int. J. Innov. Technol. Explor. Eng.* **2019**, *9*, 2278–3075. [CrossRef]
13. Yin, S.; Li, H. Hot Region Selection Based on Selective Search and Modified Fuzzy C-Means in Remote Sensing Images. *IEEE J. Sel. Top. Appl. Earth Obs. Remote Sens.* **2020**, *13*, 5862–5871. [CrossRef]
14. Xu, J.; Zhao, T.; Feng, G.; Ni, M.; Ou, S. A Fuzzy C-Means Clustering Algorithm Based on Spatial Context Model for Image Segmentation. *Int. J. Fuzzy Syst.* **2021**, *23*, 816–832. [CrossRef]
15. Ma, W.; Li, N.; Zhou, H.; Jiao, L.; Tang, X.; Guo, Y.; Hou, B. Feature Split–Merge–Enhancement Network for Remote Sensing Object Detection. *IEEE Trans. Geosci. Remote Sens.* **2022**, *60*, 5616217. [CrossRef]
16. Khamael, A.; Mustafa, R. Satellite image classification and segmentation by using JSEG segmentation algorithm. *Int. J. Image Graph. Signal Process.* **2012**, *10*, 48–53. [CrossRef]
17. Wang, C.; Shi, A.Y.; Wang, X.; Wu, F.M.; Huang, F.C.; Xu, L.Z. A novel multi-scale segmentation algorithm for high resolution remote sensing images based on wavelet transform and improved JSEG algorithm. *Optik* **2014**, *125*, 5588–5595. [CrossRef]
18. Basaeed, E.; Bhaskar, H.; Al-Mualla, M. Supervised remote sensing image segmentation using boosted convolutional neural networks. *Knowl. Based Syst.* **2016**, *99*, 19–27. [CrossRef]
19. Wu, Z.; Tang, Y.; Hong, H.; Liang, B.; Liu, Y. Enhanced Precision in Dam Crack Width Measurement: Leveraging Advanced Lightweight Network Identification for Pixel-Level Accuracy. *Int. J. Intell. Syst.* **2023**, *2023*, 9940881. [CrossRef]

20. Chen, L.; Gao, I.; Lopes, A.M.; Zhang, Z.; Chu, Z.; Wu, R. Adaptive fractional-order genetic-particle swarm optimization Otsu algorithm for image segmentation. *Appl. Intell.* **2023**, *53*, 26949–26966. [CrossRef]
21. Sharma, R.; Ravinder, M. Remote sensing image segmentation using feature-based fusion on FCM clustering algorithm. *Complex Intellelligent Syst.* **2023**, *9*, 7423–7437. [CrossRef]
22. Zheng, Y.; Jeon, B.; Xu, D.; Wu, J.Q.M.; Zhang, H. Image segmentation by generalized hierarchical fuzzy C-means algorithm. *J. Intell. Fuzzy Syst.* **2015**, *28*, 961–973. [CrossRef]
23. Qi, Y.; Zhang, A.; Wang, H.; Li, X. An efficient FCM-based method for image refinement segmentation. *Vis. Comput.* **2022**, *38*, 2499–2514. [CrossRef]
24. Cai, W.; Chen, S.C.; Zhang, D.Q. Fast and robust fuzzy c-means clustering algorithms incorporating local information for image segmentation. *Pattern Recognit.* **2007**, *40*, 825–838. [CrossRef]
25. Di Martino, F.; Loia, V.; Sessa, S. A segmentation method for images compressed by fuzzy transform. *Fuzzy Sets Syst.* **2010**, *161*, 56–74. [CrossRef]
26. Perfilieva, I. Fuzzy Transforms: Theory and Applications. *Fuzzy Sets Syst.* **2006**, *157*, 993–1023. [CrossRef]
27. Di Martino, F.; Orciuoli, F. A computational framework to support the treatment of bedsores during COVID-19 diffusion. *J. Ambient. Intell. Humaniz. Computing* **2022**, *27*, 1–11. [CrossRef] [PubMed]
28. Hu, Y.-M.; Yu, M.-Q.; Du, J. An improved image segmentation approach using FGFCM with an edges-based neighbor selection strategy and PSO. In Proceedings of the 2017 36th Chinese Control Conference (CCC), Dalian, China, 26–28 July 2017; pp. 10951–10955. [CrossRef]
29. Song, J.; Zhang, Z. A Modified Robust FCM Model with Spatial Constraints for Brain MR Image Segmentation. *Information* **2019**, *10*, 74. [CrossRef]
30. Sesadri, U.; Nagaraju, C.; Ramakrishna, M. An efficient Image Segmentation based on Generalized FCM. *Int. J. Appl. Eng. Res.* **2018**, *13*, 27.
31. Tang, Y.; Huang, J.; Pedrycz, W.; Li, B.; Ren, F. A Fuzzy Clustering Validity Index Induced by Triple Center Relation. *IEEE Trans. Cybern.* **2023**, *53*, 5024–5036. [CrossRef]
32. Dunn, J.C. A fuzzy relative of the ISODATA process and its use in detecting compact well-separated clusters. *J. Cybern.* **1973**, *3*, 32–57. [CrossRef]
33. Hamming, R.W. Error detecting and error correcting codes. *Bell Syst. Tech. J.* **1950**, *29*, 147–160. [CrossRef]

sensors

MDPI

Article

Non-Contact Face Temperature Measurement by Thermopile-Based Data Fusion

Faraz Bhatti [1,*], Grischan Engel [1], Joachim Hampel [1], Chaimae Khalil [2], Andreas Reber [1], Stefan Kray [1,*] and Thomas Greiner [1,*]

1 Department of Engineering, Pforzheim University, 75175 Pforzheim, Germany
2 Pyramid Computer GmbH, 79111 Freiburg, Germany
* Correspondence: faraz.bhatti@hs-pforzheim.de (F.B.); stefan.kray@hs-pforzheim.de (S.K.); thomas.greiner@hs-pforzheim.de (T.G.)

Abstract: Thermal imaging cameras and infrared (IR) temperature measurement devices act as state-of-the-art techniques for non-contact temperature determination of the skin surface. The former is cost-intensive in many cases for widespread application, and the latter requires manual alignment to the measuring point. Due to this background, this paper proposes a new method for automated, non-contact, and area-specific temperature measurement of the facial skin surface. It is based on the combined use of a low-cost thermopile sensor matrix and a 2D image sensor. The temperature values as well as the 2D image data are fused using a parametric affine transformation. Based on face recognition, this allows temperature values to be assigned to selected facial regions and used specifically to determine the skin surface temperature. The advantages of the proposed method are described. It is demonstrated by means of a participant study that the temperature absolute values, which are achieved without manual alignment in an automated manner, are comparable to a commercially available IR-based forehead thermometer.

Keywords: non-contact temperature measurement; thermopile sensor; data fusion; intelligent access control system

Citation: Bhatti, F.; Engel, G.; Hampel, J.; Khalil, C.; Reber, A.; Kray, S.; Greiner, T. Non-Contact Face Temperature Measurement by Thermopile-Based Data Fusion. *Sensors* **2023**, *23*, 7680. https://doi.org/10.3390/s23187680

Academic Editors: Stelios Krinidis and Christos Nikolaos E. Anagnostopoulos

Received: 27 July 2023
Revised: 3 September 2023
Accepted: 4 September 2023
Published: 6 September 2023

1. Introduction

The non-contact measurement of skin temperature enables the early detection of potential signs of illness without the need for unwanted direct interaction with individuals. Additionally, it offers a practical solution for efficiently scanning large groups of people, supporting effective screening measures in both public and private spaces. Currently, there are two state-of-the-art non-contact methods:

- measurement of the forehead skin temperature using an infrared (IR) temperature measuring device;
- deployment of a thermal imaging camera at an exposed location for measuring the skin temperature.

Existing state-of-the-art approaches have certain limitations. The majority of these systems are based on IR temperature measurement sensors [1–9]. IR thermometer-based approaches do not allow for tracking the contour of the face. This requires the person's face to be positioned within a predetermined frame, which can be error-prone and less convenient. As a result, they are unsuitable for deployment in crowded areas and are economically unviable due to the extensive need for personnel.

Systems based on IR thermal imaging [10–12] offer facial recognition and tracking, but they are significantly more expensive than conventional image sensors, rendering them economically impractical for many manufacturers of integrated systems. Current systems only provide temperature measurements based on the overall facial outline or non-specific facial regions. Specific facial areas are not considered or detected, making it impossible to determine temperature reliably and consistently in the same facial regions. This is crucial since facial skin temperature can vary significantly [13,14]. Furthermore, various capabilities for a fully automated solution are lacking. Some solutions have relay outputs which are indirectly controlled via Wi-Fi, requiring additional peripheral electronics. Current systems offer limited capabilities for remote reconfiguration, such as adjusting calibration data or individual measurement logic. Some researchers are combining RGB and thermal imagery for various applications, such as traffic monitoring and interdisciplinary inventory [15–17].

Only a few state-of-the-art approaches employ inexpensive thermopile sensors [18,19]. Thermopile-based systems currently lack facial recognition and/or tracking capabilities due to their limited resolution.

Given this context, the objective of this paper is to present an automated approach for contactless and facial area-specific skin temperature measurement. This method relies on the unique combination of an inexpensive thermopile sensor array and a 2D image sensor. Temperature and 2D image data are fused using a parametric affine transformation. A special calibration target is designed to determine this transformation. Through facial recognition, specific facial areas can be assigned with temperature values, which are then used to determine the skin surface temperature. Algorithms for detecting facial features and fusing data from the thermopile sensor array and 2D image sensor are described. Furthermore, the distributed system architecture and its components are introduced. Finally, the feasibility of the approach is demonstrated by a small participant study and the results are discussed.

2. Materials and Methods

2.1. Thermopile Sensor and Data Readout

For this study, a thermopile sensor with 60 × 40 pixels (HTPA60 × 40, from Heimann Sensor GmbH, Dresden, Germany) was chosen. Thermopiles are temperature sensors based on thermocouple elements consisting of two different conductor materials. One junction is opposed to the thermal radiation, generating a voltage signal proportional to the temperature difference to the other junction by the Seebeck effect [20].

Our sensor is controlled by a custom-programmed microcontroller. The integrated program involves reading calibration data from the sensor, capturing sensor raw data, and transmitting this data to a mobile PC via USB transfer. The calibration information is sensor-specific (e.g., sensitivity coefficients, number of defective pixels, etc.) and is required for the accurate calculation of the object temperature, as well as the configuration of the sensor's clock frequency, ADC resolution, and the common mode voltage of the preamplifier. The calibration data are stored on an electrically erasable programmable read-only memory (EEPROM) in the sensor.

Since the object temperature calculation takes place on a mobile PC, the calibration data are transmitted once at the beginning of communication, while temperature raw data and other values (e.g., thermal drift) are continuously updated during processing.

The mobile PC polls the sensor for raw data and corrects them based on the calibration information. The sensor raw data either provide a reference voltage proportional to the absolute temperature (PTAT) and the active pixels raw data or the electrical offsets, depending on the readout command. The ambient temperature is calculated from the sensor average measured PTAT value and from EEPROM calibration variables, such as the PTAT gradient and the PTAT offset. The sensor pixels voltages are subjected to different compensations before they can be used to determine the object's temperature; initially, it is necessary to deduct the sensor's thermal offset from each pixel to counteract potential thermal drift. Additionally, the outcome of the thermal gradient multiplied by the PTAT average is adjusted by the scaling coefficient for the thermal gradient stored in the EEPROM. Next, the electrical offsets are subtracted to compensate for changes in the supply voltage. Then a second supply voltage compensation (VddComp) is performed using the supply voltage of the sensor (Vdd) which is measured internally. After that, the sensitivity coefficients are calculated using EEPROM data. Finally, the sensitivity-compensated pixel values are calculated by dividing the pixels voltages by the sensitivity coefficients.

The sensitivity-compensated pixels and the ambient temperature are both needed to calculate absolute temperature for each pixel with the help of a look-up table, provided by the manufacturer. The look-up table rows represent the range of ambient temperatures supported by the sensor, and the columns represent the temperature values. When mapping the two values, a bilinear interpolation calculates the absolute object temperature for each pixel. As a result, temperature data matrices (thermopile images) of the captured scene are generated.

The measured temperature is also dependent on the emissivity [20]. Charlton et. al. have shown that the emissivity for human skin is nearly constant for all skin types of the Fitzpatrick scale [21]. Thus, in the following, the emissivity of skin is assumed to be constant with a value of $\varepsilon = 0.972$, close to an ideal black body radiator. The result scales by $\sqrt[4]{\varepsilon} = \sqrt[4]{0.972} = 0.993$ (see [20]), meaning it is only slightly influenced by the skin color. However, the measured skin temperature fluctuates due to changing ambient and physiological conditions.

2.2. Thermopile Sensor Characterization

An artificial head was built to characterize the sensor (see Figure 1). The head is constructed of sheet metal and painted with black paint to mimic a black body radiator with an emissivity close to 1. The head includes heating resistors inside at the bottom plate. External electronics allow the head to be set to a targeted temperature. The setup is used to characterize the sensor noise, the signal-to-noise ratio (SNR), and the frame rate.

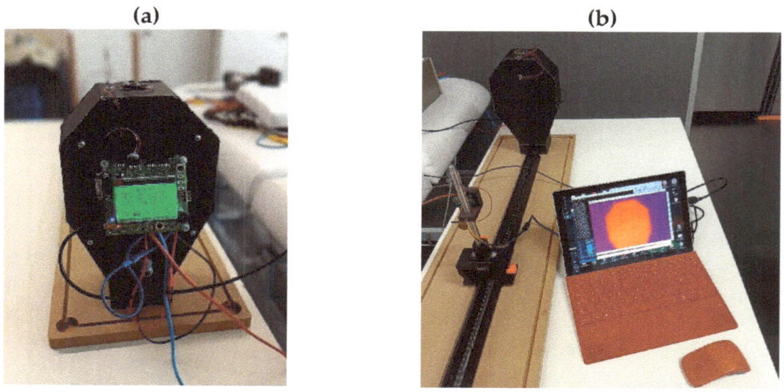

Figure 1. (**a**): backside of the artificial head showing the control electronics; (**b**): front side of the heated head, measured by the thermopile sensor. The artificial head provides a constant surface temperature and an emissivity similar to human skin.

The pixel noise measured as mean squared error (MSE) using the jth sensor image $S_j(x,y)$ with discrete co-ordinates (x,y) of a homogeneous area with constant temperature μ, is ([22]):

$$\text{MSE} = \frac{1}{MN}\sum_{x=1}^{M}\sum_{y=1}^{N}\left(S_j(x,y)-\mu\right)^2 \tag{1}$$

We extend this definition by taking a number of J temporal images into account:

$$\text{MSE} = \frac{1}{JMN}\sum_{j=1}^{J}\sum_{x=1}^{M}\sum_{y=1}^{N}\left(S_j(x,y)-\mu\right)^2 \tag{2}$$

The root mean squared error ($RMSE$) is the square root of Equation (2). For a single pixel, the $RMSE$ corresponds to the temporal standard deviation of that pixel.

The actual SNR is calculated using the signal (the constant temperature μ) and the MSE ([22]):

$$\mathrm{SNR} = 10 \cdot \log_{10} \frac{\mu^2}{MSE} = 20 \cdot \log_{10} \frac{\mu}{RMSE} \quad (3)$$

For correlation analysis, we use the definition of the correlation coefficient r for two discrete lists of values, p_i and q_i, and their respective averages, $\overline{p}$ and $\overline{q}$ [23]:

$$r = \frac{\Sigma((p_i - \overline{p}) \cdot (q_i - \overline{q}))}{\sqrt{\Sigma(p_i - \overline{p})^2 \cdot \Sigma(q_i - \overline{q})^2}} \quad (4)$$

2.3. Multimodal Sensor Setup and System Calibration

The imaging part of our system is comprised of the 60 × 40 thermopile sensor and a 2D color camera. The thermopile sensor array and the 2D image sensor are positioned closely together and are mechanically fixed. Although the camera is full-HD capable, we use a resolution of 600 × 400 in this study to have a better alignment to the thermopile resolution.

Both sensors image the same scene from a slightly different perspective. Consequently, calibration procedures used for stereo imaging might seem obvious. However, due to the completely different wavelength regions (~10 µm for the thermal sensor and 0.4 µm–0.7 µm for the visible range), commonly used methods fail. As the two sensors are based on different principles, have different resolutions, have different fields of view, and provide different types of data, feature-based algorithms cannot be applied. Intensity-based image approaches fail as the thermopile sensor provides temperature data and does not measure the scene's visible light intensity.

We propose a modified calibration approach here. We identify correspondences between the data from the two sensors based on a contrast-rich calibration scene with distinct features. A custom calibration device is developed. It is used to generate circular features that exhibit significant temperature variations compared to the surroundings and emit light at the same time. This approach allows both sensors to capture these circular features with sufficient SNR for subsequent calibration algorithms. Figure 2 illustrates the sensor and calibration target.

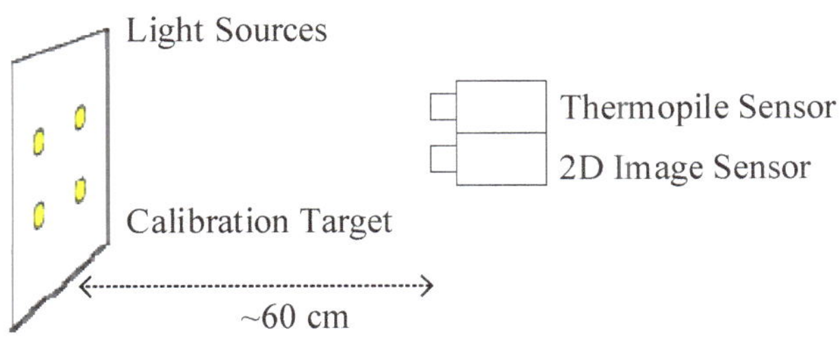

Figure 2. A thermopile sensor and a 2D color sensor image the calibration target.

The calibration target consists of a vertically oriented surface with four integrated, self-heated light sources. The two sensors are aligned to the calibration target at a distance of approximately 60 cm. Figure 3 illustrates the images captured by each respective sensor. The temperature data from the thermopile sensor are color coded.

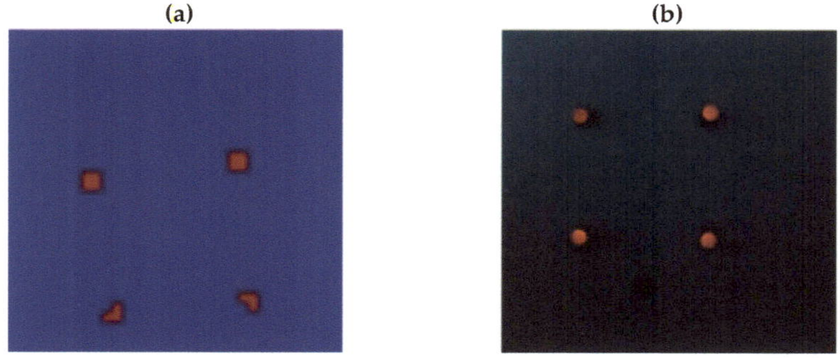

Figure 3. Image data with circular features, captured by the thermopile sensor (**a**), and by the 2D color image sensor (**b**).

The displayed circular features can be detected and matched to each other. The image from the 2D image sensor serves as reference, and corresponding circular features are searched in the image data from the thermopile sensor. Based on this information, an equation for a parametric affine transformation can be determined, taking into account scaling, rotation, shear, and translation of the data. In this manner we obtain a transformation, which allows us to match the thermopile image with the rgb image.

The steps of the algorithm for determining corresponding circular features and calculating a transformation matrix are in detail:

1. capturing the thermopile image I_t and the camera image I_c;
2. converting I_t and I_c into binary images to remove redundant information;
3. detecting the contours of the circular features in I_t and I_c;
4. determining the center points of the circular features, I_{t0}, I_{t1}, I_{t2}, and I_{t3} in I_t, as well as I_{c0}, I_{c1}, I_{c2}, and I_{c3} in I_c;
5. spatially sorting the detected center points from I_t and I_c to ensure correct correspondence;
6. calculating the real-valued coefficients a_0, a_1, a_2, b_0, b_1, and b_2 of the transformation matrix based on the centers of the circular features and the linear transformation equations:

$$x_c = a_0 x_t + a_1 y_t + a_2$$
$$y_c = b_0 x_t + b_1 y_t + b_2$$

Result: transformation matrix $T = \begin{bmatrix} a_0 & a_1 & a_2 \\ b_0 & b_1 & b_2 \\ 0 & 0 & 1 \end{bmatrix}$, which is used to transform the co-ordinates of a data point from the thermopile sensor into the co-ordinate system of the 2D image sensor.

The described algorithm is applied once before using the skin temperature measurement to determine the transformation matrix T. The calibration remains valid as long as the alignment between the two sensors is not altered. Thereafter, it can be employed within the sensor data fusion in combination with facial region detection and tracking, as described in the following section.

2.4. Skin Temperature Measurements and Signal Processing

The temperature of the facial skin can vary significantly in different areas, i.e., by more than 1 °C [13,14]. Therefore, for accurate determination of skin temperature, especially across different individuals, it is crucial to conduct targeted and consistent temperature measurements in specific facial regions.

Furthermore, during the measurement of a person's temperature, it is essential to ensure that head movements do not distort the temperature measurement result. To detect facial contour points and specific facial regions, available state-of-the-art methods can be used [24]. Therefore, facial contour tracking is employed for dynamically adjusting the temperature determination.

Figure 4 shows how this facial contour tracking is done. The face contour is detected with the help of Mediapipe Face Mesh, a machine learning framework provided by Google. Face Mesh determines characteristic landmarks within the face, making it possible to identify eyes, mouth, nose, and also forehead. The forehead region (red quadrilateral in Figure 4) is selected by using the proper Face Mesh nodes (node numbers 68, 103, 297, and 333 are used in this work). The red quadrilateral covers an area of approximately of 8 cm × 1.5 cm = 12 cm^2, measured at a working distance of 50 cm.

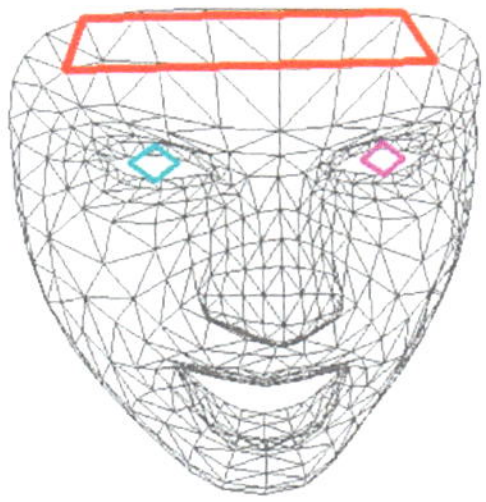

Figure 4. Mediapipe Face Mesh map of a person's face detected by the color sensor. Each line in the image connects one of the 478 nodes. The eyes are marked by colored squares. The red quadrilateral marks the measured area on the forehead.

With the help of the transformation matrix T, the thermopile image is transformed to match the rgb image. All mapped thermopile temperature values within the forehead region are averaged spatially. The landmarks are continuously tracked, even during slight face movements. Thus, it is possible to identify the same area within consecutive frames. These areas are then also averaged temporally over a time interval (e.g., 1 s, 5 s), depending on the settings of the software. All steps are performed in real time, which allows for continuous detection and tracking of the corresponding region. Ultimately, temperature data for the entire facial area are always available during the measurement process and can be evaluated accordingly.

The steps of the algorithm for combined sensor data fusion with facial region detection and tracking are, in detail:

1. capturing the thermopile image I_t and the camera image I_c;
2. determining the transformed image $I_{t'}$ from I_t. For each point (x_t and y_t) in I_t, the following transformation equation applies:

$$\begin{bmatrix} x' \\ y' \\ 1 \end{bmatrix} = T \begin{bmatrix} x_t \\ y_t \\ 1 \end{bmatrix}$$

   The transformation matrix T is obtained from the initial sensor calibration shown above; x' and y' represent the transformed points in $I_{t'}$;
3. identifying the forehead landmark points G_c based on I_c (using Face Mesh);
4. estimating the temperature data within the area enclosed by G_c using $I_{t'}$;
5. spatially averaging the temperature data from $I_{t'}$ within the respective area;
6. temporally averaging the temperature data from $I_{t'}$ within the respective area;
7. visualization of the fused image with corresponding facial features in I_c.

2.5. Distributed System Architecture and Components

This section presents the system architecture used. It is designed to be fundamentally re con figurable, enabling flexible adaptation to different requirements of various ap pli cation scenarios. The corresponding system architecture consists of five fundamental components:

- mobile PC with an integrated touchscreen [25];
- thermopile sensor: Heimann HTPA60 × 40d sensor with ARM Cortex M0 (Pyramid Computer GmbH, Freiburg, Germany);
- binocular camera with two lenses: 2MP AI dual lens camera module (1920 × 1080, RGB and IR camera, Hampo Electronic Technology, Dongguan, China);
- electronic relays which can be connected to further actuators;
- RFID reader;
- edge server system for providing configuration parameters.

In this context, the mobile PC plays a central role. Besides providing an interactive display to visualize the measurement process, it takes charge of the entire measurement and evaluation logic. The thermopile sensor with an integrated microcontroller is connected to the mobile PC via a USB serial interface. The processing of raw sensor data is conducted in real time on the mobile PC, using the Python programming language, along with the software frameworks OpenCV (v 4.6.0) and Google Mediapipe (v 0.8.2). The acquisition of the raw sensor data from the thermopile sensor on the ARM Cortex-M0 is accomplished using the C programming language. Additionally, the system architecture can be optionally incorporated into a broader cloud/edge system. This enables remote and location-independent adjustments of both (sensor-specific) configuration parameters and individual measurement logic. Furthermore, the mobile PC features an RFID reader and universal switching outputs (relays) which can be utilized to control automated processes such as actuators based on temperature measurements for access control tasks.

3. Results

3.1. Sensor Characteristics

The sensor was characterized by imaging the artificial head, which was heated to a constant temperature of ~33 °C. Figure 4 shows the mean and standard values of the thermopile sensor over a time interval of 5 s.

The averaged temperature values in Figure 5a are showing a fixed pattern noise imposed on the image. The temperature values of the head slightly increase towards the lower end where the heating resistors are located. The pixels in the upper, more homogeneous part show a pixel noise between $\sigma = 0.45\ °C - 0.7\ °C$ (Figure 5b), on average approximately $\sigma = 0.53\ °C$, which is quite high for most applications.

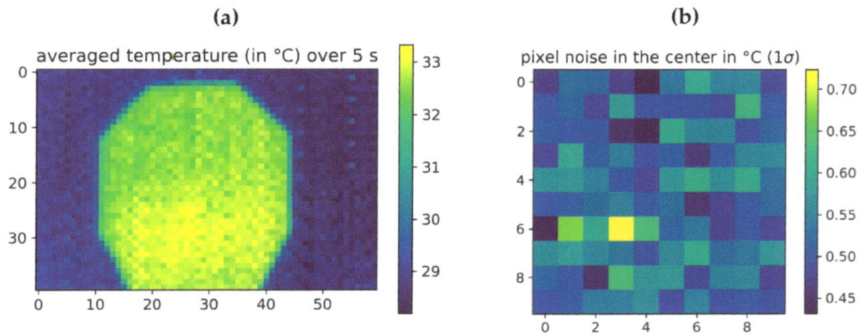

Figure 5. Averaged temperature image (**a**) and temporal standard deviation of 10 × 10 pixels in the upper homogeneous region of the image (**b**). Both images were taken over a time interval of 5 s. The frame rate of the thermopile sensor is 22 frames per second. The exposure time of the sensor is not controllable by us, it is set internally in the sensor.

Pixel noise, determined according to Equation (2), can be improved by spatial and temporal averaging, shown in Figure 6. The noise can be decreased by spatial averaging, assuming a homogeneous surface area is measured. The noise is reduced by 50% by averaging 4 pixels (see Figure 6). By averaging a few dozen pixels, the statistical noise is dramatically reduced, also eliminating variations due to fixed pattern noise. A further noise reduction is possible by additional temporal averaging. The statistical temperature noise reaches levels of $\sigma = 0.01$ °C by averaging over 64 pixels for 5 s. The SNR of a single temperature pixel is 36.6 dB. Averaging 64 pixels over 5 s increases the SNR value to ~70 dB. In this manner, spatial and temporal averaging allows for precise temperature measurements even with noisy thermopile sensors.

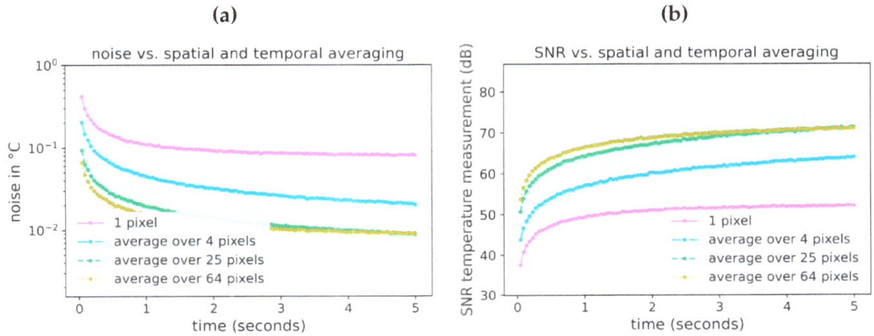

Figure 6. Effect of spatial and temporal averaging on noise (**a**) and SNR (**b**). The noise and SNR values are obtained and calculated from the measurement of the artificial head shown in Section 2.2 with a constant surface temperature μ. Noise values are calculated as RMSE for pixel areas of different sizes. Temporal averaging is done by averaging successive frames over a given time span. SNR values are calculated according to Equation (3).

3.2. System Characteristics

Figure 7a illustrates the overall implemented system. The mobile PC is equipped with an integrated camera and the thermopile sensor is added on top. The system prototype can be connected via an electronic relay to an actuated door, which served as a use-case scenario for temperature-based access control.

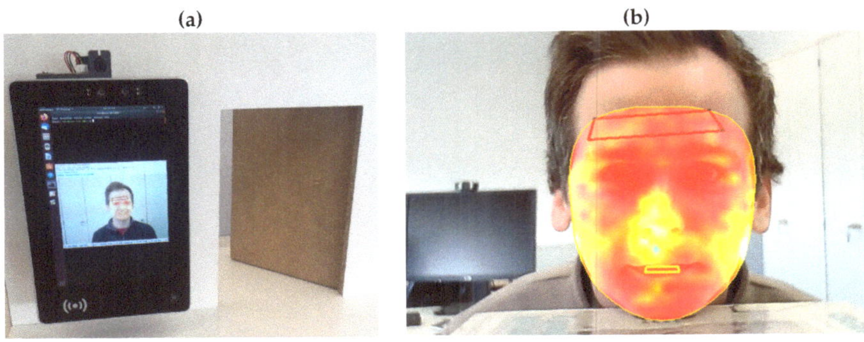

Figure 7. (**a**): System prototype; (**b**): exemplary illustration of the combined sensor data fusion with facial region detection and tracking.

The Figure 7b shows the detection of both the facial contour and specific facial regions, such as the forehead. The face is tracked continuously. The skin surface temperature is visualized through a colormap. The red quadrilateral shows the area used for determining forehead temperature. The region is identified with the help of Face Mesh and mapped to the thermopile image, consisting of roughly 25 thermopile pixels. The pixels are spatially averaged to one temperature value. Furthermore, the area is tracked during several seconds

(5 s, in this case) and averaged over this time interval to the final temperature value. The system is capable of performing all necessary calculations in real time.

Inside and outside temperature as well as physical activities influence the measured skin temperature. We performed some qualitative measurements and found that the ambient temperature has a correlation coefficient of $r \approx 0.35$ (calculated by Equation (4)). The distance to the subject also influences the result and has a correlation coefficient of $r \approx 0.27$. However, these influences are difficult to reproduce.

Foreign objects located near the face or covering specific facial areas can adversely affect the measurement of skin temperature. Examples of such objects include wearing a mask or glasses. The proposed approach enables the indirect detection of these objects by assuming that skin areas covered by a foreign object have lower temperatures than uncovered areas. In this way, foreign objects can be identified by detecting cooler temperature regions on the face. The described principle is exemplified in Figure 8.

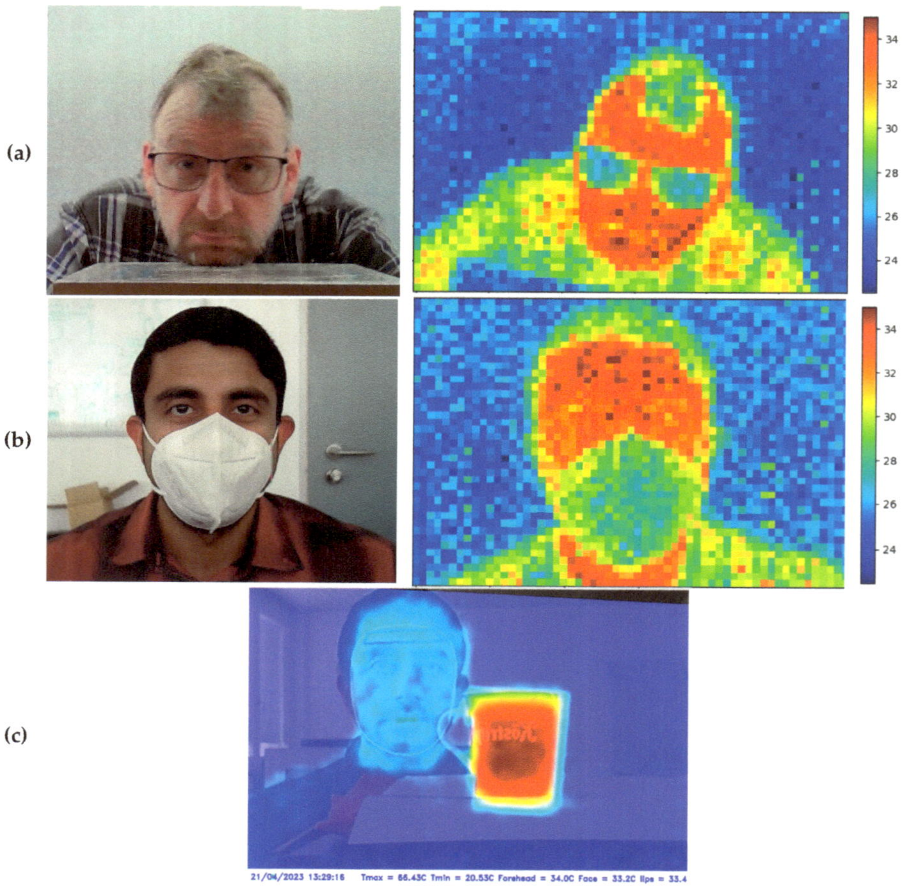

Figure 8. Temperature measurements and identification of foreign objects, such as glasses, mask, or hot objects. (**a**,**b**): Each image pair shows the rgb image and the corresponding thermopile image before registration. (**c**): The image shows the fused rgb and thermopile image after registration.

To detect specific foreign objects, certain facial regions (e.g., chin, mouth, nose, and cheek areas, corresponding to the position of a mask) can be defined. Each facial region corresponds to a set of temperature values. To identify a foreign object in a facial region, the number of temperature values is counted which deviate from the expected temperature range. If the number is too high, the person is asked to remove any objects present in the facial region. Otherwise, temperature values below the defined threshold are filtered out and are not considered in the temperature measurement.

Even larger objects do not disturb the temperature measurement. The last example of Figure 8 shows that despite the presence of a hot cup with a temperature of approximately 66 °C, the face is recognized, and the surface temperature of the forehead (34 °C) is accurately determined.

3.3. Participant Study and Quantitative Measurement Results

To validate the temperature measurement approach, especially regarding the surface forehead temperature, a participant study was conducted. The system was calibrated as described in Section 2.3. The calibrated system was validated by holding a hot object (e.g., a hand or a cup) at different points in the image and ensuring that both image sources matched in the fused image. The temperature data were compared to those produced by a commercially available, manually operated forehead thermometer (Medisana TM A79). The study involved five participants with skin types I, II, and V according to the Fitzpatrick scale. Five measurements were conducted for each individual under room temperature conditions (22 °C). For better repeatability, all measurements were conducted without any physical activities of the participants. All subjects had approximately the same distance of 50 cm–60 cm to the sensor. The participants remained immobile to provide more stable results. Approximately 50 images from the thermopile sensor were captured and subsequently spatially as well as temporally averaged over a five-second period. The reference measurements with the forehead thermometer were performed manually.

For each participant, the time, reference temperature as well as thermopile temperature were measured. The five participants were measured one after each other, and the procedure was repeated five times. The raw data of the thermopile system were corrected by the emissivity factor of 0.99, as mentioned in Section 2.1. The average values and the standard deviations have been calculated for each participant, as shown in Table 1. The standard deviation fluctuates approximately between 0.07 °C and 0.24 °C for the thermopile and reference measurement. The mean 1σ deviation of the reference measurement is $\sigma_{ref} = 0.14\ ^\circ\text{C}\ (2\sigma_{ref} = 0.28\ ^\circ\text{C})$ and $\sigma_{tp} = 0.12\ ^\circ\text{C}\ (2\sigma_{tp} = 0.24\ ^\circ\text{C})$ for the thermopile sensor.

Table 1. Average and standard deviation for all study participants. For each person, five measurements were performed and used for calculating mean and standard deviation (1σ).

Participants	μ_{ref} (°C)	σ_{ref} (°C)	μ_{tp} (°C)	σ_{tp} (°C)
1	34.06	0.174	33.64	0.116
2	33.9	0.237	33.71	0.242
3	35.1	0.071	34.81	0.116
4	33.82	0.133	33.83	0.074
5	33.36	0.08	33.45	0.07

The results of the measurements are presented in Figure 9, illustrating a linear correlation with a slope of approximately 1 (after correction for skin emissivity) between the data from the commercially available forehead thermometer and the temperature data by our setup. The correlation coefficient between the reference and thermopile measurements is calculated to be $r = 0.92$, indicating a very strong correlation between those values. The RMSE (compare Equation (2)) between the corrected dataset and reference measurement is calculated to be RMSE = 0.22 °C, indicating that the measured absolute values are comparable to those of the commercial reference system.

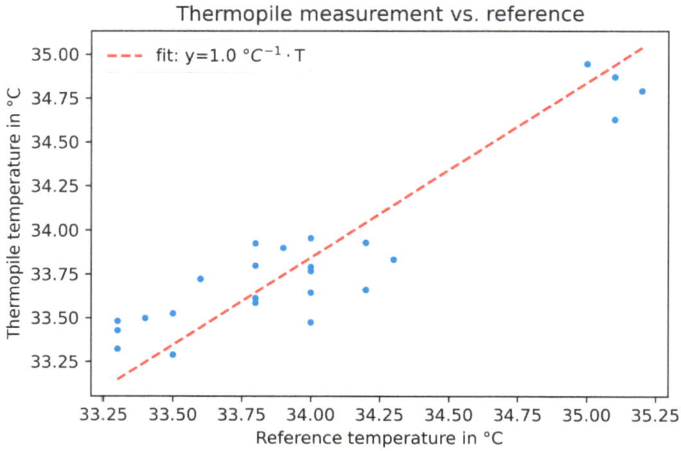

Figure 9. Measurements of the forehead temperature with the Medisana TM reference device compared to our thermopile sensor. A linear correlation between the measurements was found with a fitted slope of 1.0. The emissivity of skin was assumed to be $\varepsilon = 0.972$.

4. Discussion

Our thermopile sensor has a native pixel noise of approximately 0.53 °C. This noise value is higher than most manually operated temperature devices. However, through spatial and temporal averaging, the noise value can be brought into a subordinate range of <<0.1 °C. Averaging also increases the SNR accordingly.

While the ratio between thermopile measurements and reference measurements closely approximates 1 in our scenario, we recommend conducting an initial validation of this factor. A low-cost generic thermometer produced results that differed slightly from those obtained with the Medisana reference thermometer. According to our findings, a sample size of 3–5 individuals is sufficient to check the quality and the scaling factor of the thermometer.

In the participant study, a temperature noise of $\sigma = 0.12$ °C was measured on average for the individual participants. The temperature noise observed in this study is higher than statistically anticipated. This means, that this noise is probably caused by systematic deviations due to, e.g., fluctuations in skin temperature, blood perfusion, varying environmental conditions, or movements of the person which are not corrected properly by the face tracking. We did not observe relevant differences in the temperature measurements for different skin phototypes according to the Fitzpatrick scale. This finding is also supported by other studies [21].

In our study, we mostly used controlled conditions. For example, the distance was not varied, the subjects were staying at room temperature, and did not move excessively. Prior to the participant study, we qualitatively investigated skin temperature changes. We found that temperature slightly increases over distance and is also dependent on ambient temperature conditions, e.g., coming from the cold outside. This problem affects not only our measurement methodology, but all non-contact methods based on thermal imaging.

Our system can be used, for example, for contactless temperature measurements in combination with an access control system. It can be connected directly to further actuators, such as electronic doors, via an electronic relay interface. We were able to implement such a temperature-based access control connected to a door as a use case scenario in the lab. The covid pandemic has shown that such systems are needed at the entrance to critical areas, like airports, hospitals, nursing homes, or large buildings, such as residential complexes.

However, further studies are necessary to determine the temperature under more difficult conditions, such as wearing reflective glasses, face masks, headpieces, scarfs or under changing ambient conditions. Moreover, the accuracy of the measurement can decrease as the 2D image sensor and thermopile sensor move farther away from the face.

This is because the captured forehead area in the 2D image covers a smaller area with fewer pixels due to the increased distance. As a consequence, spatial and temporal averaging might yield less accurate results. A higher spatial resolution of the thermopile sensor could potentially reduce this effect and enable better registration results. Additionally, further noise suppressing techniques, such as bandpass filtering, 2D image filters, etc., might be helpful to reduce noise. Furthermore, our system features a second, independent infrared camera, which could be beneficial for operating the system under low light conditions.

Additional work is also needed to understand external influences to skin temperature. This is a challenging task as it is necessary to understand and monitor more parameters in the system. A further aspect might be to include heart rate monitoring [26], remote photoplethysmography [26], or blood perfusion detection [27]. Integrating these parameters into the analysis, along with temperature measurements from extremities like arms, hands, and toes, could lead to a holistic understanding of the body's response to different external stimuli. This knowledge might be used to infer core body temperature, which is an important vital sign and used commonly for diagnosing fever. Ultimately, leveraging inverse heat transfer methods [28–30] in conjunction with these physiological indicators could significantly enhance the prediction of the thermal state of the human subject. Inverse heat transfer relies on simulation models, working backward from temperature measurements to the boundary conditions or heat sources that could have caused those measurements. This might offer insights into inferring the core body temperature accurately.

5. Conclusions

This paper presents an approach for automated, contactless, and region-specific measurement of skin surface temperature on the face. The method is based on data fusion from a thermopile sensor and a 2D image sensor. By capturing and tracking the facial outline, specific temperature values are assigned to selected facial areas, which are used to determine the skin surface temperature accurately.

The application of the proposed approach was demonstrated. In qualitative terms, the participation study has shown that facial capture and tracking reduce the susceptibility to errors in temperature measurement, particularly when foreign objects are in close proximity to the face. In quantitative terms, the subject study demonstrated that the measured temperature absolute values have an RMSE of 0.22 °C, rendering them comparable to those of a commercially available, manually operated reference system. As such, our system holds the potential to become a valuable tool in the future for accurate and automated non-contact temperature measurements.

Author Contributions: Conceptualization, F.B., G.E., S.K. and T.G.; methodology, F.B., G.E., J.H., C.K., A.R., S.K. and T.G.; software, F.B., G.E., J.H. and C.K.; validation, F.B., G.E., J.H., C.K., A.R., S.K. and T.G.; formal analysis, F.B., G.E. and S.K.; investigation, F.B., G.E., J.H., C.K., A.R., S.K. and T.G.; resources, F.B., G.E., J.H., A.R., S.K. and T.G.; data curation, F.B., G.E. and S.K.; writing—original draft preparation, F.B., G.E., C.K. and S.K.; writing—review and editing, F.B., G.E., C.K, S.K. and T.G.; visualization, F.B., G.E., J.H., A.R., S.K. and T.G.; supervision, S.K. and T.G.; project administration, S.K. and T.G.; funding acquisition, S.K. and T.G. All authors have read and agreed to the published version of the manuscript.

Funding: This research was funded under the Central Innovation Programme for Small and Medium-Sized Enterprises (ZIM) by the Federal Ministry for Economic Affairs and Climate Action (grant number KK5314702CR1).

Institutional Review Board Statement: Ethical review and approval were waived for this study because it solely addresses data protection issues, for which we are responsible, aligning with Pforzheim University's ethical guidelines.

Informed Consent Statement: Informed consent was obtained from all subjects involved in the study.

Data Availability Statement: The data are not publicly available due to confidentiality rules.

Conflicts of Interest: The authors declare no conflict of interest.

References

1. Putri, N.K.; Ogi, D. Implementation of PRESENT Algorithm on Contactless Access Control Using Raspberry Pi. In Proceedings of the 2022 International Conference on Informatics, Multimedia, Cyber and Information System (ICIMCIS), Jakarta, Indonesia, 16–17 November 2022; pp. 213–218. [CrossRef]
2. Jadhav, A.; Khadse, V. Real-Time Face Mask Detection and Contactless Body Temperature Measurement on Edge Device Using Deep Learning. In Proceedings of the 2022 4th International Conference on Advances in Computing, Communication Control and Networking (ICAC3N), Greater Noida, India, 16–17 December 2022; pp. 1111–1116. [CrossRef]
3. Bhogal, R.K.; Potharaju, S.; Kanagala, C.; Polla, S.; Jampani, R.V.; Yennem, V.B.R. Corona Virus Disinfectant Tunnel Using Face Mask Detection and Temperature Monitoring. In Proceedings of the 2021 5th International Conference on Intelligent Computing and Control Systems (ICICCS), Madurai, India, 6–8 May 2021; pp. 1704–1709. [CrossRef]
4. Girinath, N.; Swathi, R.; Swedha, N.; Nisarga, V. Face Mask Detection with Contactless Temperature Using MATLAB. In Proceedings of the 2022 3rd International Conference on Smart Electronics and Communication (ICOSEC), Trichy, India, 20–22 October 2022; pp. 1536–1539. [CrossRef]
5. Ampex. Soluzioni Tecnologiche per La Misurazione Temperatura Corporea. Available online: http://www.ampex.tn.it/covid-19/brochure_termoscanner.pdf (accessed on 3 September 2023).
6. Grupocartroni. Sistema IA de Reconocimiento Facial y Detección de Temperatura. Available online: https://grupocartronic.com/wp-content/uploads/2020/07/Datasheet-Sistema-de-medici%C3%B3n-Infinilink-INF-MBPC-K304.pdf (accessed on 3 September 2023).
7. Tyalux. Access Control Dispaly. Available online: https://tyalux.com/access-control-dispaly-8-inch/ (accessed on 3 September 2023).
8. ADM Electronic. Corona Zugangskontrolle. Available online: https://www.adm-electronic.de/zutrittskontrolle-corona/ (accessed on 3 September 2023).
9. Elotouch. Einfachere Zugangskontrolle Und Besucherverwaltung. Available online: https://www.elotouch.de/elo-access (accessed on 3 September 2023).
10. Bhowmik, T.; Mojumder, R.; Banerjee, I.; Das, G.; Bhattacharya, A. IoT Based Non-Contact Portable Thermal Scanner for COVID Patient Screening. In Proceedings of the 2020 IEEE 17th India Council International Conference (INDICON), New Delhi, India, 10–13 December 2020; pp. 1–6. [CrossRef]
11. Paramasivam, S.; Shen, C.H.; Zourmand, A.; Ibrahim, A.K.; Alhassan, A.M.; Eltirifl, A.F. Design and Modeling of IoT IR Thermal Temperature Screening and UV Disinfection Sterilization System for Commercial Application Using Blockchain Technology. In Proceedings of the 2020 IEEE 10th International Conference on System Engineering and Technology (ICSET), Shah Alam, Malaysia, 9 November 2020; pp. 250–255. [CrossRef]
12. Abirami, M.; Saundariya, K.; Senthil Kumaran, R.; Yamuna, I. Contactless Temperature Detection of Multiple People and Detection of Possible Corona Virus Affected Persons Using AI Enabled IR Sensor Camera. In Proceedings of the 2021 Sixth International Conference on Wireless Communications, Signal Processing and Networking (WiSPNET), Chennai, India, 25–27 March 2021; pp. 166–170. [CrossRef]
13. Lee, C.M.; Jin, S.P.; Doh, E.J.; Lee, D.H.; Chung, J.H. Regional Variation of Human Skin Surface Temperature. *Ann. Dermatol.* **2019**, *31*, 349–352. [CrossRef] [PubMed]
14. Kim, J.-H.; Seo, Y.; Quinn, T.; Yorio, P.; Roberge, R. Intersegmental differences in facial warmth sensitivity during rest, passive heat and exercise. *Int. J. Hyperth. Off. J. Eur. Soc. Hyperthermic Oncol. N. Am. Hyperth. Group* **2019**, *36*, 653–658. [CrossRef] [PubMed]
15. Ben-Shoushan, R.; Brook, A. Fused Thermal and RGB Imagery for Robust Detection and Classification of Dynamic Objects in Mixed Datasets via Pre-Trained High-Level CNN. *Remote. Sens.* **2023**, *15*, 723. [CrossRef]
16. Alldieck, T.; Bahnsen, C.H.; Moeslund, T.B. Context-Aware Fusion of RGB and Thermal Imagery for Traffic Monitoring. *Sensors* **2016**, *16*, 1947. [CrossRef] [PubMed]
17. Paziewska, J.; Rzonca, A. Integration of Thermal and RGB Data Obtained by Means of a Drone for Interdisciplinary Inventory. *Energies* **2022**, *15*, 4971. [CrossRef]
18. Tang, H.-F.; Hung, K. Design of a Non-Contact Body Temperature Measurement System for Smart Campus. In Proceedings of the 2016 IEEE International Conference on Consumer Electronics-China (ICCE-China), Guangzhou, China, 19–21 December 2016; pp. 1–4. [CrossRef]
19. Moisello, E.; Vaiana, M.; Castagna, M.E.; Bruno, G.; Malcovati, P.; Bonizzoni, E. An Integrated Micromachined Thermopile Sensor with a Chopper Interface Circuit for Contact-Less Temperature Measurements. *IEEE Trans. Circuits Syst. I Regul. Pap.* **2019**, *66*, 3402–3413. [CrossRef]
20. Moisello, E.; Malcovati, P.; Bonizzoni, E. Thermal Sensors for Contactless Temperature Measurements, Occupancy Detection, and Automatic Operation of Appliances during the COVID-19 Pandemic: A Review. *Micromachines* **2021**, *12*, 148. [CrossRef] [PubMed]
21. Charlton, M.; Stanley, S.A.; Whitman, Z.; Wenn, V.; Coats, T.J.; Sims, M.; Thompson, J.P. The effect of constitutive pigmentation on the measured emissivity of human skin. *PLoS ONE* **2020**, *15*, e0241843. [CrossRef] [PubMed]
22. Sohag, S.A.; Islam, M.K.; Islam, M.B. A Novel Approach for Image Steganography Using Dynamic Substitution and Secret Key. *Am. J. Eng. Res. (AJER)* **2013**, *2*, 118–126. Available online: https://www.ajer.org/papers/v2(9)/Q029118126.pdf (accessed on 3 September 2023).

23. Al-Fahoum, A.; Harb, B. A Combined Fractal and Wavelet Angiography Image Compression Approach. *Open Med. Imaging J.* **2013**, *7*, 9–18. [CrossRef]
24. Khabarlak, K.; Koriashkina, L. Fast Facial Landmark Detection and Applications: A Survey. *J. Comput. Sci. Technol.* **2022**, *22*, e02. [CrossRef]
25. Faytech. 10.1″ Thermal Temperature Testing PC. Available online: https://www.faytech.com/product/special-products/10-1-thermal-temperature-testing-pc/ (accessed on 3 September 2023).
26. Premkumar, S.; Hemanth, D.J. Intelligent Remote Photoplethysmography-Based Methods for Heart Rate Estimation from Face Videos: A Survey. *Informatics* **2022**, *9*, 57. [CrossRef]
27. Kumar, M.; Suliburk, J.W.; Veeraraghavan, A.; Sabharwal, A. PulseCam: A camera-based, motion-robust and highly sensitive blood perfusion imaging modality. *Sci. Rep.* **2020**, *10*, 4825. [CrossRef] [PubMed]
28. Singh, M.; Flores, H.; Ma, R.; Zhu, L. Extraction of Baseline Blood Perfusion Rates in Mouse Body and Implanted PC3 Tumor Using Infrared Images and Theoretical Simulation. In Proceedings of the Summer Biomechanics, Bioengineering and Biotransport Conference, Virtual, 17–20 June 2020. Available online: http://hdl.handle.net/11603/25295 (accessed on 3 September 2023).
29. Singh, M.; Turnbaugh, B.; Ma, R.; Zhu, L. Managing Cooling Penetration and Minimizing Systemic Hypothermia after Surgery Using a Cooling Pad–Whole Body Heat Transfer Simulation. In Proceedings of the Summer Biomechanics, Bioengineering and Biotransport Conference, Virtual, 17–20 June 2020. Available online: http://hdl.handle.net/11603/25296 (accessed on 3 September 2023).
30. Caporale, A.; Lombardo, J.; Singh, M.; Zhu, L. Development of an Empirical Correlation to Predict Cooling Penetration into Tissue for Practical Use. In Proceedings of the Summer Biomechanics, Bioengineering and Biotransport Conference, Cambridge, MD, USA, 20–23 June 2022. Available online: http://hdl.handle.net/11603/25595 (accessed on 3 September 2023).

sensors

MDPI

Article

Real-Time Embedded Eye Image Defocus Estimation for Iris Biometrics

Camilo A. Ruiz-Beltrán, Adrián Romero-Garcés, Martín González-García, Rebeca Marfil and Antonio Bandera *

Departamento Tecnologia Electronica, ETSI Telecomunicacion, University of Málaga, 29071 Málaga, Spain; camilo@uma.es (C.A.R.-B.); argarces@uma.es (A.R.-G.); martin@uma.es (M.G.-G.); rebeca@uma.es (R.M.)
* Correspondence: ajbandera@uma.es

Abstract: One of the main challenges faced by iris recognition systems is to be able to work with people in motion, where the sensor is at an increasing distance (more than 1 m) from the person. The ultimate goal is to make the system less and less intrusive and require less cooperation from the person. When this scenario is implemented using a single static sensor, it will be necessary for the sensor to have a wide field of view and for the system to process a large number of frames per second (fps). In such a scenario, many of the captured eye images will not have adequate quality (contrast or resolution). This paper describes the implementation in an MPSoC (multiprocessor system-on-chip) of an eye image detection system that integrates, in the programmable logic (PL) part, a functional block to evaluate the level of defocus blur of the captured images. In this way, the system will be able to discard images that do not have the required focus quality in the subsequent processing steps. The proposals were successfully designed using Vitis High Level Synthesis (VHLS) and integrated into an eye detection framework capable of processing over 57 fps working with a 16 Mpixel sensor. Using, for validation, an extended version of the CASIA-Iris-distance V4 database, the experimental evaluation shows that the proposed framework is able to successfully discard unfocused eye images. But what is more relevant is that, in a real implementation, this proposal allows discarding up to 97% of out-of-focus eye images, which will not have to be processed by the segmentation and normalised iris pattern extraction blocks.

Keywords: eye detection; Haar-like features; convolution kernels; defocus test; Ultrascale+ MP SoC

Citation: Ruiz-Beltrán, C.A.; Romero-Garcés, A.; González-García, M.; Marfil, R.; Bandera, A. Real-Time Embedded Eye Image Defocus Estimation for Iris Biometrics. *Sensors* **2023**, *23*, 7491. https://doi.org/10.3390/s23177491

Academic Editor: Wataru Sato

Received: 22 July 2023
Revised: 14 August 2023
Accepted: 25 August 2023
Published: 29 August 2023

1. Introduction

Biometric identification by iris recognition is based on the analysis of the iris pattern using mathematical techniques. Although it is a relatively recent technique (the first automatic identification system was developed and patented by John Daugman in the last decade of the 20th century), its excellent identification characteristics have led to its rapid evolution. Thus, using the most recent developments, it has become a mature technique. The challenge now is to use it in a scenario where the cooperation of the person is not required to obtain a focused image of the eye but where the person can be allowed to continue walking, keeping the image sensor at a distance of more than 1 m from the person's face. In iris recognition at a distance (IAAD) systems [1], it is common to use a camera that, thanks to its large field of view (FoV), detects and tracks the face of people approaching the system, and to have several high-resolution iris cameras, with a narrower FoV, that move according to what is determined by the first camera, in order to capture the image that will have the irises to be processed. These systems will therefore employ pan–tilt and control units. As the person is moving, capturing an image of the iris with the appropriate quality in contrast and resolution will require predicting where the person's face will be at each instant to capture a quality image [2]. However, in cases where the field of view to be covered is not too large, such as an access controlled point, a static system, in which a single high-resolution camera is located, can be used. In this situation, the system shall be able to process these high-resolution images at high speed [3]. The need to capture high-resolution images is imposed by the maintenance of a certain FoV, which would otherwise be too low. On the other hand, having to capture a large number of frames per second is due to the fact that, under normal circumstances, the depth of field of the camera (i.e., the distance around the image plane for which the image sensor is focused [4]) will be very shallow [1]. Because the person to be identified is moving, it is difficult to time the camera shutter release to coincide with the moment when the iris is in the depth of field and therefore in focus. Only by processing many images per second can the system try to ensure that one of the images has captured this moment.

Capturing and pre-processing a large volume of input images requires the use of a powerful edge-computing device. Currently, the options are mainly in the form of graphics processing units (GPUs), application-specific integrated circuits (ASICs), or field-programmable gate arrays (FPGAs). Due to the high power consumption and size of GPUs, and the low flexibility of ASICs, FPGAs are often the most interesting option [5,6]. Moreover, if the traditional development approach of FPGAs using low-level hardware languages (such as Verilog and VHDL) is usually time-consuming and very inefficient, the use of high-level language synthesis (HLS) tools allows developers to program hardware solutions using C/C++ and OpenCL. This significantly improves the efficiency in FPGA developments [7,8]. Finally, FPGAs are nowadays integrated in multi-processor system-on-chips (MPSoC), in which computer and embedded logic elements are combined. These MPSoCs thus offer the acceleration capabilities of the FPGA and the computational capabilities that allow it to work as an independent stand-alone system, which does not have to be connected to an external computer/controller.

Using a MPSoC as the hardware basis (the AMD/Xilinx Zynq™ UltraScale+™ XCZU4EV), we recently proposed a real-time eye detection system, which was able to process the 47 frames per second (fps) provided by a EMERALD 16MP image sensor from Teledyne e2v [3]. In the actual deployment of this system, the eye images (640 × 480 pixels in size) were sent to an Intel i9 computer for processing and final identification of the user. The problem is that, as there is about 2–3 s of recording per user in which eyes are detected, the number of eye images sent to the external computer can exceed 250 images. The external computer cannot process this volume of information before the user has left the access point at a normal pace. The solution to this problem is to filter out the large number of images in which the iris is not in focus. In our case, this means discarding almost 97% of the eye images captured by the system.

The main contribution of this work is to describe the implementation, in the programmable logic (PL) part of the MPSoC, of a module that evaluates which points of an image are in focus. Integrated together with an eye image detection system, this module allows to discard the detections that do not pass this out-of-focus test, thus preventing them from being processed by the next stages of an iris recognition identification system. The whole framework was mainly built using Vitis HLS and synthesised in the aforementioned AMD/Xilinx UltraScale+ MPSoC (multiprocessor system-on-chip). Given the characteristics of an FPGA, the design option selected for this defocus blur evaluation module was based on convolution kernels [7,9].

The rest of the paper is organised as follows: The state of the art in the topic is briefly revised in Section 2. Section 3 provides an overview of the whole proposed framework for eye detection implemented in the processing system (PS) and programmable logic (PL) parts of the MPSoC and details about the implementation of the defocus estimation core, synthesised as a functional block in the PL of the MPSoC. Experimental results are presented in Section 4. Finally, the conclusions and future work are drawn in Section 5.

2. Related Work

Defocus blur is the result of an out-of-focus optical imaging system [4]. When an object is not in the focal plane of the camera, the rays associated with it do not converge on the same spot on the sensor but to a region called the circle of confusion (CoC). The CoC can be characterised using a point spread function (PSF), such as Gaussian, defined by a radius/scale parameter [10]. This radius increases as this object gets farther away from the focal plane. In practice, it is assumed that there is a range of distances from the camera, associated with the focal plane, at which an object is considered to be in focus. This is the so-called depth of field of the camera.

The detection of blurred image regions is a relevant task in computer vision. Significantly, defocus blur is considered by several authors to be one of the main sources of degradation in the quality of iris images [2,11–14]. Many single-image defocus blur estimation approaches have been proposed [4,10]. They can be roughly classified into two categories: edge-based and region-based approaches [10]. The edge-based approach models the blurry edges to estimate a sparse defocus map. Then, the blur information at the edge points can be propagated to the rest of the image to provide a dense blur map [15]. Edge blur estimation models typically consider that the radius of the CoC is roughly constant, and define the edge model using this parameter [16,17]. However, other more complex models of defocused edges can be used [18]. Although edge detection and blur estimation can be performed simultaneously [10], the problem with these approaches is that obtaining the dense defocus map can be a time-consuming step. To alleviate this problem, Chen et al. [16] proposed to divide the image into superpixels, and consider the level of defocus blur in them to be uniform. The method needs a first step in which this division into superpixels is generated. One additional problem with edge-based defocus map estimation is that they usually suffer from textures of the input image [19]. It should be noted that our region of interest, the iris, is primarily a texture region.

Region-based approaches avoid the propagation procedure to obtain dense defocus maps in edge-based approaches, dividing up the image into patches and providing local defocus blur estimation values. They are free of textures [19]. Some of these approaches work in the frequency domain, as the defocus blur has a frequency response with a known parametric form [20]. Oliveira et al. [21] proposed to assume that the power spectrum of the blurred images is approximately isotropic, with a power-law decay with the spatial frequency. A circular Radon transform was designed to estimate the defocus amount. Zhu et al. [22] proposed to measure the probability of the local defocus scale in the continuous domain, analysing the Fourier spectrum and taking into consideration the smoothness and colour edge information. In the proposal by Ma et al. [23], the power of the high, middle and low frequencies of the Fourier transform is used. Briefly, the ratio of the middle-frequency power to the other frequency powers is estimated. This ratio should be larger for the clear images than for the defocused and motion blurred images. For classifying the images into valid or invalid ones, a support vector machine (SVM) approach is used. In all cases, the approach is simple and fast. These approaches take advantage of the fact that convolution corresponds to a product in the Fourier domain. Also assuming spatially invariant defocus blur, Yan et al. [24] proposed a general regression neural network (GRNN) for defocus blur estimation.

To avoid the computation of the Fourier transform, other researchers prefer to directly work in the image domain. As J.G. Daugman pointed out [25], defocus can be represented, in the image domain, as the convolution of an in-focus image with a PSF of the defocused optics. For simplicity, this function can be modelled as an isotropic Gaussian one, its width being proportional to the degree of defocus [21]. Then, in the Fourier domain, this convolution can be represented as

$$D_\sigma(\mu,\nu) = e^{-\left(\frac{\mu^2+\nu^2}{\sigma^2}\right)} F(\mu,\nu) \tag{1}$$

$D_\sigma(\mu,\nu)$ and $F(\mu,\nu)$ are the 2D Fourier transforms of the image defocused to degree σ and the in-focus image, respectively. Significantly, for low (μ,ν) values, the exponential term approaches unity, and both Fourier transforms are very similar. The effect of defocus is mainly to attenuate the highest frequencies in the image [25]. As the computational complexity of estimating the Fourier transform is relatively high, Daugman suggested to take into consideration Parseval's theorem

$$\iint |I(x,y)|^2 dxdy = \iint |F(\mu,\nu)|^2 d\mu d\nu \quad (2)$$

and to not estimate the total power at high frequencies in the Fourier domain but in the image domain. Thus, the idea is to filter the image with a high pass (or a band-pass filtering within a ring of high spatial frequency). After filtering the low-frequency part of the image, the total power in the filtered image is computed using the equation

$$P = \frac{1}{M \cdot N} \sum_{i=0}^{M-1} \sum_{j=0}^{N-1} |C(i,j)|^2 \quad (3)$$

where $C(i,j)$ is the filtered image of $M \times N$ dimension. In order to reduce the computational complexity of the Fourier transform, Daugman proposed to obtain the high frequency of the image using a 8×8 convolution kernel (Figure 1a). Briefly, this kernel is equivalent to superposing two centred square box functions with a size of 8×8 (and amplitude -1) and 4×4 (and amplitude +4). The 2D Fourier transform of this kernel can be expressed as the equation

$$K(\mu,\nu) = \frac{sin(\mu)sin(\nu)}{\pi^2\mu\nu} - \frac{sin(2\mu)sin(2\nu)}{4\pi^2\mu\nu} \quad (4)$$

In short, the result is a band-pass filter with a central frequency close to 0.28125 and with a bandwidth of 0.1875. The Fourier spectrum of this kernel is shown in Figure 2a.

It must be noted that, although there is no reference image, to obtain a normalised score between 0 and 100, Daugman proposed that the obtained spectral power x be passed through a compressive non-linearity of the form

$$f(x) = 100 \times \frac{x^2}{x^2 + c^2} \quad (5)$$

where c is the half power of a focus score corresponding to 50%. This last normalisation step presupposes the existence of a canonical (reference) iris image [26].

Finally, once the signal power value associated with the image (or a sub-image within the image) has been obtained, a threshold value can be set for determining whether the image is clear or out-of-focus [14].

Since, in our scenario, it is possible to assume the isotropic behaviour of the PSF and that the blur is due to bad focusing, this approach based on convolution kernels applied in the image domain is fully valid [21,25]. Furthermore, the convolution filtering of digital images can be efficiently addressed using FPGA devices [7,27,28].

Similar to Daugman's filter, the proposal by Wei et al. [11] is a band-pass filter, but it selects higher frequencies (central frequency around 0.4375 and a bandwidth of 0.3125). The convolution kernel is shown in Figure 1b. It is a 5×5 kernel that superposes three centred square box functions. The frequency response is shown in Figure 2b. Kang and Park [29] also proposed a kernel with a size of 5×5 pixels:

$$K(\mu,\nu) = \frac{sin(\frac{3}{2}\mu)sin(\frac{3}{2}\nu)}{\frac{9}{4}\pi^2\mu\nu} - \frac{sin(\frac{5}{2}\mu)sin(\frac{5}{2}\nu)}{\frac{25}{4}\pi^2\mu\nu} - 4 \cdot \frac{sin(\frac{1}{2}\mu)sin(\frac{1}{2}\nu)}{\frac{1}{4}\pi^2\mu\nu} \quad (6)$$

This band-pass filter has a central frequency close to 0.2144 and a bandwidth of 0.6076. Thus, the shape of the Fourier spectrum is similar to the one of Daugman's proposal but with a significantly higher bandwidth (see Figure 2c). Figure 1c shows that it combines three square box functions (one of size 5×5 and amplitude -1, one of size 3×3 and amplitude +5, and other of size 1×1 and amplitude -5).

(a)

–1	–1	–1	–1	–1	–1	–1	–1
–1	–1	–1	–1	–1	–1	–1	–1
–1	–1	3	3	3	3	–1	–1
–1	–1	3	3	3	3	–1	–1
–1	–1	3	3	3	3	–1	–1
–1	–1	3	3	3	3	–1	–1
–1	–1	–1	–1	–1	–1	–1	–1
–1	–1	–1	–1	–1	–1	–1	–1

(b)

–1	–1	–1	–1	–1
–1	2	2	2	–1
–1	2	0	2	–1
–1	2	2	2	–1
–1	–1	–1	–1	–1

(c)

–1	–1	–1	–1	–1
–1	–1	4	–1	–1
–1	4	4	4	–1
–1	–1	4	–1	–1
–1	–1	–1	–1	–1

(d)

0	1	1	2	2	2	1	1	0
1	2	4	5	5	5	4	2	1
1	2	4	5	5	5	4	2	1
2	5	3	–12	–24	–12	3	5	2
2	5	0	–24	–40	–24	0	5	2
2	5	3	–12	–24	–12	3	5	2
1	2	4	5	5	5	4	2	1
1	2	4	5	5	5	4	2	1
0	1	1	2	2	2	1	1	0

(e)

–1	–1	–1
–1	8	–1
–1	–1	–1

Figure 1. Convolution kernels proposed by (**a**) Daugman [25], (**b**) Wei et al. [11], (**c**) Kang and Park [29] and (**d**,**e**) Wan et al. [30].

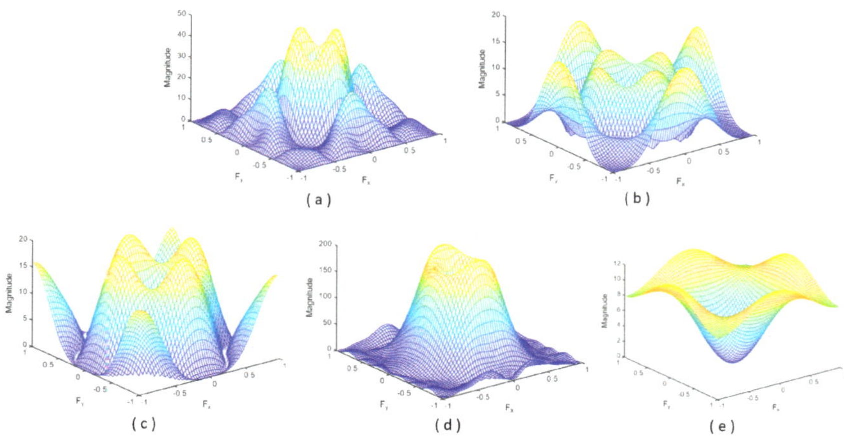

Figure 2. Fourier spectrum for the convolution kernels (Figure 1) proposed by (**a**) Daugman [25], (**b**) Wei et al. [11], (**c**) Kang and Park [29], and (**d**,**e**) Wan et al. [30].

Since high frequency is associated with sharp changes in intensity, one way to estimate its presence in the image would be to use the Laplacian [30]. The Laplacian $L(x, y)$ of an image $I(x, y)$ is given by

$$L(x, y) = \frac{\partial^2 I}{\partial x^2} + \frac{\partial^2 I}{\partial y^2} \quad (7)$$

This operator is the basis for the convolution kernel proposed by Wan et al. [30]. To reduce the sensitivity to noise, the Laplacian is applied to an image that is first smoothed by a Gaussian smooth filter. The 2D LoG (Laplacian-of-Gaussian) function has the form

$$LoG(x,y) = -\frac{1}{\pi\sigma^4}\left[1 - \frac{x^2+y^2}{2\sigma^2}\right]e^{\frac{x^2+y^2}{2\sigma^2}} \quad (8)$$

with σ being the Gaussian standard deviation. For σ equal to 1.4, the convolution kernel takes the form shown in Figure 1d. The associated Fourier spectrum is illustrated in Figure 2d. However, to simplify the computation, the authors propose an alternative 3×3 Laplace operator, combining two square box functions (one of size 3×3 and amplitude -1, and other of size 1×1 and amplitude +9) (Figure 1e).

3. Implementation

3.1. Overview of the Proposed Framework

Figure 3 shows the schematic of the proposed logic architecture for detecting iris images, in which the core for evaluating defocus is integrated. The figure shows how the images are captured from a Teledyne e2v EMERALD sensor. This sensor can provide 16 Mpx images at a speed of 47 fps, using 16 low voltage differential signalling (LVDS) lines. The first of the cores in the architecture, EMERALD core, is responsible for deserialising the signals encoded on these 16 LVDS lines, generating the input video stream. A first video direct memory access (VDMA) channel allows up to 8 frames to be stored in the DDR3 RAM available on the hardware platform.

The size of the frames received in the input video stream is modified by the Resizer core to a size of 128×128, which is also stored in the RAM using a second VDMA channel. In addition, the input stream is processed by the DEFOCUS core to generate a binary image stream, in which out-of-focus areas are marked with 0 and in-focus areas with 1. These images are stored in RAM using a third VDMA channel. The algorithm implemented in the DEFOCUS core is discussed in detail in Section 3.2.

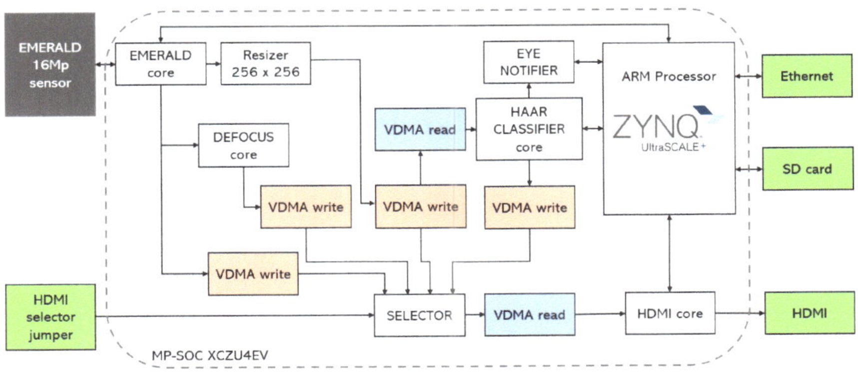

Figure 3. Overview of the proposed framework.

The detection of the eyes present in the rescaled image is carried out by the HAAR CLASSIFIER core. The core reads from the VDMA that manages this stream of rescaled images and implements a parallelised version of the popular classifier proposed by Viola and Jones [31]. Briefly, this detector uses a set of Haar-like features to characterise an image region and a supervised learning scheme (AdaBoost) to boost the classification performance of a simple learning algorithm. The result is an ensemble of weak classifiers, each of which internally computes a Haar-like feature and uses a threshold value to determine whether the region can be the desired object or not. As described in [3], the classifier is not organised as a sequence of weak classifiers but as a decision tree [32]. Thus, instead of having to use hundreds of classifiers (which will have to be evaluated almost entirely if the region to be studied has similarity to the person's eye), the system evaluates 120 Haar features in parallel, which form five stages, each of which has three trees with 8 nodes each ($5 \times 3 \times 8$ features) [3]. All five stages are executed in parallel.

Figure 4 shows the internal scheme of the classifier. First, the integral image, the tilted integral image, and the standard deviation of the grey level of the image are calculated [31,32]. These parameters allow the parallel evaluation of the 120 Haar features which, compared to a threshold, generate 120 binary values (node 0 to 119). With these values, 15 vectors of 10 components are formed. In each vector, the first two bits encode the tree within the stage, and the next eight are taken from the evaluation of eight nodes. The vectors are grouped in threes, and each group of three vectors is used to address a look-up table (LUT). This allows three values per LUT to be obtained, which are summed. Theoretically, the final value of each tree should be computed by multiplying each node by a weight and summing the results. In order to accelerate the execution, the LUT is implemented with the results of each 256 possibilities. This allows us to remove the hardware employed for math computations and have the results in a single clock. Finally, the resulting value is compared to a threshold. This comparison generates a Boolean value, which determines whether that stage evaluates the region as an eye or not. Figure 5 provides the output images from the five stages in the classifier for a given input image. If the output Boolean values computed by the five stages are true, the evaluated region is marked as true (results mixer core). Figure 6 (middle) shows the raw detection image associated to the five output images in Figure 5. Significantly, the whole process is executed in only four steps [3]. The raw detection image, the output of the results mixer core, is slightly smaller than the original rescaled one (if the size of the rescaled image and evaluated region are $M \times N$ and $m \times n$, respectively, the size of this image will be $(M - m + 1) \times (N - n + 1)$).

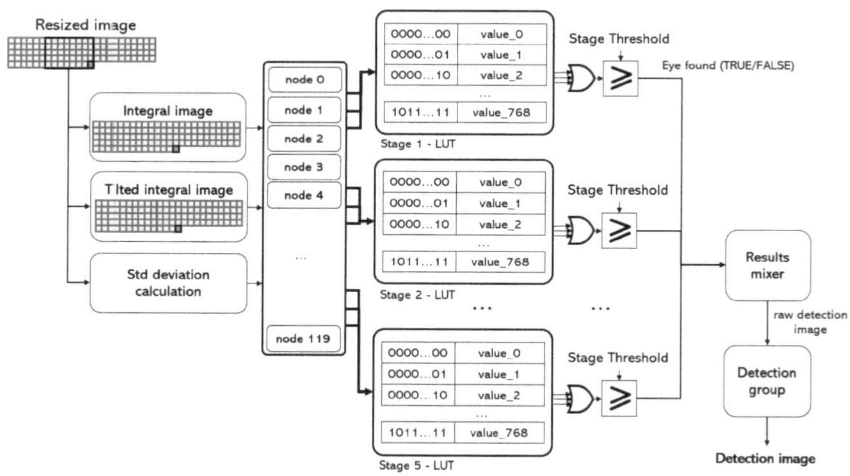

Figure 4. The classification structure.

The problem with this scheme is that the jump between the regions being evaluated is one pixel, so the overlap between regions is very large. This results in very close positive detections, which generates many images that are associated with a single iris image. The detection group core filters the raw detection image, adding the values within a sliding window and thresholding the obtained sum value to mark the pixel as a positive detection (its value will be equal to the sum value) or not (0). Figure 6 shows the raw detection image associated to an input image, and the filtered version obtained by the detection group core. The sliding window is 20×15 pixels, and the threshold value is set to 7 (it can be modified if desired). Both images in Figure 6 are inverted in colour to help visualise the points obtained. It can be noted how the detection group core filters the isolated dots of the raw detection image, and groups the dots into higher entities. In fact, three major entities are identified (both eyes and one false positive detection).

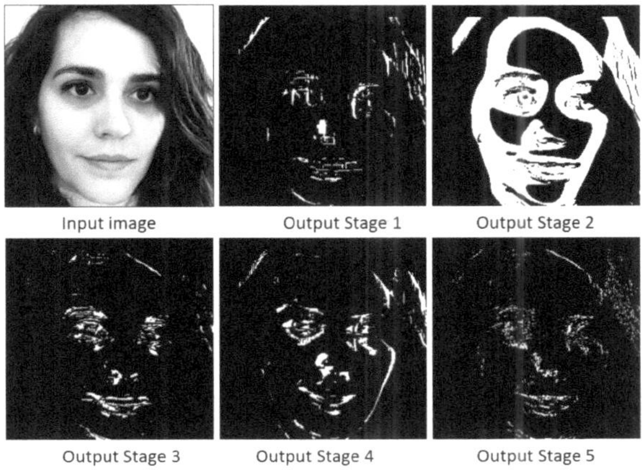

Figure 5. Input image and outputs provided by the five stages in the classifier.

Figure 6. (**Left**) Input image; (**middle**) raw detection image; and (**right**) filtered detection image. Both images are inverted in colour to help visualise the points obtained.

The eye notifier core is responsible for providing high-resolution image cropping (640 × 480) for each entity detected in the image. A sliding window is now used to detect the maximum values inside the entities provided by the detection group core. For each entity, a 640 × 480 pixel image is cropped from the high-resolution image. This cropping is centred on the position, transposed from the scaled image to the original input image, provided by the maximum detected value.

3.2. Defocus Estimation

The basis of the designed defocus blur estimation core is convolution. A 2D convolution can be mathematically represented by the equation

$$C(i,j) = \sum_{u=-U}^{U} \sum_{v=-V}^{V} h(u,v) \cdot I(i-u, j-v) \tag{9}$$

with $I(i,j)$ and $C(i,j)$ being the input and the filtered images, respectively. $h(u,v)$ is the convolution kernel, with size $(2 \cdot U + 1) \times (2 \cdot V + 1)$. If the size of the convolution kernel is 8×8, the expansion of Equation (9) results in 64 multiplications and 63 summations to be computed for each pixel.

If a large storage capacity is available (as is the case when using CPU or GPU), it is possible to store the complete rows read from the image sensor, and apply convolution when the complete dataset is available. This is not the situation when working with an FPGA. If the input image and convolution kernel are small in value (e.g., Wei et al. [7] work with 640 × 480 pixel images and a convolution kernel of 3 × 3), FIFO (first in–first out) memories can be implemented to store the necessary image rows (3 FIFO memories with a depth of 637 each in the case of Wei et al.'s implementation). In our case, the EMERALD 16MP sensor provides images of 4096 × 4096 pixels, and the design will need 8 FIFO memories with a depth of 4088 (4096 − 8) each. Storing these data in order to apply the convolution kernel when all the data are available would increase the data usage and latency.

However, it is important to note that, in our case, the defocus map obtained must evaluate the pixels of the detection image (the output of the eye notifier core, see Figure 4). The size of this detection image is much smaller than that of the input image. Therefore, an initial data reduction step is to apply the convolution kernel not to each pixel of the input image but in steps of S pixels as proposed by Daugman [25]. In our case, as the sensor data are read in blocks of 8 pixels, a value of S equal to 8 is used. Applying the convolution kernel in 8 pixel steps allows, in one clock cycle, to have the 8 multiplications and 7 additions of that kernel row. Using a memory of $4096/8 \cdot 32$ bits, it is possible to store, without losing resolution, the summation and to accumulate results until the results of the last row of the kernel are obtained. The result is a defocus map of size 512 × 512, storing 32-bit values. Given the size of the scaled image (128 × 128), we further reduce the size of the defocus map by adding in 4 × 4 blocks and obtaining a 128 × 128 map, with 32-bit values. Figure 7 schematizes the procedure for obtaining the defocus map. The first and second steps involve convolving the kernel with the input data (in batches of 8 pixels per clock) and performing 4 × 4 compression. Both steps require only two line buffers: one of 512 and one of 128 values (i.e., whole images are not managed). The data are stored as 32-bit values.

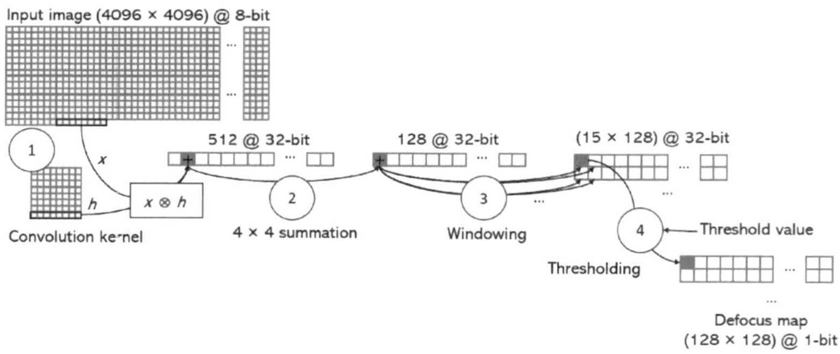

Figure 7. Graphical representation of the defocus blur map estimation. The first step implies the generation of a line buffer of 512 positions with 32-bit values. Each position in this buffer stores the convolution of a 8 × 8 block of the input image with the convolution kernel. The second step is a 4 × 4 summation for generating a line buffer of 128 positions with 32-bit values. A sliding window of 20 × 15 size is used in this map to accumulate the values (Step 3). A final thresholding process permits the core to obtain the final defocus map.

The third step averages the defocus map using the same sliding window used to detect eyes (of size 20 × 15 in our implementation). Actually, to reduce computations, the values in the window are not averaged but simply accumulated. The result obtained is thresholded using a configurable value from the ARM (step 4 in Figure 7). In this way, the final result is a defocus map with the same size as the detection image obtained by the classifier and which stores binary values (0 if the value is not in focus and 1 otherwise). Using a VDMA channel, the map is stored to be used for validating each positive eye detection.

In order to speed up the design of this core, as well as to optimise its performance characteristics, the Vitis HLS tool (https://docs.xilinx.com/r/en-US/ug1399-vitis-hls/Introduction (accessed on 15 August 2023)) from Xilinx is used. Vitis HLS allows the developer not to have to generate RTL using a hardware HDL language but to use a high/medium level language (C/C++, System C) and obtain, from these sources, the core IP in RTL. In our case, the design is coded using C/C++. From the set of directives that Vitis HLS provides to help developers optimise a hardware design, we implement the PIPELINE directive to parallelise the execution of the multiple computations in the convolution operation. In addition, in step 4 (thresholding), the UNROLL directive is used to transform loops by creating multiple copies of the loop body in the RTL design. In this way, some or all iterations of the loop can occur in parallel. Finally, we use HLS directives to establish when Block RAMs (BRAMs) or UltraRAM (URAMs) should be used in the design. BRAMs are required to be dual ported. URAM blocks have a fixed width of 72 bits, so two 32-bit values are joined together to be stored at each location.

The convolution kernels described in Section 2 are implemented and tested in the DEFOCUS core of the proposed framework. Table 1 shows the resource usage for the DEFOCUS core implementing Daugman's proposal. As a detail, regarding memory, everything is optimised to fit in one URAM and two BRAM, as the rest of the design demands a lot of BRAM (which is faster but has been used for the HAAR CLASSIFIER). The total resource usage for the whole system is shown in Table 2.

Table 1. Total resource utilisation for the DEFOCUS core (Daugman's kernel). FF stands for flip flops, LUT for look-up tables, and DSP48E for digital signal processing elements. The DSP48E combines an 18-bit by 25-bit signed multiplier with a 48-bit adder and programmable mux to select the adder input.

Name	BRAM_18K	DSP48E	FF	LUT	URAM
DSP	–	–	–	–	–
Expression	–	–	0	2	–
FIFO	0	–	65	332	–
Instance	2	1	2606	4193	1
Memory	–	–	–	–	–
Multiplexer	–	–	–	–	–
Register	–	–	–	–	–
Total	2	1	2671	4527	1
Available	256	728	175,680	87,840	48
Utilisation (%)	0	0	1	5	2

Table 2. Total resource usage for the whole system.

Name	BRAM_18K	DSP48E	FF	LUT	URAM
Classifier	81	582	34,645	25,740	0
Defocus	2	1	2671	4527	1
Total	83	583	37,316	30,267	1
Available	256	728	175,680	87,840	48
Usage (%)	32	80	21	34	2

4. Experimental Evaluation

4.1. Experimental Setting

The system is built as a portable device. The computational core is the TE0820-03-4DE21FA micromodule from Trenz Electronic. This micromodule is an industrial-grade 4 × 5 cm MPSoC System on Module (SoM) integrating an AMD/Xilinx Zynq™ UltraScale+™ XCZU4EV. Moreover, the micromodule includes 2 GByte DDR4 SDRAM, 128 MByte Flash memory for configuration and operation, and powerful switch-mode power supplies for all on-board voltages. A large number of configurable I/Os is provided via rugged high-speed stacking connections. The TE0820-03-4DE21FA micromodule is mounted on a compatible carrier board that provides physical connections between the module and peripherals. On this first version of the hardware system, the carrier board for the Trenz Electronic 7 Series was used. This carrier provides us with all the components needed during the development phase, such as power delivery, Ethernet connection, debugging interface, UART to USB bridge, HDMI, PMOD connectors, and a FMC (FPGA Mezzanine Card) connector. The image sensor is the EMERALD 16MP from Teledyne e2v, which is mounted on a sensor board. Apart from other I/O connections, the sensor board provides as outputs the LVDS lines. An adaptation board was designed to mount the sensor board and to interface with the FMC connector in the carrier board. Figure 8 provides a snapshot of this first version of the camera.

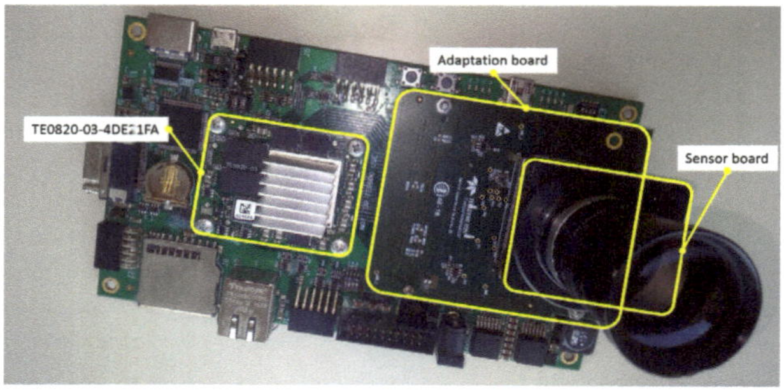

Figure 8. The first version of the system, mounting the TE0820-03-4DE21FA and the adaptation board on a carrier board for Trenz Electronic 7 Series. The adaptation board in turn mounts the sensor board (with image sensor and optics).

The carrier board for Trenz Electronic 7 Series provides peripherals that are not strictly necessary. In order to customize this carrier board, and also to remove the adaptation board from the scheme, a specific carrier board was designed and tested. Thus, in this final version, the carrier board includes only the peripherals needed by the proposed system. Parts that were initially used for debugging (HDMI, debugging interface, and UART USB bridge) were also left out. This resulted in an smaller and cheaper board (see Figures 9 and 10).

Figure 9. Custom PCB design of the carrier board.

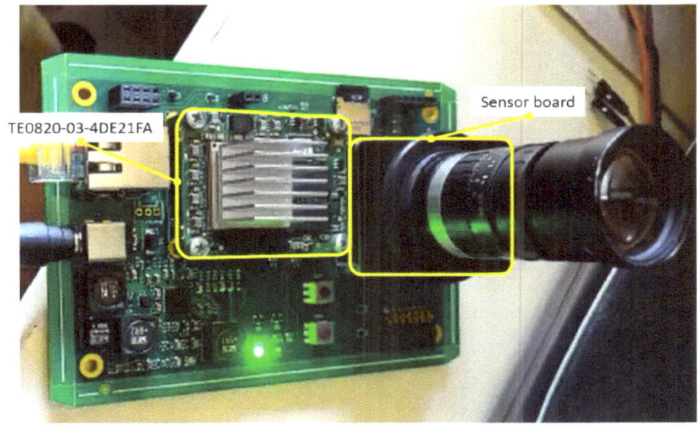

Figure 10. The final version of the system, mounting the TE0820-03-4DE21FA and the sensor board on a customised carrier board.

With respect to the illumination subsystem, the EMERALD 16MP features a very small true global shutter pixel (2.8 µm). Moreover, the sensor was designed to exhibit a very reduced dark signal nonuniformity (DSNU) value. Both properties allow the sensor to correctly work in a low-light scenario. In any case, the first hardware design employs a VARIO2 IPPoE infrared lamp from Raytec (see Figure 11 (left)). Similar to the proposal of Dong et al. [33], this lamp cannot be synchronised with the trigger of the camera and provides a very wide angle when compared with the FoV covered by the EMERALD sensor.

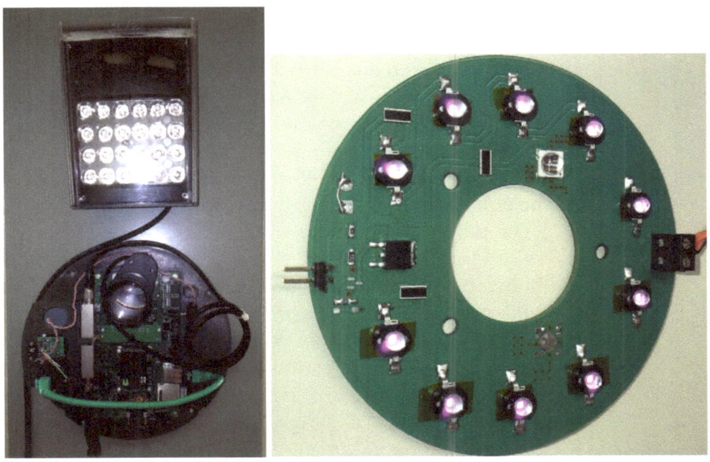

Figure 11. (**left**) The previous version of the hardware system, showing the VARIO2 IPPoE infrared lamp from Raytec, and (**right**) the board design of the new illumination module.

The Raytec lamp provides 51W continuously. To avoid this continuous irradiation, a plate with high power LEDs (3W) was designed and synchronised with the triggering of the camera. The board design is shown in Figure 11 (right). In total, 10 LEDs are mounted on the board. The new system ensures sufficient brightness and allows shortening the exposure time of the sensor, reducing motion blur caused by the subject walking through the FoV of the camera. The new design also ensures a more homogeneous illumination of the person's face in the capture position (about 1.7 m from the camera).

4.2. Hardware Implementation

In the design of the cores that are integrated in the programmable logic (PL) region of the AMD/Xilinx device, intensive use is made of Vitis High-Level Synthesis (HLS). Specifically, the two main streams described in this manuscript are synthesised in two cores generated from C++ files using HLS. The first one includes all the algorithms necessary to detect the eye regions: calculation of the integral and tilted integral images, calculation of the standard deviation, image processing using the HAAR features in the five stages described, and the subsequent filtering that generates the dot image with the positive detections. The second includes the focused region detection algorithm. Synthesising and merging all modules into only two cores allows the employed synthesis and implementation tool to further optimize and share resources. To achieve a faster runtime and allow the tool to optimise resources, source C++ files are subsequently modified by adding specific directives (see Section 3.2).

In the proposed design, all the hardware implemented on the programmable logic (PL) part is initialised and controlled from the processing system (PS) part (in this case, in the Cortex-A53 ARM processor available in the SoC). Initialisation follows a series of stages. First of all, the peripherals are tested. Then, the EMERALD sensor is initialised. After that, the onboard memory is configured and synchronised with the VDMA cores. It can be noted that there are four VDMA write cores (see Figure 3). These video streams (input images, resized images, contrast images, and detection images) can be viewed in real time using the HDMI interface. This allows easy real-time debugging. Finally, the EMERALD sensor is configured to produce 47 fps at 16MP, and the video stream is started. Alternatively, for debugging purposes, the video stream can be generated from a still image stored on an SD card.

When the desired object (an eye) is found on the video stream, the eye notifier interrupts the ARM processor and saves the frame number and coordinates. When the processor attends to the interrupt, it reads the data, checks the defocus map to determine if the detection is in focus, and, if so, it is considered a valid detection. If the detection is a valid one, the processor goes to the referenced frame temporary stored on the frame buffer used by the VDMA, crops the region, and stores it into another buffer that will be sent through Ethernet. By dividing the tasks between the two ARM cores, the system can send the detected eyes (in 640×480 images) with minimal detection delay (variable depending on the number of eyes detected per input image).

The processing is made around a video stream implemented using an AXI Stream interface. As a throughput of at least 47 fps at 16MP resolution is required (to get the most out of the image sensor), the design choice was to use a data width of 8 bytes per clock at 150 MHz. As described in Section 3.1, this video stream is buffered in the onboard DDR3 memory by means of VDMA cores. This allows the ARM processor to access the frames to accomplish the cropping and sending steps. The hardware resource usage is shown in Table 2. The intensive use of block RAMs (BRAMs) in the classifier module is due to the need to store pixels already read from the sensor but needed to estimate the integral and rotated integral images, and to calculate the standard deviation (see Figure 4). It is also due to the storage needs of the result mixer and detection group modules. They are then used to implement memories and FIFOs. One UltraRAM (URAM) is employed in the defocus module. The DSP48 blocks are intensively used for implementing the arithmetic required in the classifier module.

4.3. Obtained Results

The set of tests carried out to check the validity of the proposal is divided into three blocks. In the first, a publicly available database, the Clarkson dataset LivDet2013 [34], is used to evaluate the ability of the convolution kernel as an estimator of the level of defocus blur. This database includes sets of images with different levels of defocus blur. They are images of eyes, so they do not allow evaluation of the eye detection system. In the second test block, the CASIA-Iris-Distance V4 database (http://biometrics.idealtest.org/

(accessed on 15 August 2023)) is extended by incorporating slightly blurred versions of the component images. In this case, it is possible to evaluate the system's ability to detect eyes and to discard those detections that are out of focus. Finally, a third block of tests is carried out in a real environment, using the hardware described above. In addition to evaluating the system's ability to detect eyes in a real environment (using the EMERALD 16MP sensor and with people in motion), these tests allow us to determine that the system is capable of discarding a large volume of out-of-focus detections. The following sections provide details on these three test blocks.

4.3.1. Evaluation of the Defocus Blur Estimation in Eye Subimages

Although the actual validation of the proposal must be performed with the system described in Section 4.1, in order to evaluate the performance for estimating the iris image defocus of the four convolution kernels described in Section 3.2, the Clarkson dataset LivDet2013 [34] is used. In this dataset, images are collected through the use of video capture of 100 frames at 25 fps using a Dalsa camera. The sequence is started out of focus and is moved through the focus range across full focus and back to being out of focus. Images are grouped according to their blur level in five categories:

- Group #1–10% blur in frame before least blurry image;
- Group #2–5% blur in frame before least blurry image;
- Group #3 least blurry image;
- Group #4–5% blur in frame after least blurry image;
- Group #5–10% blur in frame after least blurry image.

A total of 270 images are available for training. In Figure 12, the blue curve joins the response values for applying the convolution kernel proposed by Daugman [25] to these images. A higher value indicates that the image has, on average, a higher focus. The images are sorted by groups (the first 53 belong to group #1, the next 55 to group #2, and so on). The mean values of each group are shown in black (the curve forms five steps). The values cannot be clearly delimited into a range per group, as there are images with values much higher (or lower) than the group mean. The results obtained by applying the convolution kernels proposed by Wei et al. [11] and Wan et al. [30] are very similar, varying only mainly in the absolute value of the results.

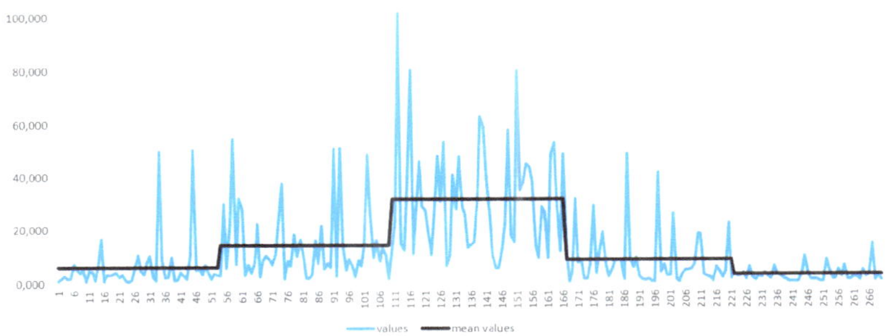

Figure 12. Mean filter responses (in blue colour) obtained from applying Daugman's [25] filtering to the 270 images in the training set of the Clarkson dataset LivDet2013 [34]. The five steps that form the black line are associated with the five categories in the database. The value of each step is the average value of the responses obtained on the images in each category.

The problem when evaluating an image is that the value averages those associated with each pixel. As shown in Figure 13, the contrast value of the image is not really the one associated with the iris. In this figure, the three images shown belong to group #1. However, while the first one has a very high average value (4201.04 using Kang and Park's proposal), the third one has a very low value (92.55). In both cases, the iris area is clearly out of focus (blue values in the filter response images), but the presence of the eyebrows, in the first and also in the second image (978.43 average value), makes the average value high. Eyebrows and eyelashes offer higher values of focus if, in addition, the skin area of the face is saturated by the infrared illumination (for example, in the third image, where the eyebrow is also visible, the pixels associated with it are barely in focus).

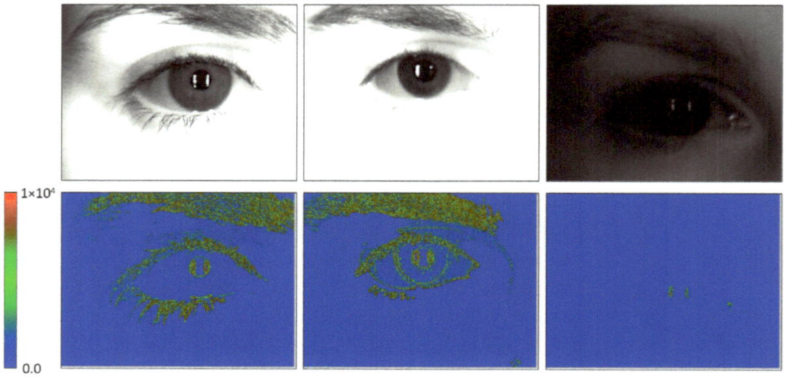

Figure 13. (**Top**) Examples of images with low contrast from the Clarkson dataset LivDet2013 [34]. The three images are included in the group #1 (10% blur in frame before least blurry image). (**Down**) Filter responses obtained when applied Kang and Park's [29] filtering to the images in the top row. The scale values is shown at the left (responses greater than 1×10^4 are drawn in red colour).

Joining groups 1 and 5, and 2 and 4 (where the level of blur is the same), you have three groups (10% blur, 5% blur, and in focus). In Table 3, the results of the evaluated kernel approaches are summarised. The first column in the table identifies the kernel, and the rest of the columns provide the mean and standard deviations of the total power at high-frequency bands processing each kernel with the images in the three groups. For Wan et al.'s proposal, the 3×3 convolution kernel is used (Figure 1e). Given the dispersion shown in Figure 12, the Z-score (68% confidence interval) is employed to remove outliers before estimating the parameters for each group. The last column in the table illustrates the threshold values for the discrimination of defocus and in-focus images. ROC (receiver operation characteristic) curves are used to obtain these threshold values. Using these schemes, the 246 images available for testing in the Clarkson dataset LivDet2013 [34] are evaluated. Using the threshold values shown in Table 3, the number of rejected iris images (out-of-focus images) that are actually focused images (false rejection rate, FRR) is less than 1% for all tested kernels. This is because the problems present in the images mostly force an increase in the power obtained. This causes them to be falsely taken as in-focus images (increasing the false accept rate (FAR)). It is important to note that the FRR is the percentage of error valid for us, since what the system cannot do in any case is to discard any image with the iris that really is in focus.

Table 3. Results obtained when evaluating convolution kernels using the Clarkson LivDet2013 [34].

Convolution Kernel	Groups 1–5		Groups 2–4		Group 3		Threshold Value
	Mean	Std	Mean	Std	Mean	Std	
Daugman [25]	2284.74	952.20	7044.57	5794.04	27,411.10	12,113.73	14,067.99
Wei et al. [11]	108.70	75.61	244.81	211.72	1206.34	599.93	531.47
Kang and Park [29]	194.31	131.00	450.28	399.91	2222.65	1132.99	969.92
Wan et al. [30]	51.39	16.67	67.79	29.95	186.91	83.30	100.68

4.3.2. Quantitative Evaluation Using an Extended CASIA-Iris-Distance V4 Database

Significantly, when working with real images, the defocus test does not apply to eye images (such as those shown in Figure 13). In contrast, when the test is applied, there is no estimation of where the eye is, and what is done is to compute the convolution kernel values and threshold them using the previously obtained values. The kernel allows regions that are correctly focused (and have edges) to be marked with a value of 1 on a background value of 0. For obtaining quantitative results, the proposed approach is evaluated using an extended version of the CASIA-Iris-Distance, version 4.0 database. The original database contains 2567 images of 142 people, most of them graduate students of the Chinese Academy of Sciences' Institute of Automation (CASIA). The database is captured indoors, with a distance of more than 2 m, and using a self-developed long-range multi-modal biometric image acquisition and recognition system (LMBS). In our extended version, the dataset is doubled, incorporating, for each image in the database, a version affected by defocus blur, simulated using a Gaussian of radius 2 pixels. The effect of this filtering is virtually unnoticeable. For example, Figure 14 shows two original database images and, to their right, the generated versions. Only by zooming in can the smoothing effect be seen.

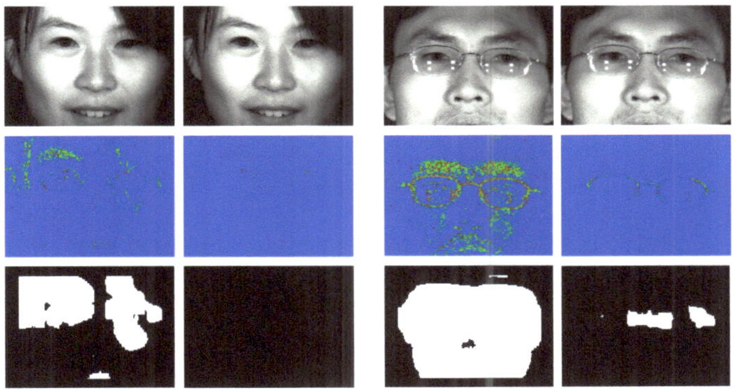

Figure 14. (**Top**) Original images from the CASIA-Iris-Distance V4 database and, on the right, defocused versions. (**Middle**) 147 × 108 pixel images obtained after applying the first two steps (see Figure 7) of the defocus blur estimation process. The blue tones are associated with out-of-focus pixels, while the green and red tones are associated with increasingly focused pixels. (**Bottom**) Defocus maps associated with the images in the top row. When the person wears glasses, these defocus maps are not completely zeroed, which causes certain eye detections to be falsely considered to be in focus (see text).

The images have a size of 2352 × 1728 pixels. As in the real deployment, the convolution kernel is applied on a block of 8 × 8 pixels without overlapping, resulting in an image of 294 × 216 pixels. In our case, an array of 294 positions would be sufficient to store 32-bit data. The next step applies a compression that, instead of using a 4 × 4, uses a 2 × 2 block. Thus, the result is an image of 147 × 108. Although it would be sufficient to use an array of 147 positions, complete images are generated to show them as intermediate results in Figure 14 (middle). In the images, the defocus blur values are shown on a scale ranging from pure blue tone (out of focus) to green and, finally, red tones (higher level of focus). Using a sliding window of 23 × 17 pixels and a threshold value proportional to the ones in Table 3, the system provides the defocus maps shown in Figure 14 (bottom). It can be seen how the presence of glasses or reflections generates strong edges, which are not considered to be out of focus, and which generate defocus maps that are not completely zeroed.

The eye detection results in this database are 100%, matching those obtained when only the original version is used [3]. Of the total set of 10,268 eyes present in the entire extended database, 5016 images are correctly discarded for being out of focus. Of the total set of 10,268 eyes present in the entire extended database, 5016 images are correctly discarded for being out of focus. In total, 118 eye images are erroneously categorised as being in focus. All of them belong to images of people wearing glasses.

4.3.3. Evaluation in a Real Scenario

The whole framework is tested in a real deployment. Figure 15 schematically illustrates how it works. Given an input frame, the framework simultaneously generates two masks. The upper image is generated by detecting the in-focus areas (in the example, they are marked in white colour; planar areas are also marked as defocus regions). The second mask is generated by the HAAR CLASSIFIER kernel. In this case, the white dots correspond to positive detection values. Both masks are merged by the ARM to generate a final mask that will be used to crop the eyes from the 16MP input image (white dots in the right image in Figure 15 (right)).

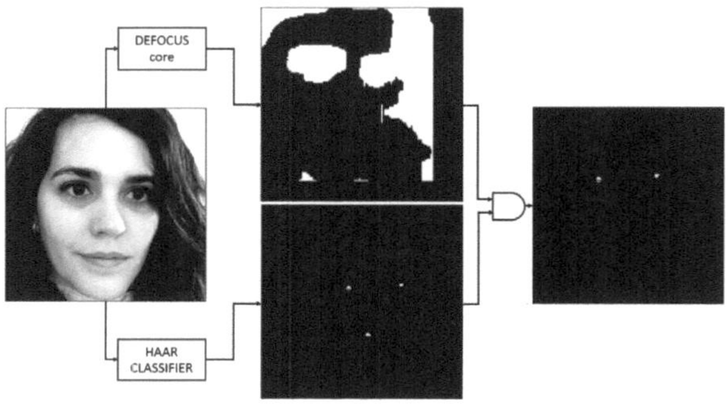

Figure 15. Generation of the contrast and eye detection masks from an input image, and combination for obtaining the final detection points. Both masks are of size 128 × 128 pixels.

Figure 16 shows an example of zooming in on the sensor. In the first frames on the left, the face is not in the depth of field (in the second frame part of the image is already in focus). In the next two frames, the face is in focus. In the middle row, the white areas (mask values equal to 1), which are associated with the eyes or hair (fringe, moustache, or goatee), are visible. In the areas where there are no borders, the mask does not return positive values (1). The bottom row shows how the left eye regions are better defined in these two frames. When the person's face leaves the depth of field, the whole face is out of focus again.

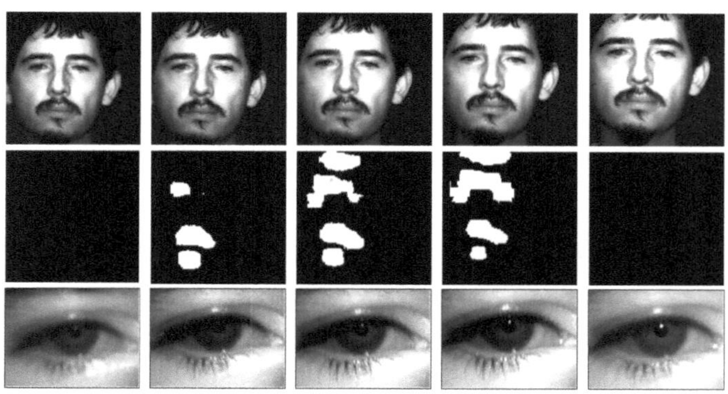

Figure 16. (**Top row**) Frames of a sequence of a person passing in front of the sensor. (**Middle row**) Masks obtained by the DEFOCUS core (white values are associated with regions in focus). (**Bottom row**) Images of the left eye of the images in the top row. It can be seen how the third and fourth images are in focus, while the rest are not.

In tests with the system, it always detects eyes that are correctly focused, discarding a large number of unfocused eyes. Several image sequences are recorded and used for validation. Of these sequences, 97% of the eye detections are correctly discarded because they are out of focus. The proposed design aims to ensure that no eye that may be in focus is discarded, so the selection of the threshold can be considered conservative. This is the reason why, occasionally, images of out-of-focus eyes are allowed to pass to the iris pattern segmentation and normalisation phase. These images are discarded by the iris recognition module.

5. Conclusions and Future Work

This paper details how to design a defocus blur estimation core on the PL part of an MPSoC to allow an eye detection system to discard in real-time those eye images that are out of focus. The design is implemented on an AMD/Xilinx Zynq™ UltraScale+™ XCZU4EV platform by converting the C/C++ code to hardware logic core through Vitis HLS. The core design is optimised by using different optimisation directives to reduce latency. Thus, the proposed real-time eye detection system can correctly discard out-of-focus detected eye images but also process 57 images with 4096 × 4096 size per second. Significantly, when integrated in an IAAD recognition system, the defocus blur estimation core allows the system to discard 97% of the detected eye images.

Future work focuses on (1) implementing the next steps of the iris recognition system (iris pattern segmentation and normalisation) in the MPSoC, using the resources that are not yet being used (in the PL, but also in the PS, such as the GPU or the dual-core Cortex-R5); (2) further exploiting the use of the optimisation directives provided by Vitis HLS to improve the current proposed framework; and (3) adding the framework with the cores to implement an iris presentation attack detection system (iPAD).

Author Contributions: Conceptualization , C.A.R.-B., A.R.-G., M.G.-G. and A.B.; Data curation, C.A.R.-G. and A.B.; Formal analysis, C.A.R.-G., A.R.-G., M.G.-G. and R.M.; Funding acquisition, R.M. and A.B.; Investigation, C.A.R.-B., A.R.-G., M.G.-G. and R.M.; Methodology, A.R.-G. and M.G.-G.; Software, C.A.R.-G. and A.R.-G.; Supervision, M.G.-G.; Validation, C.A.R.-B. and A.B.; Writing—original draft, A.R.-G. and A.B.; Writing—review & editing, C.A.R.-B., A.R.-G. and A.B. All authors have read and agreed to the published version of the manuscript.

Funding: This work has been supported by grants CPP2021-008931, PDC2022-133597-C42, TED2021-131739B-C21 and PID2022-137344OB-C32, funded by MCIN/AEI/10.13039/501100011033 and by the European Union NextGenerationEU/PRTR (for the first three grants), and "ERDF A way of making Europe" (for the fourth grant).

Institutional Review Board Statement: Not applicable.

Informed Consent Statement: Not applicable.

Data Availability Statement: Not applicable.

Acknowledgments: Portions of the research in this paper use the CASIA-Iris V4 collected by the Chinese Academy of Sciences—Institute of Automation (CASIA).

Conflicts of Interest: The authors declare no conflict of interest.

References

1. Nguyen, K.; Fookes, C.; Jillela, R.; Sridharan, S.; Ross, A. Long range iris recognition: A survey. *Pattern Recognit.* **2017**, *72*, 123–143. [CrossRef]
2. Tan, C.W.; Kumar, A. Accurate Iris Recognition at a Distance Using Stabilized Iris Encoding and Zernike Moments Phase Features. *IEEE Trans. Image Process.* **2014**, *23*, 3962–3974. [CrossRef] [PubMed]
3. Ruiz-Beltrán, C.A.; Romero-Garcés, A.; González, M.; Pedraza, A.S.; Rodríguez-Fernández, J.A.; Bandera, A. Real-time embedded eye detection system. *Expert Syst. Appl.* **2022**, *194*, 116505. [CrossRef]
4. Zeng, K.; Wang, Y.; Mao, J.; Liu, J.; Peng, W.; Chen, N. A Local Metric for Defocus Blur Detection Based on CNN Feature Learning. *IEEE Trans. Image Process.* **2019**, *28*, 2107–2115. [CrossRef] [PubMed]

5. Li, J.; Un, K.F.; Yu, W.H.; Mak, P.I.; Martins, R.P. An FPGA-Based Energy-Efficient Reconfigurable Convolutional Neural Network Accelerator for Object Recognition Applications. *IEEE Trans. Circuits Syst. II Express Briefs* **2021**, *68*, 3143–3147. [CrossRef]
6. Zhu, J.; Wang, L.; Liu, H.; Tian, S.; Deng, Q.; Li, J. An Efficient Task Assignment Framework to Accelerate DPU-Based Convolutional Neural Network Inference on FPGAs. *IEEE Access* **2020**, *8*, 83224–83237. [CrossRef]
7. Wei, C.; Chen, R.; Xin, Q. FPGA Design of Real-Time MDFD System Using High Level Synthesis. *IEEE Access* **2019**, *7*, 83664–83672. [CrossRef]
8. Kerdjidj, O.; Amara, K.; Harizi, F.; Boumridja, H. Implementing Hand Gesture Recognition Using EMG on the Zynq Circuit. *IEEE Sens. J.* **2023**, *23*, 10054–10061. [CrossRef]
9. Javier Toledo-Moreo, F.; Javier Martínez-Alvarez, J.; Garrigós-Guerrero, J.; Manuel Ferrández-Vicente, J. FPGA-based architecture for the real-time computation of 2-D convolution with large kernel size. *J. Syst. Archit.* **2012**, *58*, 277–285. [CrossRef]
10. Karaali, A.; Jung, C.R. Edge-Based Defocus Blur Estimation with Adaptive Scale Selection. *IEEE Trans. Image Process.* **2018**, *27*, 1126–1137. [CrossRef] [PubMed]
11. Wei, Z.; Tan, T.; Sun, Z.; Cui, J. Robust and Fast Assessment of Iris Image Quality. In *Advances in Biometrics*; Zhang, D., Jain, A.K., Eds.; Springer: Berlin/Heidelberg, Germany, 2005; pp. 464–471.
12. Belcher, C.; Du, Y. A Selective Feature Information Approach for Iris Image-Quality Measure. *IEEE Trans. Inf. Forensics Secur.* **2008**, *3*, 572–577. [CrossRef]
13. Li, X.; Sun, Z.; Tan, T. Comprehensive assessment of iris image quality. In Proceedings of the 2011 18th IEEE International Conference on Image Processing, Brussels, Belgium, 11–14 September 2011; pp. 3117–3120. [CrossRef]
14. Colores, J.M.; García-Vázquez, M.; Ramírez-Acosta, A.; Pérez-Meana, H. Iris Image Evaluation for Non-cooperative Biometric Iris Recognition System. In *Advances in Soft Computing*; Batyrshin, I., Sidorov, G., Eds.; Springer: Berlin/Heidelberg, Germany, 2011; pp. 499–509.
15. Zhuo, S.; Sim, T. Defocus map estimation from a single image. *Pattern Recognit.* **2011**, *44*, 1852–1858.
16. Chen, D.J.; Chen, H.T.; Chang, L.W. Fast defocus map estimation. In Proceedings of the 2016 IEEE International Conference on Image Processing (ICIP), Phoenix, AZ, USA, 25–28 September 2016; pp. 3962–3966. [CrossRef]
17. Ma, H.; Liu, S.; Liao, Q.; Zhang, J.; Xue, J.H. Defocus Image Deblurring Network with Defocus Map Estimation as Auxiliary Task. *IEEE Trans. Image Process.* **2022**, *31*, 216–226. [CrossRef] [PubMed]
18. Liu, S.; Zhou, F.; Liao, Q. Defocus Map Estimation From a Single Image Based on Two-Parameter Defocus Model. *IEEE Trans. Image Process.* **2016**, *25*, 5943–5956. [CrossRef] [PubMed]
19. Liu, S.; Liao, Q.; Xue, J.H.; Zhou, F. Defocus map estimation from a single image using improved likelihood feature and edge-based basis. *Pattern Recognit.* **2020**, *107*, 107485. [CrossRef]
20. Bertero, M.; Boccacci, P. *Introduction to Inverse Problems in Imaging*; IOP Publishing: London, UK, 1998.
21. Oliveira, J.P.; Figueiredo, M.A.T.; Bioucas-Dias, J.M. Parametric Blur Estimation for Blind Restoration of Natural Images: Linear Motion and Out-of-Focus. *IEEE Trans. Image Process.* **2014**, *23*, 466–477. [CrossRef] [PubMed]
22. Zhu, X.; Cohen, S.; Schiller, S.; Milanfar, P. Estimating Spatially Varying Defocus Blur from A Single Image. *IEEE Trans. Image Process.* **2013**, *22*, 4879–4891. [CrossRef] [PubMed]
23. Ma, L.; Tan, T.; Wang, Y.; Zhang, D. Personal identification based on iris texture analysis. *IEEE Trans. Pattern Anal. Mach. Intell.* **2003**, *25*, 1519–1533. [CrossRef]
24. Yan, R.; Shao, L. Blind Image Blur Estimation via Deep Learning. *IEEE Trans. Image Process.* **2016**, *25*, 1910–1921. [CrossRef]
25. Daugman, J. How iris recognition works. *IEEE Trans. Circuits Syst. Video Technol.* **2004**, *14*, 21–30. [CrossRef]
26. Kalka, N.D.; Zuo, J.; Schmid, N.A.; Cukic, B. Image quality assessment for iris biometric. In *Biometric Technology for Human Identification III, Proceedings of the Defense and Security Symposium, Orlando, FL, USA, 17–21 April 2006*; Flynn, P.J., Pankanti, S., Eds.; International Society for Optics and Photonics (SPIE): Bellingham, WA, USA, 2006; Volume 6202, p. 62020D.
27. Mohammad, K.; Agaian, S. Efficient FPGA implementation of convolution. In Proceedings of the 2009 IEEE International Conference on Systems, Man and Cybernetics, San Antonio, TX, USA, 11–14 October 2009; pp. 3478–3483. [CrossRef]
28. Sreenivasulu, M.; Meenpal, T. Efficient Hardware Implementation of 2D Convolution on FPGA for Image Processing Application. In Proceedings of the 2019 IEEE International Conference on Electrical, Computer and Communication Technologies (ICECCT), Coimbatore, India, 20–22 February 2019; pp. 1–5. [CrossRef]
29. Kang, B.J.; Park, K.R. A Study on Iris Image Restoration. In *Audio- and Video-Based Biometric Person Authentication*; Kanade, T., Jain, A., Ratha, N.K., Eds.; Springer: Berlin/Heidelberg, Germany, 2005; pp. 31–40.
30. Wan, J.; He, X.; Shi, P. An Iris Image Quality Assessment Method Based on Laplacian of Gaussian Operation. In Proceedings of the IAPR International Workshop on Machine Vision Applications, Tokyo, Japan, 16–18 May 2007.
31. Viola, P.; Jones, M. Rapid object detection using a boosted cascade of simple features. In Proceedings of the 2001 IEEE Computer Society Conference on Computer Vision and Pattern Recognition. CVPR 2001, Kauai, HI, USA, 8–14 December 2001; Volume 1, pp. 511–518. [CrossRef]
32. Lienhart, R.; Liang, L.; Kuranov, A. A detector tree of boosted classifiers for real-time object detection and tracking. In Proceedings of the 2003 International Conference on Multimedia and Expo. ICME '03. Proceedings (Cat. No.03TH8698), Baltimore, MD, USA, 6–9 July 2003; Volume 2. [CrossRef]

33. Dong, W.; Sun, Z.; Tan, T. A Design of Iris Recognition System at a Distance. In Proceedings of the 2009 Chinese Conference on Pattern Recognition, Nanjing, China, 4–6 November 2009; pp. 1–5. [CrossRef]
34. Yambay, D.; Doyle, J.S.; Bowyer, K.W.; Czajka, A.; Schuckers, S. LivDet-iris 2013—Iris Liveness Detection Competition 2013. In Proceedings of the IEEE International Joint Conference on Biometrics, Clearwater, FL, USA, 29 September–2 October 2014; pp. 1–8. [CrossRef]

sensors

MDPI

Article

Automated Identification of Hidden Corrosion Based on the D-Sight Technique: A Case Study on a Military Helicopter

Andrzej Katunin [1,*], Piotr Synaszko [2] and Krzysztof Dragan [2]

1 Department of Fundamentals of Machinery Design, Faculty of Mechanical Engineering, Silesian University of Technology, Konarskiego 18A, 44-100 Gliwice, Poland
2 Airworthiness Division, Air Force Institute of Technology, Ks. Bolesława 6, 01-494 Warsaw, Poland; piotr.synaszko@itwl.pl (P.S.); krzysztof.dragan@itwl.pl (K.D.)
* Correspondence: andrzej.katunin@polsl.pl; Tel.: +48-32-237-1069

Abstract: Hidden corrosion remains a significant problem during aircraft service, primarily because of difficulties in its detection and assessment. The non-destructive D-Sight testing technique is characterized by high sensitivity to this type of damage and is an effective sensing tool for qualitative assessments of hidden corrosion in aircraft structures used by numerous ground service entities. In this paper, the authors demonstrated a new approach to the automatic quantification of hidden corrosion based on image processing D-Sight images during periodic inspections. The performance of the developed processing algorithm was demonstrated based on the results of the inspection of a Mi family military helicopter. The nondimensional quantitative measurement introduced in this study confirmed the effectiveness of this evaluation of corrosion progression, which was in agreement with the results of qualitative analysis of D-Sight images made by inspectors. This allows for the automation of the inspection process and supports inspectors in evaluating the extent and progression of hidden corrosion.

Keywords: D-Sight; hidden corrosion; damage identification; DAIS; non-destructive testing; aircraft structures

Citation: Katunin, A.; Synaszko, P.; Dragan, K. Automated Identification of Hidden Corrosion Based on the D-Sight Technique: A Case Study on a Military Helicopter. *Sensors* **2023**, *23*, 7131. https://doi.org/10.3390/s23167131

Academic Editors: Christos Nikolaos E. Anagnostopoulos, Stelios Krinidis and Karim Benzarti

Received: 17 July 2023
Revised: 25 July 2023
Accepted: 8 August 2023
Published: 11 August 2023

1. Introduction

Widely applied regulations for periodic inspections of aircraft structures using non-destructive testing (NDT) techniques are one of the crucial tasks within the damage tolerance philosophy. According to this philosophy, such inspections ensure the timely detection and identification of damage and the further monitoring of its growth in defined limits, which makes it possible to guarantee appropriate reliability and, therefore, the structural integrity and safety of aircraft. The need for effective inspection methods for such structures drives the development of new NDT techniques and the enhancement of existing ones to improve their sensitivity to various types of damage, as well as the smaller and smaller sizes of damage sites that are possible to detect.

Regardless of the wide application of composite materials in aircraft structures, there are still numerous elements made of metallic alloys in both new and older aircraft, which means adjusting the inspection approaches to these elements and structures. Under the umbrella of structural aging, besides fatigue, one of the most widespread and costly types of damage in such structures is corrosion. The cost of aircraft corrosion is on the level of billions of dollars annually in the United States alone, according to various reports [1–4]. Moreover, as reported in [5], corrosion is a primary cause of structural issues (80% of all issues) related to the aging of aircraft, resulting in tens of incidents annually caused by this phenomenon. This demonstrates the need for effective inspection techniques to detect and identify corrosion in a timely manner. For this purpose, corrosion prevention and control programs that regulate inspection frequency, applied techniques, and corrosion identification, have been introduced based on the recommendations of the Federal Aviation

Administration (FAA) advisory circular [6] and manufacturer recommendations for specific aircraft.

A variety of corrosion types are described in detail, e.g., in [2,6–8], and often require various approaches for their detection and evaluation. The most widely used approach for the inspection of corrosion in aircraft structures is visual inspection, which is used to detect corrosion spots. However, in some cases, visual inspection can be ineffective, especially in cases of so-called hidden corrosion, which appears primarily in rivet joints because of moisture penetration in lap joints and results in the initiation of electrochemical processes. These processes cause the appearance of corrosion products, which, given their much higher stiffness compared with the surrounding material—usually aluminum alloys—imply internal stresses, resulting in the appearance of surface deformations [9]. Because of this, the process is called the pillowing effect. Because hidden corrosion, even at high severity, is barely visible to the naked eye, NDT-based approaches are necessary for sensing and detection. According to the literature [2,3,6,10,11], the following NDT techniques are used most often in such inspections: X-ray radiography/tomography, ultrasonography, eddy current testing, thermography, and shearography. However, their application is often biased by a certain amount of uncertainty in detecting hidden corrosion spots because of the low magnitude of the resulting deformations and the specificity of the application of particular techniques. For example, the application of the eddy current NDT technique is difficult in such cases because of the allowable tolerances for manufactured sheets in the aviation industry, and therefore, the detected spots of hidden corrosion are on a level of measurement uncertainty, as reported by Komorowski et al. [12]. Moreover, all of the above-mentioned NDT techniques are characterized by high labor intensity, a relatively long period of testing, the need to dismantle tested elements in numerous situations, and comparatively high inspection costs since specially trained personnel are needed [6]. Considering this, effective, fast, accurate, and inexpensive techniques for inspection of this type of damage are favorable.

The technique that meets all these requirements is double-pass retroreflection, also known also by its commercial name—D-Sight—developed in 1983 by Diffracto Ltd. (Windsor, ON, Canada) in Canada for hidden corrosion detection in aircraft structures and successfully implemented in 1988 by the Canadian Institute for Aerospace Research National Research Council. D-Sight is an optical technique based on the evaluation of surface deformations. In this way, the mentioned problems of measurement uncertainties do not take place. The principle of operation for the D-Sight technique is based on imaging the tested surface at an oblique angle. This surface is illuminated with a light source shifted from the camera, and the light is reflected from the surface onto a special retroreflective screen, which disperses this light and reflects it back on the test surface. This makes it possible to highlight tiny deformations caused by the pillowing effect, and this image is then captured by the camera. The mathematical background for the principle of operation of this technique can be found, e.g., in [13]. The attempts of the inventors and the first working group that used the D-Sight technique for the evaluation of hidden corrosion can be found in [14–16]. The construction of the test device used to sense the hidden corrosion based on the D-Sight technique, known as the D-Sight Aircraft Inspection System (DAIS), is simple, which makes inspection comparatively inexpensive. Moreover, the inspection of large areas of aircraft is possible in a short testing time, which introduces savings in labor intensity. This makes it possible to perform inspections in a fast and reliable way. Nevertheless, in addition to the mentioned advantages, the D-Sight technique is mostly used as a qualitative technique, and it cannot provide enough information for the evaluation of the extent of corrosion or its growth. Several attempts have been made in the past to improve this technique by incorporating supporting finite element models and analysis of profiles based on grayscale images resulting from inspections [12,17–19]; however, their applicability under routine inspection conditions was still limited. Moreover, the authors of this paper identified additional difficulties during previous studies [20], namely that the angle of observation of a tested structure and the illumination should be strictly repeatable, which is especially important during the comparison of historical data from inspections. Additionally, the determination of corroded areas, visible as darker regions in D-Sight images, is also a challenging problem because of small differences in the colors of healthy and corroded areas, as well as the nonuniformity of color distribution in the vicinity of rivets, where the hidden corrosion appears. However, these areas are observable by the naked eye in D-Sight images; therefore, it is reasonable to apply perceptual color contrast measures. Numerous approaches have been developed for this purpose, which can be found in the literature. They include numerous contrast measures reviewed, e.g., in [21–24]. Most of them, however, use sophisticated algorithms, which can extend processing runtimes. The selection of an effective and fast algorithm to evaluate hidden corrosion remains an open question in the processing of D-Sight images.

Recently, numerous steps have been taken toward improving the D-Sight technique to become a quantitative system. Brandoli et al. [25] demonstrated the application of deep neural networks (DNNs) for the detection of hidden corrosion in aircraft fuselage structures. A similar approach was presented by Zuchniak et al. [26], where the authors used machine learning to detect hidden corrosion spots. Nevertheless, the problem of the quantitative evaluation of hidden corrosion remains of great importance from the point of view of supporting inspections and ground maintenance for aircraft. Some steps toward solving the mentioned disadvantages of D-Sight inspections and the quantification of hidden corrosion based on D-Sight images were undertaken by a team of authors in the following study. In [27], the authors proposed a method of image processing that includes procedures to reduce the influence of the angle of observation and non-uniform illumination and detect corroded areas, and it included the first attempts at their quantification. Furthermore, laboratory tests were performed on specimens with simulated hidden corrosion to evaluate the sensitivity of the D-Sight technique and find a correlation between the true dimensions of the corrosion spots obtained using reference methods, both planar and in the direction normal to the surface of the tested structures, as well as those estimated based on image processing of D-Sight images [28].

The current study is motivated by the need to develop a computationally efficient method for quantifying hidden corrosion in aircraft structures, which will allow for the automation of the evaluation process and support inspectors in the evaluation of the extent of corrosion and its growth over the years of an aircraft's operation. Such evaluation is possible using subsequent analyses of D-Sight images collected during periodic inspections using this technique. The D-Sight technique is used as a routine approach for inspections at the Air Force Institute of Technology in Warsaw, which performs maintenance on aircraft for the Polish Armed Forces. The study was carried out on the inspection results of selected structures of the Mi family military helicopters to demonstrate the performance of the proposed approach using realistic inspection results. It highlights the difficulties and open questions in the process of evaluating hidden corrosion and demonstrates the processing algorithm, which can be used as a supporting tool for inspectors using this technique.

2. Inspections and Data Acquisition

2.1. Pillowing Phenomenon and Challenges of D-Sight Inspections

In practice, hidden corrosion occurs most often in the multilayer lap joints of aluminum skins joined by rivets. The increase in the volume of aluminum oxides is greater than the decrease in volume caused by the loss of layer thickness, which results in growing deformations in the skin between the rivets (see Figure 1).

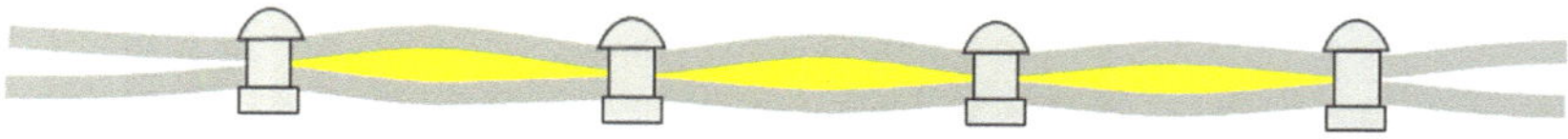

Figure 1. Schematic representation of pillowing phenomenon in multilayered lap joints.

The observation of deformations allows us to detect corrosion, which may also occur in the second and subsequent layers of the lap joint. The stiffness of the aluminum oxides is larger than that of the aluminum alloys, which leads to skin deformation as well as layer thickness decreases. In this process, trihydrate oxides have a dominant influence on the resulting deformations, which is additionally amplified by the appearance of monohydrate oxides [12]. An increase in the volume of the oxides leads to not only material loss but also to pillowing deflection in the skin, which also increases the shear stress in the lap joints. This may lead to critical failures, such as multiple-site fatigue cracking, which, for example, was the main cause of cracked elements that led to failure in the infamous Aloha Airline Flight 243 accident in 1988 [29].

As discussed above, because of the small magnitude of deformation resulting from the pillowing effect, hidden corrosion is difficult to detect with numerous NDT techniques, especially in the early stages of its development. The results of previous studies [28] have shown that the lowest detectable magnitude is at the level of 30 μm. This result was obtained using the D-Sight NDT technique, which demonstrated a high sensitivity to such deformations.

Inspections using D-Sight techniques have been implemented with hardware created by Diffracto Ltd. in a system known as DAIS. The principle of operation for this testing device is based on the above overview of the D-Sight testing approach and can be found in numerous previous publications; see, e.g., [27,28]. However, during inspections, numerous factors influencing the quality of the resulting D-Sight images need to be taken into consideration, such as the reflectivity of the tested surface and the position of the testing device, which has a direct influence on the angle of observation and illumination. Considering the curvatures present in aircraft fuselage structures, ensuring these conditions is not always a trivial task, as can be seen in the example photograph from the inspection (see Figure 2). Because of this, the D-Sight technique is currently used in the practice of aircraft inspections mostly as a qualitative approach, which allows for the evaluation of the severity of corrosion based on the subjective opinion of a single inspector.

Figure 2. The inspection of an aircraft using DAIS.

2.2. Inspections and Acquisition of D-Sight Images

The current study focuses on the improvement of the D-Sight technique for the purpose of quantitatively evaluating D-Sight images and automating the evaluation process based on real aircraft structures after successfully testing the developed approach on specimens with artificially introduced deformations that simulate hidden corrosion [28]. For this purpose, historical data from periodic inspections of the Mi family of military helicopters (see Figure 3), which are a part of the arsenal of the Polish Armed Forces, were considered as a case study. According to FAA, European, and national recommendations, inspection data should be collected and analyzed throughout the service life of helicopters, especially for riveted lap joints. Some examples of the tested structures are presented in Figure 4.

Figure 3. The Mi-type helicopter.

(a) (b)

Figure 4. Examples of riveted lap joints in the tested helicopter: views from the outer (**a**) and inner (**b**) sides.

The inspections were carried out by the Air Force Institute of Technology, Warsaw, using the DAIS 250C scanning system (Diffracto Ltd., Windsor, ON, USA) (Figure 5).

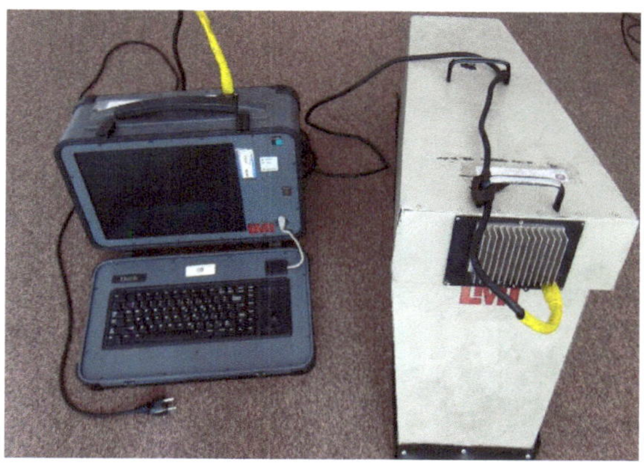

Figure 5. The DAIS 250C scanning system.

Before testing, the surface is covered with Ecolink Electron® antireflective agent (Tucker, GA, USA) to maximize light reflection. The role of the hood, visible in Figure 5, is to isolate the measurement system from ambient light. Images with spatial resolutions of 640 × 480 pixels are collected during inspections in accordance with a scheme of a given element, which is recoded in the settings file in the DAIS system. An example of such a schematic for the tested helicopter is presented in Figure 6. The cyan rectangles represent the areas marked for the expected appearance of hidden corrosion. The inspector reads information about the subsequent positions of the measuring device from the file, which allows him to conduct the measurements in an orderly manner.

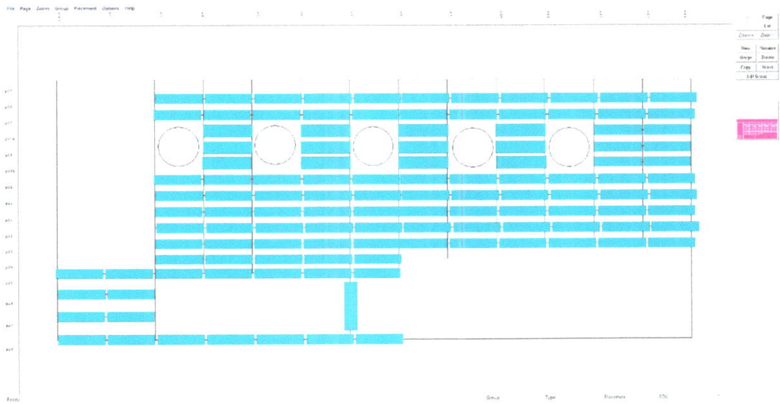

Figure 6. Scheme of the grid for the tested helicopter, indicating the location of interest.

For the following case study, a single location of interest was selected to demonstrate the performance of the developed quantification procedure. Images acquired from the same location were collected in an inspection period of 13 years of operation for the considered helicopter. During this period, five inspections were performed. The resulting D-Sight images from these inspections are presented in Figure 7. The hidden corrosion in these images manifests in local color changes around the rivets. The corrosion severity for the tested area was classified by an inspector as small for the images collected in the period of 2009–2014 (see Figure 7a–c) and moderate for the subsequent period (see Figure 7d,e). The presented results of the inspection demonstrate the mentioned challenges in the quantitative evaluation of hidden corrosion: all of the D-Sight images have different angles of observation and illumination. Moreover, one can observe inaccuracies in the spatial positioning of the testing system, resulting in an offset observable in these images, which is a common situation in inspection practices due to the performance of inspections by various inspectors in various conditions as well as a lack of positioning systems for testing.

(a) (b) (c)

Figure 7. *Cont.*

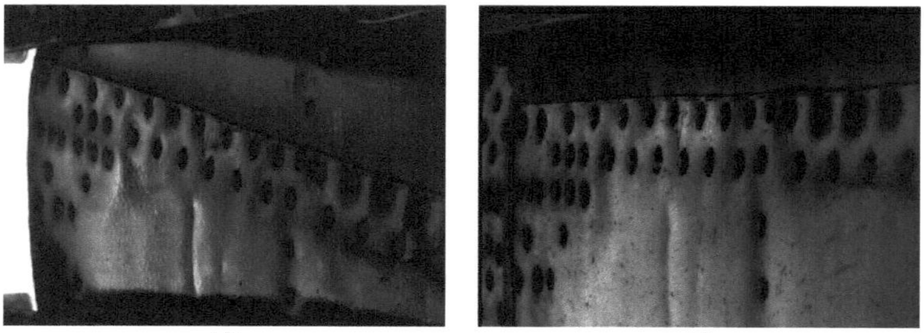

(**d**) (**e**)

Figure 7. The collected D-Sight images from the inspections performed in (**a**) 2009, (**b**) 2012, (**c**) 2014, (**d**) 2017, and (**e**) 2022 for the same area of the inspected helicopter.

3. Processing and Evaluation of D-Sight Images

3.1. Data Preparation and Processing Algorithm

To evaluate the extent of corrosion in the tested area and its evolution, the acquired D-Sight images required preprocessing to rotate them to a planar view. The algorithms described below were implemented using the MATLAB 2022a (MathWorks®, Natick, MA, USA) environment with the Image Processing Toolbox. The calculations were performed on a Windows 10 laptop equipped with an Intel® Core™ i7 quad-core processor and 16 GB of RAM. The preprocessing algorithm, developed previously and in [28], consists of three main steps: image alignment, orthonormalization, and illumination equalization. The parameters of the algorithm can be found in [28]. In the first step, the edge detection procedure was applied, and then, the images were subjected to the application of Hough and shearing transforms. In the next step, orthonormalization was performed to rotate the images to a planar view, which allows for the evaluation of the dimensions of the corrosion spots. In the last step, the contrast of the images was improved to highlight the color differences in the hidden corrosion spots in the vicinity of rivets. The preprocessed D-Sight images of the analyzed sequence are presented in Figure 8. The color difference in the vicinity of rivets, which represents hidden corrosion spots, is still well visible in the preprocessed images.

Based on an analysis of the literature, the local ΔE^* metric was selected to evaluate perceptual color differences. To detect and quantify the corrosion spots, the preprocessed images were converted into the CIELAB color space, known also as the L*a*b* color space, according to the CIE76 standard proposed by the International Commission of Illumination (Commission internationale de l'éclairage) in 1976. The conversion into this color space is due to its perceptual uniformity within the human eye, which makes it possible to measure color differences. Next, the ΔE^* metric, defined as

$$\Delta E^* = \sqrt{\left(L_2^* - L_1^*\right)^2 + \left(a_2^* - a_1^*\right)^2 + \left(b_2^* - b_1^*\right)^2} \tag{1}$$

where $\left(L_1^*, a_1^*, b_1^*\right)$ and $\left(L_2^*, a_2^*, b_2^*\right)$ represent two colors defined in the L*a*b* color space, was applied to converted images to quantify color differences. This metric is based on the calculation of the Euclidean distance in particular channels of the L*a*b* color space. To limit this approach to local color changes, the square window with a 20 times smaller size with respect to the height of the analyzed image was applied. This value was determined empirically and adjusted to typical dimensions of the corrosion spots in the analyzed images. The introduction of such a dependency allowed for a reduction in differences in the dimensions of the images obtained after preprocessing (Figure 8). Using this window, the local mean values $\left(\overline{L}^*, \overline{a}^*, \overline{b}^*\right)$ in each channel of the L*a*b* color space were calculated, and then, the original values of the analyzed image in the specified window were subtracted from this local mean value. In this way, the terms of the Euclidean distance formula were obtained, being a modification of (1):

$$\Delta E^* = \sqrt{\left(\overline{L}^* - L_1^*\right)^2 + \left(\overline{a}^* - a_1^*\right)^2 + \left(\overline{b}^* - b_1^*\right)^2}. \quad (2)$$

The results presented in [21] show that the applied approach is the fastest among similar and more advanced algorithms. The example of an image obtained after these operations for the case shown in Figure 8a is presented in Figure 9a. Then, a quantization procedure was performed on an analyzed image using thresholding based on Otsu's method. The thresholds determined during this procedure were adjusted to the color intensity of an analyzed image, which additionally reduced the problem of non-uniform illumination within the sequence of D-Sight images being analyzed. The result of this procedure is presented in Figure 9b.

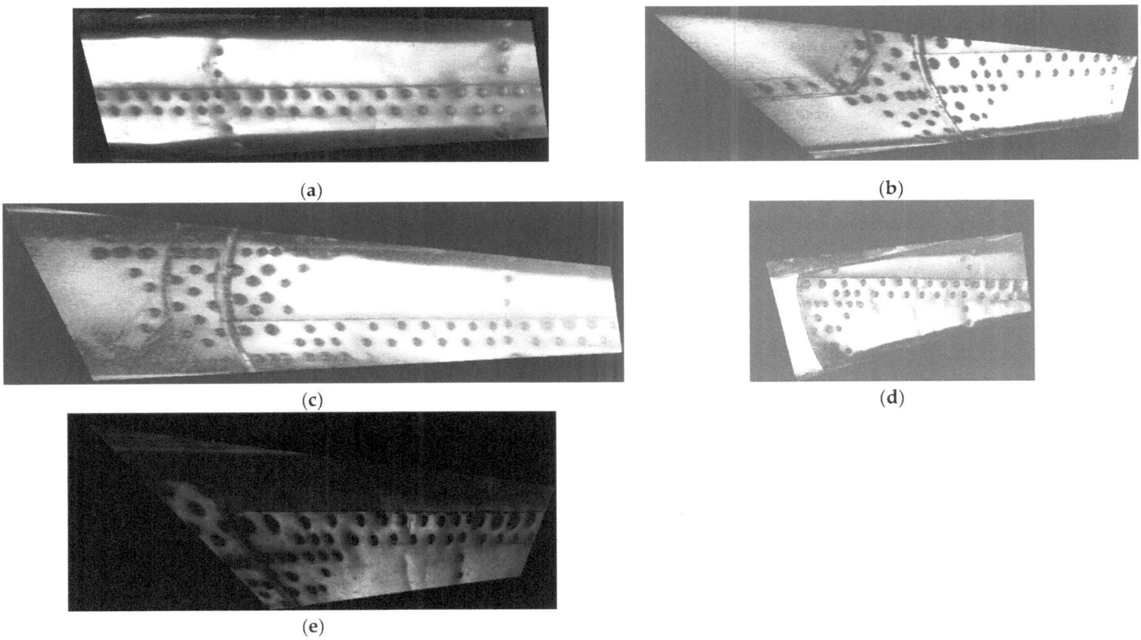

Figure 8. The preprocessed D-Sight images of the analyzed sequence for the tested area for the inspections performed in (**a**) 2009, (**b**) 2012, (**c**) 2014, (**d**) 2017, and (**e**) 2022.

Further, to identify corroded areas and rivets, morphology operations were applied. In the beginning, very large and very small objects that correspond to the surrounding frames of the tested structures and the long edges of overlapping sheets, and noise, respectively, were removed using a morphological opening. The threshold for very large objects was set to 100,000 px, while the threshold for very small objects was calculated as the total area of an image in px divided by 12,000; these thresholds were selected empirically by analyzing the considered images and were further applied within XOR logical operation. In the next step, to classify corrosion spots and rivets, the following criteria were applied. The rivets were classified based on the criterion of the roundness of the convex areas of the remaining objects, while the corrosion spots were classified based on the aspect ratio of the bounding boxes of the remaining objects with a threshold set at five. All objects for which this threshold was exceeded were removed from the image. The latter operation allowed for the removal of residues from the surrounding frames and shadows inappropriately classified as corrosion spots after the initial cleaning. The resulting image after the application of morphological operations is presented in Figure 9c. In the last step, the corrosion spots were visualized as a mask on the preprocessed D-Sight image (see Figure 9d), and the nondimensional corrosion extent was calculated using the following quantitative measure. To determine its value, the determined areas of the convex areas of the detected rivets (the number of considered rivets was in a range of 20–40) were averaged and then used as a divisor for the total area of the detected corrosion spots. Given the size of the D-Sight images after preprocessing, the runtimes for each case were ca. 30 min. For clarity, the processing algorithm is presented in the form of a flowchart in Figure 10.

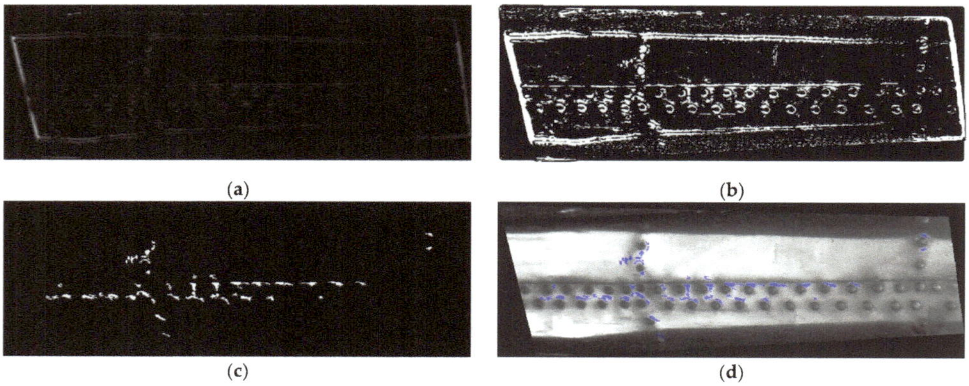

Figure 9. The results of subsequent operations during the processing of the preprocessed D-Sight images: (**a**) determination of the local ΔE^* metric; (**b**) quantization; (**c**) morphological operations; (**d**) visualization of the detected corrosion spots.

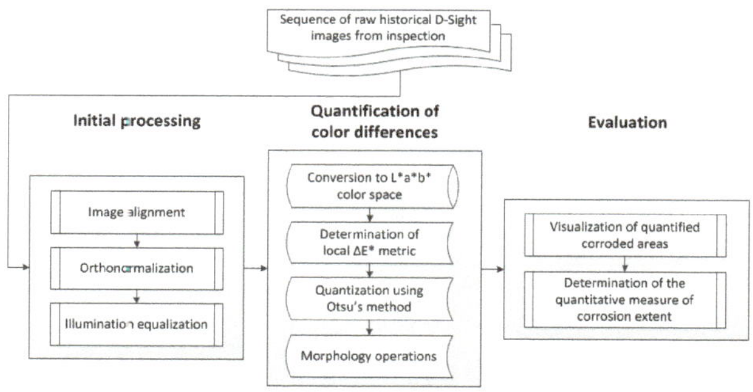

Figure 10. The flowchart of the processing algorithm.

3.2. Evaluation of Hidden Corrosion

The sequence of D-Sight images analyzed in this study (see Figure 7) was processed using the algorithm presented in Section 3.1. The results of processing are presented in Figures 9d and 11a–d.

From the presented sequence, the progression of the hidden corrosion that developed over the tested helicopter's years of operation is clearly visible; i.e., the corroded area around the rivets, labeled with a blue mask in the preprocessed D-Sight images, significantly increased. It can be seen that, in some locations, hidden corrosion was not properly detected (see, e.g., the vertical line of rivets on the bottom right in Figure 11d); however, the vast majority of the corrosion spots were well detected and identified. The corroded areas were also assessed quantitatively based on the measure introduced in Section 3.1. The selection of this measure was based on the simplicity of its implementation and runtime efficiency, which is a critical parameter during the automated evaluation of D-Sight images. The obtained results are presented in Table 1.

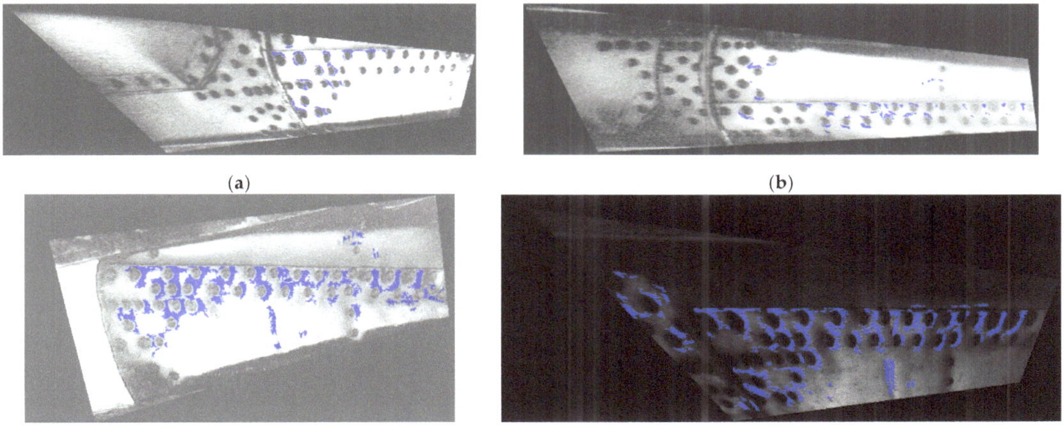

(**a**) (**b**) (**c**) (**d**)

Figure 11. The results of processing the preprocessed D-Sight images based on the analyzed historical sequence, which were captured in (**a**) 2012, (**b**) 2014, (**c**) 2017, and (**d**) 2022.

Table 1. The results of the quantitative analysis of corroded areas for the analyzed sequence of D-Sight images.

Year of Inspection	**2009**	**2012**	**2014**	**2017**	**2022**
Value of quantitative measure	13.1842	12.6146	14.0894	39.7338	52.5461

The determined quantitative measure of the analyzed sequence of D-Sight images demonstrates the increasing trend of the extent of the corrosion, which is in agreement with the results of the qualitative assessment of corrosion severity made by the inspector (see Section 2.2), despite the mismatches in the observed corroded areas (the spatial offset between particular D-Sight images during subsequent inspections). These mismatches are the reason for the decrease in the value of the quantitative measure of the inspection in 2012. Nevertheless, the obtained quantitative results confirm the trend in corrosion progression in the analyzed period of operation for the tested helicopter and can be used as an automated supporting tool for quantitative assessments of the extent and progression of hidden corrosion.

4. Discussion and Conclusions

The presented case study demonstrates an approach to processing D-Sight images collected during periodic inspections of a selected area of a fuselage of a Mi-family military helicopter, which allows one to quantify the extent of hidden corrosion and its progression in an automatic way. The results of the case study show numerous challenges during the quantitative analysis of corroded areas (variable angle of observation, inhomogeneous and non-repeatable illumination, spatial offsets of the acquired D-Sight images), which have the precedent in the repeatable conditions of inspections, being difficult to maintain over long service periods. An improvement in the repeatability of the performed inspections may significantly improve the analysis of the acquired inspection results. As demonstrated in this paper, this is especially important for quantitative evaluations of corrosion extent and progression. Despite the mentioned challenges, the presented processing algorithm, which is based on the determination of the local ΔE^{*} metric, demonstrates high sensitivity to changes in colors in the vicinity of rivets and allows for the successful identification of corroded areas. The obtained quantitative results for the analyzed sequence of D-Sight images demonstrated the increasing trend in the total area of the corrosion spots and are in agreement with the results of the qualitative analysis performed by the inspector, despite the aforementioned challenges and uncertainties, which consist of, among others, spatial offset from inspection to inspection. To further automatize the process of tracking the hidden corrosion growth, the problem of this offset needs to be solved, which is currently one of the limitations of the algorithm and is planned to be resolved during further studies. Moreover, currently the algorithm requires a lot of time to process D-Sight images, mainly due to the significant extension of the size of images after the pre-processing step (see Figure 10). From the point of view of further analysis, such resolution seems to be unnecessary; therefore, optimal scaling needs to be applied in further steps to reduce runtimes and retain the detectability of corrosion spots. Finally, as observed in Figure 11c,d, some dents present on the surface of the tested element were classified as hidden corrosion, since the deformations were similar to those of the appearance of hidden corrosion. This requires checking the results by an inspector to detect such cases, which is another limitation on the way of automating of this approach to be solved in future studies.

The automation of the damage extent and severity assessment process, based on the processing of acquired D-Sight images, makes it possible to effectively support inspectors in the evaluation of the extent, severity, and progression of hidden corrosion during service periods. Moreover, the possibility of quantitative analysis opens up new perspectives that were not available when the technique had a qualitative character, e.g., the possibility of predicting hidden corrosion progression and the development of maintenance programs. Such an approach may influence cost reductions in maintenance and increase the availability of helicopters. This requires further deep studies, which will consolidate knowledge about electrochemical processes during the appearance of this type of corrosion with the obtained results in the following and preceding studies; analyses of loading and environmental factors, which also have an influence on the process; and numerous other factors. The development of such models should be considered in future authors' studies. Finally, the automatic evaluation of the extent of corrosion using D-Sight images makes it possible to prepare large datasets that can be used to train DNNs to further improve the effectiveness of the identification of hidden corrosion spots and avoid their incorrect identification in the case of the presence of other deformed areas.

Author Contributions: Conceptualization, A.K., P.S. and K.D.; methodology, A.K.; software, A.K.; validation, A.K. and P.S.; formal analysis, A.K.; investigation, A.K. and P.S.; resources, P.S. and K.D.; data curation, A.K.; writing—original draft preparation, A.K.; writing—review and editing, A.K., P.S. and K.D.; visualization, A.K.; supervision, A.K.; project administration, A.K.; funding acquisition, A.K. All authors have read and agreed to the published version of the manuscript.

Funding: The publication is supported under the Excellence Initiative—Research University Program of the Silesian University of Technology, grant no. 10/060/SDU/10-21-01, year 2021.

Institutional Review Board Statement: Not applicable.

Informed Consent Statement: Not applicable.

Data Availability Statement: The data that support the findings of this study were taken from the Air Force Institute of Technology. Restrictions apply to the availability of these data, which were used with permission for this study.

Conflicts of Interest: The authors declare no conflict of interest. The funders had no role in the design of the study; in the collection, analyses, or interpretation of data; in the writing of the manuscript; or in the decision to publish the results.

References

1. Komorowski, J.P.; Forsyth, D.S. The role of enhanced visual inspections in the new strategy for corrosion management. *Aircr. Eng. Aerosp. Technol.* **2000**, *72*, 5–13. [CrossRef]
2. Liao, M. Corrosion damage atlas for aircraft corrosion management and structural integrity assessment. In Proceedings of the NATO RTO STO-MP-AVT-303 Corrosion Management, Athens, Greece, 10–14 December 2018.

3. Li, L.; Chakik, M.; Prakash, R. A review of corrosion in aircraft structures and graphene-based sensors for advanced corrosion monitoring. *Sensors* **2021**, *21*, 2908. [CrossRef] [PubMed]
4. Nusser, J.K.; Herzberg, E.; Stimson, D.J.; Babish, C.A. Data-driven corrosion prevention and control decisions for the USAF. In Proceedings of the Society for Protective Coatings 2017 (SSPC 2017) Conference, Tampa, FL, USA, 30 January–2 February 2017.
5. Cassidy, M.; Waldie, J.; Palanisamy, S. A method to estimate the cost of corrosion for Australian Defense Force aircraft. In Proceedings of the AIAC16 Sixteenth Australian International Aerospace Congress, Melbourne, Australia, 23–24 February 2015.
6. *Corrosion Control for Aircraft, Advisory Circular no. 43-4B*; U.S. Department of Transportation, Federal Aviation Administration: Cambridge, MA, USA, 2018.
7. Cole, G.K.; Clark, G.; Sharp, P.K. *The Implications of Corrosion with Respect to Aircraft Structural Integrity*; DSTO Aeronautical and Maritime Research Laboratory: Melbourne, Australia, 1997.
8. Benavides, S. (Ed.) *Corrosion Control in the Aerospace Industry*; Woodhead Publishing: Cambridge, UK, 2009.
9. Cardinal, J.W.; Burnside, H. Damage tolerance risk assessment of T-38 wing skin cracks. In Proceedings of the 2005 USAF Structural Integrity Program (ASIP) Conference, Memphis, TN, USA, 29 November–1 December 2005.
10. Knight, S.P.; Salagaras, M.; Trueman, A.R. The study of intergranular corrosion in aircraft aluminum alloys using X-ray tomography. *Corros. Sci.* **2011**, *53*, 727–734. [CrossRef]
11. Ciężak, P.; Rdzanek, A. Corrosion monitoring of aircraft based on the corrosion prognostic health management (CPHM) system. *J. CONBiN* **2020**, *50*, 205–216. [CrossRef]
12. Komorowski, J.P.; Bellinger, N.C.; Gould, R.W. The role of corrosion pillowing in NDI and in the structural integrity of fuselage joints. Fatigue in New and Aeging Aircraft. In Proceedings of the 19th Symposium of the International Committee on Aeronautical Fatigue, Edinburgh, UK, 18–20 June 1997; pp. 251–266.
13. Heida, J.; Bruinsma, A. D-Sight technique for rapid impact damage detection on composite aircraft structures. In Proceedings of the 7th European Conference on Non-Destructive Testing, Copenhagen, Denmark, 26–29 May 1998; NDT.net: Copenhagen, Denmark, 1998. Volume 4.
14. Reynolds, R.L.; Karpala, F.; Clarke, D.A.; Hageniers, O.L. Theory and applications of a surface inspection technique using double-pass retroreflection. *Opt. Eng.* **1993**, *32*, 2122–2129. [CrossRef]
15. Hageniers, O.L. D Sight for large area aircraft inspection. *Proc. SPIE* **2001**, *1993*, 248–256.
16. Bellinger, N.; Komorowski, J.; Benak, T. Residual life predictions of corroded fuselage lap joints. *Int. J. Fatigue* **2001**, *23*, 349–356. [CrossRef]
17. Komorowski, J.P.; Bellinger, N.C.; Gould, R.W.; Marincak, A.; Reynolds, R. Quantification of corrosion in aircraft structures with double pass retroreflection. *Can. Aeronaut. Space J.* **1996**, *42*, 76–82.
18. Bellinger, N.C.; Komorowski, J.P. Corrosion pillowing stresses in fuselage lap joints. *AIAA J.* **1997**, *35*, 317–320. [CrossRef]
19. Forsyth, D.S.; Komorowski, J.P.; Gould, R.W. Use of solid film highlighter in automation of D sight image interpretation. *Proc. SPIE* **1998**, *3397*, 50–56.
20. Katunin, A.; Dragan, K. Qualitative to quantitative non-destructive evaluation: A concept for D-Sight inspections of aircraft structures. *Appl. Mech. Mater.* **2022**, *909*, 69–74. [CrossRef]
21. Xiong, G.; Li, X.; Gong, J.; Chen, H.; Lee, D.-J. Color rank and census transforms using perceptual color contrast. In Proceedings of the 2010 11th International Conference on Control Automation Robotics & Vision, Singapore, 7–10 December 2010; pp. 1225–1230.
22. Simone, G.; Pedersen, M.; Hardeberg, J.Y. Measuring perceptual contrast in digital images. *J. Vis. Commun. Image Represent.* **2012**, *23*, 491–506. [CrossRef]
23. Palma-Amestoy, R.; Provenzi, E.; Bertalmío, E.; Caselles, V. A perceptually inspired variational framework for color enhancement. *IEEE Trans. Pattern Anal.* **2009**, *31*, 458–474. [CrossRef] [PubMed]
24. Beghdadi, A.; Larabi, M.-C.; Bouzerdoum, A.; Iftekharuddin, K.M. A survey of perceptual image processing methods. *Signal Process. Image Commun.* **2013**, *28*, 811–831. [CrossRef]
25. Brandoli, B.; de Geus, A.; Souza, J.; Spadon, G.; Soares, A.; Rodrigues, J., Jr.; Komorowski, J.; Matwin, S. Aircraft fuselage corrosion detection using artificial intelligence. *Sensors* **2021**, *21*, 4026. [CrossRef] [PubMed]
26. Zuchniak, K.; Dzwinel, W.; Majerz, E.; Pasternak, A.; Dragan, K. Corrosion detection on aircraft fuselage with multi-teacher knowledge distillation. In *Computational Science–ICCS 2021*; Lecture Notes in Computer Science; Paszynski, M., Kranzlmüller, D., Krzhizhanovskaya, V., Dongarra, J., Sloot, P., Eds.; Springer: Cham, Germany, 2021; Volume 12747, pp. 318–332.
27. Katunin, A.; Nagode, M.; Oman, S.; Cholewa, A.; Dragan, K. Monitoring of hidden corrosion growth in aircraft structures based on D-Sight inspections and image processing. *Sensors* **2022**, *22*, 7616. [CrossRef] [PubMed]
28. Katunin, A.; Lis, K.; Joszko, K.; Żak, P.; Dragan, K. Quantification of hidden corrosion in aircraft structures using enhanced D-Sight NDT technique. *Measurement* **2023**, *216*, 112977. [CrossRef]
29. *Aircraft Accident Report–Aloha Airlines, Flight 243, Boeing 737-200, N73711, near Maui, Hawaii, April 28, 1988, NTSB/AAR-89/03*; National Transportation Safety Board Bureau of Accident Investigation: Washington, DC, USA, 1989.

sensors

MDPI

Review

Object Detection, Recognition, and Tracking Algorithms for ADASs—A Study on Recent Trends

Vinay Malligere Shivanna [1,*] and Jiun-In Guo [1,2,3]

1 Department of Electrical Engineering, Institute of Electronics, National Yang-Ming Chiao Tung University, Hsinchu City 30010, Taiwan; jiguo@nycu.edu.tw
2 Pervasive Artificial Intelligence Research (PAIR) Labs, National Yang Ming Chiao Tung University, Hsinchu City 30010, Taiwan
3 eNeural Technologies Inc., Hsinchu City 30010, Taiwan
* Correspondence: vinay.ms23@gmail.com

Abstract: Advanced driver assistance systems (ADASs) are becoming increasingly common in modern-day vehicles, as they not only improve safety and reduce accidents but also aid in smoother and easier driving. ADASs rely on a variety of sensors such as cameras, radars, lidars, and a combination of sensors, to perceive their surroundings and identify and track objects on the road. The key components of ADASs are object detection, recognition, and tracking algorithms that allow vehicles to identify and track other objects on the road, such as other vehicles, pedestrians, cyclists, obstacles, traffic signs, traffic lights, etc. This information is then used to warn the driver of potential hazards or used by the ADAS itself to take corrective actions to avoid an accident. This paper provides a review of prominent state-of-the-art object detection, recognition, and tracking algorithms used in different functionalities of ADASs. The paper begins by introducing the history and fundamentals of ADASs followed by reviewing recent trends in various ADAS algorithms and their functionalities, along with the datasets employed. The paper concludes by discussing the future of object detection, recognition, and tracking algorithms for ADASs. The paper also discusses the need for more research on object detection, recognition, and tracking in challenging environments, such as those with low visibility or high traffic density.

Keywords: object detection; object tracking; advanced driver assistance system (ADAS); deep learning

Citation: Malligere Shivanna, V.; Guo, J.-I. Object Detection, Recognition, and Tracking Algorithms for ADASs—A Study on Recent Trends. *Sensors* **2024**, *24*, 249. https://doi.org/10.3390/s24010249

Academic Editors: Stelios Krinidis and Christos Nikolaos E. Anagnostopoulos

Received: 28 September 2023
Revised: 13 December 2023
Accepted: 20 December 2023
Published: 31 December 2023

Correction Statement: This article has been republished with a minor change. The change does not affect the scientific content of the article and further details are available within the backmatter of the website version of this article.

1. Introduction

Advanced driver assistance systems (ADASs) are a group of electronic technologies that assist drivers in driving and parking functions. Through a safe human–machine interface, ADASs increase car and road safety. They use automated technology, such as sensors and cameras, to detect nearby obstacles or driver errors, and respond or issue alerts accordingly. They can enable various levels of autonomous driving, depending on the features installed in the car.

ADASs use a variety of sensors such as cameras, radar, lidar, and a combination of these, to detect objects and conditions around the vehicle. The sensors send data to a computing system, which then analyzes the data and determines the best course of action based on the algorithmic design. For instance, if a camera detects a pedestrian in the vehicle's path, the computing system may trigger the ADAS to sound an alarm or apply the brakes.

The chronicles of ADAS date back to the 1970s [1,2] with the development of the first anti-lock braking system (ABS). Following a slow and steady evolution, additional features such as the lane departure warning system (LDWS) and electronic stability control (ESC) emerged in the 1990s. In recent years, there has been a rapid development of numerous ADASs, with new functionalities being introduced every other day and becoming increasingly prevalent in modern vehicles, as they offer a variety of safety features that aid in preventing accidents, relying on the aforementioned variety of sensors that have made the ADAS a potential system with which to significantly reduce the number of traffic accidents and fatalities. A study by the Insurance Institute for Highway Safety [3] found that different uses of ADASs can reduce the risk of a fatal crash by up to 20–25%. Therefore, ADASs are becoming increasingly common in cars. In 2021, 33% of new cars sold in the United States had ADAS features. This number is expected to grow to 50% by 2030, as ADASs are expected to play a major role in the future of transportation [4]. By helping to prevent accidents and collisions, reducing drivers' fatigue and stress [5,6], improving fuel efficiency [7,8], making parking easier and more convenient [9] and thereby providing peace of mind to drivers and passengers [5,6], ADASs can save lives and make our roads safer.

Additionally, various features of ADASs, as shown in Figure 1, are a crucial part of the development of autonomous driving; in other words, self-driving cars, as autonomous vehicles, rely on the performance and efficiency of ADASs to detect objects and conditions in their surroundings in real-world scenarios. Self-driving cars use a combination of ADASs and artificial intelligence to drive themselves. Therefore, ADASs are continuing to play an important role in the development of autonomous driving as the technology matures.

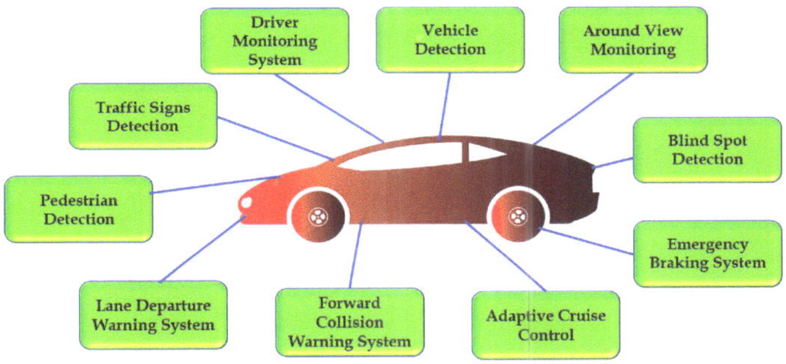

Figure 1. Different features of ADASs.

The basic functionalities of ADASs are object detection, recognition, and tracking. Numerous algorithms allow vehicles to detect and recognize—in other words, to identify and then track—other objects on the road, such as vehicles, pedestrians, cyclists, traffic signs, lanes, probable obstacles on the road, and more; warn the driver of potential hazards; and/or take evasive action automatically.

There are a number of different object detection, recognition, and tracking algorithms that have been developed for ADASs. These algorithms can be broadly classified into two main categories: traditional methods and deep learning (DL) methods, as discussed in detail in Section 1.3.

This paper attempts to provide a comprehensive review of recent trends in different algorithms for various ADAS functions. The paper begins by discussing the challenges of object detection, recognition, and tracking in ADAS applications. The paper then discusses the different types of sensors used in ADASs and different types of object detection, recognition, and tracking algorithms that have been developed for various ADAS methodologies and datasets used to train and test the methods. The paper concludes by discussing the future trends in object detection, recognition, and tracking for ADASs.

1.1. Basic Terminologies

Before diving into the main objective of the paper, the section below introduces some of the basic terminologies commonly used in the field of ADAS research:

a. Image processing is the process of manipulating digital images to improve their quality or extract useful information from them. Image processing techniques are commonly used in ADASs for object detection, recognition, and tracking tasks;
b. Object detection is the task of identifying and locating objects in a scene, such as vehicles, pedestrians, traffic signs, and other objects that could pose a hazard to the driver;
c. Object tracking involves following the movement of vehicles, pedestrians, and other objects over time to predict their future trajectories;
d. Image segmentation is the task of dividing an image into different regions, each of which corresponds to a different object or part of an object such as the bumper, hood, and wheels and other objects such as pedestrians, traffic signs, lanes, forward objects, and so on;
e. Feature extraction is the extraction of features like shape, size, color, and so on from an image or a video; these features are used to identify objects or track their movements.
f. Classification is the task of assigning a label such as vehicles, pedestrians, traffic signs, or others to an object or several images to categorize the objects;
g. Recognition is the task of identifying an object or a region in an image by its name or other attributes.

1.2. An Overview of ADASs

The history of ADAS technology can be traced back to the 1970s with the adoption of the anti-lock braking system [10,11]. Early ADASs including electronic stability control, anti-lock brakes, blind spot information systems, lane departure warning, adaptive cruise control, and traction control emerged in the 1900s and 2000s [12,13]. These systems can be affected by mechanical alignment adjustments or damage from a collision requiring automatic reset for these systems after a mechanical alignment is performed.

1.2.1. The Scope of ADASs

ADASs perform a variety of tasks using object detection, recognition, and tracking algorithms which are deemed as falling within the scope of ADASs; namely, (i) vehicle detection, (ii) pedestrian detection, (iii) traffic signs detection (TSD), (iv) driver monitoring system (DMS), (v) lane departure warning system (LDWS), (vi) forward collision warning system (FCWS), (vii) blind-spot detection (BSD), (viii) emergency braking system (EBS), (ix) adaptive cruise control (ACC), and (x) around view monitoring (AVM).

These are some of the most important of the many ADAS features that rely on detection, recognition, and tracking algorithms. These algorithms are constantly being improved as the demand for safer vehicles continues to grow.

1.2.2. The Objectives of Object Detection, Recognition, and Tracking in ADASs

An ADAS system has various functions with different objectives that can be listed as:

a. Improving road safety: ADASs can aid in improving road safety by reducing the number of accidents; this is achieved by warning drivers of potential hazards and by taking corrective actions to avoid collisions. For example, a LDWS can warn the driver if they are about to drift out of their lane, while a forward collision warning system can warn the driver if they are about to collide with another vehicle;
b. Reducing driver workload: ADASs can help to reduce driver workload by automating some of their driving tasks. This can help to make driving safer and more enjoyable. For example, ACC can automatically maintain a safe distance between the vehicle and the vehicle in front of it, and lane-keeping assist can automatically keep the vehicle centered in its lane;
c. Increasing fuel efficiency: ADASs can help to increase fuel efficiency by reducing the need for the driver to brake and accelerate, which is achieved by maintaining a constant speed and by avoiding sudden speed changes. For example, ACC can

automatically adjust the speed of the vehicle to maintain a safe distance from the vehicle in front of it, which can help to reduce fuel consumption;

d. Providing information about the road environment: ADASs can provide drivers with more information about the road environment, such as the speed of other vehicles, the distance to the nearest object, traffic signs, and the presence of pedestrians or cyclists. This information can help drivers to make better decisions about how to drive and can help to reduce the risk of accidents;
e. Assisting drivers with difficult driving tasks: ADASs can assist drivers with difficult driving tasks, such as parking, merging onto a highway, and driving in bad weather conditions, thereby reducing driver workload and enabling safer driving;
f. Ensuring a comfortable and enjoyable driving experience: ADASs can provide a more comfortable and enjoyable driving experience by reducing stress and fatigue that drivers experience which can be achieved by automating some of the tasks involved in driving, such as maintaining a constant speed and avoiding sudden changes in speed.

The ADAS algorithms are designed to achieve these objectives by using sensors, such as cameras, radar, lidar, and now a combination of these, to collect data about the road environment. The data thus obtained are processed by the algorithms as per their design to identify and track objects, predict the future movement of objects, and warn the driver of potential hazards. These ADAS algorithms are constantly being improved as new technologies are being developed. Continuous and consistent advancements in these technologies are making ADASs even more capable of improving road safety and reducing drivers' workloads.

1.2.3. The Challenges of ADASs

The task of the essential functions of ADASs, namely object detection, recognition, and tracking, is to allow ADASs to identify and track objects in the vehicle's surroundings, such as other vehicles, pedestrians, cyclists, and sometimes random objects and obstacles, using which ADASs can prevent accidents, keep the vehicle in its lane, and provide other driver assistance features. However, there are various challenges associated with object detection, recognition, and tracking in ADASs, such as:

a. Varying environmental conditions: ADASs must be able to operate in a variety of environmental conditions, including different lighting conditions like bright sunlight, dark shadows, fog, daytime, nighttime, etc., different weather conditions such as drizzle, rain, snow, and so on, along with various road conditions including dirt, gravel, etc.;
b. Occlusion: objects on the road in real scenarios can often be occluded by other objects, such as other vehicles, pedestrians, or trees, making it difficult for ADASs to detect and track objects;
c. Deformation: objects on the road can often be deformed, such as when a vehicle is turning or when a pedestrian is walking, causing difficulties for ADASs in detecting and tracking objects;
d. Scale: objects on the road can vary greatly in size, from small pedestrians to large trucks, inducing difficulties for ADASs in detecting and tracking objects of all sizes;
e. Multi-object tracking: ADASs must be able to track multiple objects simultaneously, and this can be challenging as objects move and interact with each other in complex ways in real-world scenarios;
f. Real-time performance: most importantly, ADASs must be able to detect, recognize, and track objects in real time, which is essential for safety-critical applications, as delays in detection or tracking can lead to accidents and make them unreliable.

Researchers are working on developing newer algorithms and improving the existing algorithms and techniques to address these challenges. Due to this, ADASs are becoming increasingly capable of detecting and tracking objects in a variety of challenging conditions.

1.2.4. The Essentials of ADASs

The above section discusses the challenges of different ADAS methods, whereas in this section, we discuss the numerous requirements of [14,15] ADASs, which must be tackled before the aforementioned issues can be resolved. In other words, ADAS algorithms are facing numerous additional predicaments while working on overcoming the challenges discussed in the previous section:

a. The need for accurate sensors: ADASs rely on a variety of sensors to detect and track objects on the road. These sensors must be accurate and reliable to provide accurate information to the ADAS. Nevertheless, sensors are usually affected by factors such as weather, lighting, and the environment, causing difficulties for sensors in providing accurate information, and thus leading to errors in the ADASs;
b. The need for reliable algorithms: ADASs also rely on a variety of algorithms to process the data from the sensors and make decisions about how to respond to objects on the road. These algorithms must be reliable to make accurate and timely decisions. However, these algorithms can also be affected by factors such as the complexity of the environment and the number of objects on the road. This makes it difficult for algorithms to make accurate decisions, leading to errors in the ADAS;
c. The need for integration with other systems: ADASs must be integrated with different systems in the vehicle, such as the braking system and the steering system. This integration is necessary in order for the ADAS system to take action to avoid probable accidents. Nonetheless, integration is complex and time-consuming, resulting in deployment delays of ADASs;
d. The cost of ADASs: ADASs are expensive to develop and deploy, making it difficult for some manufacturers to offer ADASs as standard features in their vehicles. As a result, ADASs are often only available as optional features, which can make them less accessible to all drivers;
e. The acceptance of ADASs by drivers: Some drivers may still be hesitant to adopt ADASs because they worry about the technology or they do not trust the technology. This will result in difficulties persuading drivers to opt for vehicles with ADASs.

Despite these challenges, ADASs have the potential to significantly improve road safety. As the technology continues to improve, ADASs are likely to become more affordable and more widely accepted by drivers. This will help to make roads safer for everyone.

1.3. ADAS Algorithms: Traditional vs. Deep Learning

There are two main types of algorithms used in ADASs: traditional algorithms and DL algorithms. In this section, we discuss the advantages and disadvantages of traditional and DL algorithms for ADASs and also some of the challenges involved in developing and deploying ADASs.

1.3.1. Traditional Algorithms

Traditional methods for object detection, recognition, and tracking are typically the most common type of algorithms used in ADASs, based on hand-crafted, rule-based features, and heuristics designed to capture the distinctive characteristics of different objects. That is, a feature for detecting vehicles might be the presence of four wheels and a windshield. This means that these algorithms use a set of pre-defined rules to determine what objects are present in the environment and how to respond to them. For instance, a traditional lane-keeping algorithm might use a rule that says, 'If the vehicle is drifting out of its lane, then turn the steering wheel in the opposite direction' or 'a rule might state that if a vehicle is detected in the vehicle's blind spot, then the driver should be warned'.

Traditional methods are less complex than DL algorithms, making them easier to develop, and are very effective in certain cases, but they are difficult to generalize to new objects or situations because they are limited by the rules that are hard-coded into them. If a new object, obstacle, or hazard is not covered by a rule, then the algorithm may not be able to detect it. Some of the basic traditional methods-based algorithms are:

a. Object detection: Traditional object detection algorithms typically use a two-step approach:
   i. The region proposal step identifies potential regions in an image that may contain objects, which is typically carried out by using a sliding window approach, where a small window is moved across the image and features are extracted from each window;
   ii. The classification step classifies each region as an object or background. This is typically carried out by using a machine learning (ML) algorithm, such as a support vector machine (SVM) [16] or a random forest [17];
b. Object recognition: Traditional object recognition algorithms typically use a feature-based approach:
   i. The feature extraction step extracts features from an image that are relevant to the object class, which is typically carried out by using hand-crafted features, such as color histograms [18], edge features [19], or shape features [20];
   ii. The classification step classifies the object class by using a ML algorithm, such as a SVM [16] or random forest [17];
c. Object tracking: Traditional object-tracking algorithms typically use a Kalman filter [21]:
   i. The state estimation step estimates the state of the object, such as its position, velocity, and acceleration;
   ii. The measurement update step updates the state estimate based on new measurements of the object.

These traditional object detection, recognition, and tracking algorithm are effective for a variety of ADAS applications. However, they can be computationally expensive and may not be able to handle challenging conditions, such as occlusion or low lighting.

In recent years, there has been a trend towards using DL algorithms for object detection, recognition, and tracking in ADASs. DL algorithms have been shown to be more accurate than traditional algorithms, and they can handle challenging conditions more effectively.

1.3.2. Deep Learning Algorithms

Inspired by the human brain, DL methods for object detection, recognition, and tracking use artificial neural networks (ANNs) to learn the features that are important for identifying different objects. They are composed of layers of interconnected nodes. Each node performs a simple calculation, and the output of each node is used as the input to the next node.

DL algorithms can learn to detect objects, obstacles, and hazards from large datasets of labeled data usually collected using a variety of sensors. The algorithm is trained to associate specific patterns in the data with specific objects or hazards. DL algorithms are generally more complex than traditional algorithms, but they can achieve higher accuracy as they are not limited by hand-crafted rules, they can learn to detect objects and hazards not covered by any rules, and they are also able to handle challenging conditions, such as occlusion or low lighting, more effectively. Some of the standard DL method-based algorithms are discussed below:

a. Object detection: DL object detection algorithms commonly use a convolutional neural network (CNN) to extract features from an image. The CNN is then trained on a dataset of images that have been labeled with the objects that they contain. Once the CNN is trained, it can be used to detect objects in new images;
b. Object recognition: DL object recognition algorithms also conventionally use a CNN to extract features from an image. However, the CNN is trained on a dataset of images that have been labeled with the class of each object. The trained CNN can be used to recognize the class of objects in new images;

c. Object tracking: DL object tracking algorithms typically use a combination of CNNs and Kalman filters [21]. The CNN is used to extract features from an image and the Kalman filter is used to track the state of the object over time.

2. Sensors Used in Object Detection, Recognition, and Tracking Algorithms of ADASs

Several sensors can be used for object detection, recognition, and tracking in ADASs. The most common sensors include cameras, radars, and lidars. In addition to these sensors, some other sensors can also be used, such as:

a. Ultrasonic sensors: used to detect objects that are close to the vehicle, aiding in preventing collisions with pedestrians or other vehicles;
b. Inertial measurement units (IMUs): employed to track the movement of the vehicle using which the accuracy of object detection and tracking can be improved;
c. GPS sensors: used to determine the position of the vehicle and are utilized to track the movement of the vehicle and to identify objects that are in the vehicle's path;
d. Gyroscope sensors: used to track the orientation of the vehicle and employed to improve the accuracy of object detection and tracking algorithms.

The choice of sensors for object detection, recognition, and tracking in ADASs depends on the specific application. For instance, a system that is designed to detect pedestrians may use a combination of cameras and radar, while a system that is designed to track the movement of other vehicles may use a combination of radar and lidar.

The combination of multiple sensors is mostly used in more recent state-of-the-art methods, as this improves the accuracy of object detection, recognition, and tracking algorithms. The combination of sensors combines the strengths of the sensors and overcomes the weaknesses of the other sensors.

2.1. Cameras, Radar, and Lidar

Cameras, radar, and lidar are the most common types of sensors used in ADASs. While there are two main types of cameras—monocular cameras are the most common type used in ADASs, which have a single lens and can only see in two dimensions, while stereo cameras have two lenses and can see in three dimensions—there are no distinctive types of radars and lidars. These sensors are used in ADASs in a variety of ways, including:

a. Object detection: the sensors are used to detect objects in the road environment such as pedestrians, vehicles, cyclists, and traffic signs, and then warn the driver of potential hazards or take corrective actions like braking or steering control using the gathered information;
b. Object recognition: the sensors are used to recognize the class of an object, such as a pedestrian, a vehicle, a cyclist, or a traffic sign. This information can be used to provide the driver with more information about the hazard, such as the type of vehicle, the type of traffic sign and the road condition ahead, or the speed of a pedestrian;
c. Object tracking: the sensors can be used to track the movement of an object over time, which is then used to predict the future position of an object, which can be used to warn the driver of potential collisions.

The advantages of cameras are their low cost, ease of installation, wide field of view (FOV), and high resolution, but they are easily impacted by weather conditions, occlusion of objects, and varying light conditions. On the other hand, both radars and lidars are resistant to varying weather conditions such as rain, snow, fog, and so on. While radars are occlusion-resistant and provide a longer range than cameras, they fail to provide as many details as cameras and are more expensive than cameras. Compared to both cameras and radars, lidars provide very accurate information about the distance and shape of objects, even in difficult conditions, and possess 3D capabilities, enabling them to create a 3D map of the road environment that makes it easier and more efficient to identify and track objects that are occluded by other objects. Nonetheless, lidars are more expensive than cameras and radars, and lidar systems are more complex, making them more challenging to install and maintain. Cameras are used in almost all ADAS functions, while radars and lidars are used in FCWS, LDWS, BSD, and ACC, with lidars having an additional application in autonomous driving.

All the above features allow these versatile sensors to be used for a variety of object detection, recognition, and tracking tasks in ADASs. However, some challenges need to be addressed before they can be used effectively in all conditions. Hence, some researchers have attempted to use a combination of these sensors, as discussed in the following section.

2.2. Sensor Fusion

Sensor fusion is the process of combining data from multiple sensors to create a more complete and accurate picture of the world. This can be used to improve the performance of object detection, recognition, and tracking algorithms in ADASs.

Numerous different sensor fusion techniques can be used for ADASs, namely:

a. Data-level fusion: a technique that combines data from different sensors at the data level by averaging the data from different sensors, or by using more sophisticated techniques such as Kalman filtering [21,22];
b. Feature-level fusion: combines features extracted from data from different sensors by combining the features, or by using more sophisticated techniques such as Bayesian fusion [23,24];
c. Decision-level fusion: combines decisions made by different sensors by taking the majority vote, or by using more sophisticated techniques such as the Dempster–Shafer theory [25–27].

The choice of sensor fusion technique is application-specific. A data-level fusion may be a good choice for applications where accuracy is critical, whereas a decision-level fusion may be a good choice for applications where speed is critical.

The benefits of using sensor fusion for object detection, recognition, and tracking in ADASs can be listed as [15,28–31]:

a. Improved accuracy: sensor fusion improves the accuracy of object detection, recognition, and tracking algorithms by combining the strengths of different sensors;
b. Improved robustness: sensor fusion also improves the robustness of object detection, recognition, and tracking algorithms by making them less susceptible to noise and other disturbances;
c. Reduced computational complexity: sensor fusion also reduces the computational complexity of object detection, recognition, and tracking algorithms, as the data from multiple sensors can be processed together, resulting in saved time and processing power.

Overall, sensor fusion is a promising, powerful technique that has the potential to make ADAS object detection, recognition, and tracking algorithms much safer and more reliable. Although sensor fusion is advantageous, it has some challenges [15,32], such as:

a. Data compatibility: the data from different sensors must be compatible to be fused, implying the data must be in the same format and have the same resolution;
b. Sensor calibration: the sensors must be calibrated to ensure that they are providing accurate data, which can be challenging, especially if the sensors are in motion;
c. Computational complexity: Sensor fusion is computationally expensive, especially if a large number of sensors are being fused. This can limit the use of sensor fusion in real-time applications.

Despite these challenges, sensor fusion is emerging with greater potential to improve the performance of ADAS object detection, recognition, and tracking algorithms. As sensor technology continues to improve, a fusion of sensors will become even more powerful and efficient, and it will likely become a standard feature in ADASs.

The following section discusses the most commonly fused sensors in ADASs.

2.2.1. Camera–Radar Fusion

Camera–radar fusion is a technique that combines data from cameras and radar sensors to improve the performance of object detection, recognition, and tracking algorithms in ADASs. As cameras are good at providing good image quality but are susceptible to weather conditions, radar sensors compensate by seeing through weather conditions. Data-level fusion and decision-level fusion are the two main approaches to camera–radar fusion.

2.2.2. Camera–Lidar Fusion

Camera–lidar fusion is a technique that combines data from cameras and lidar sensors to improve the performance of object detection, recognition, and tracking algorithms in ADASs. Cameras are good at providing detailed information about the appearance of objects, while lidar sensors are good at providing information about the distance and shape of objects. By combining data from these two sensors, it is feasible to create a complete and accurate picture of the object, leading to improved accuracy in object detection and tracking.

2.2.3. Radar–Lidar Fusion

Radar–lidar fusion is a technique that combines the data from radar and lidar sensors, improving the performance of ADAS algorithms. Radar sensors use radio waves to detect objects at long distances, while lidar sensors use lasers to detect objects in detail. By fusing the data from the two sensors, the system can obtain a more complete and accurate view of the environment.

2.2.4. Lidar–Lidar Fusion

Lidar–lidar fusion is a technique that combines data from two or more lidar sensors, improving the performance of object detection, recognition, and tracking algorithms in ADASs. Lidar sensors are good at providing information about the distance and shape of objects, but they can be limited in their ability to detect objects that are close to the vehicle or that are occluded by other objects. By fusing data from multiple lidar sensors, it is possible to create a complete and accurate picture of the environment, which can lead to improved accuracy in object detection and tracking.The above discussed advantages and disadvantages of various ADASs sensors are listed in the Table 1.

Table 1. Summary of the advantages and disadvantages of each sensor and combinations used in ADAS applications.

Sensors	Advantages	Disadvantages
Camera	i. Relatively inexpensive; ii. Easy to use; iii. High-resolution images.	i. Affected by environmental factors (lighting, weather); ii. Difficult to interpret images in low-visibility conditions; iii. Can be fooled by glare and reflections; iv. Can only detect objects in the visible spectrum.
Radar	i. Can detect objects at a longer range than cameras, even in poor visibility; ii. Less affected by weather conditions; iii. Can be used to estimate the speed of objects.	i. Lower resolution than cameras; ii. More expensive than cameras; iii. Can be complex to integrate into vehicles.
Lidar	i. Not affected by environmental factors; ii. Accurate measurement of distance, speed, and shape of objects.	i. Expensive; ii. Difficult to mount on vehicles; iii. Can produce sparse point clouds; iv. Can be limited in field of view (FOV).

Table 1. *Cont.*

Sensors	Advantages	Disadvantages
Camera–Radar Fusion	i. Combines the strengths of cameras and lidar sensors; ii. Can be used in challenging weather conditions.	i. More expensive than using a single sensor; ii. Can be complex to implement.
Camera–Lidar Fusion	i. Combines the strengths of cameras and lidar; ii. Can provide accurate 3D measurements of objects; iii. Robust object detection and tracking system; iv. Can be used in challenging weather conditions.	i. More expensive than a camera or lidar alone; ii. Can be computationally complex.
Radar–Lidar Fusion	i. Combines the strengths of radar and lidar sensors; ii. Improves accuracy of object detection and tracking in challenging weather conditions.	i. More expensive than a camera or lidar alone; ii. Can be computationally complex.
Lidar–Lidar Fusion	i. Combines data from multiple lidar sensors; ii. Can improve the accuracy of 3D mapping and object detection; iii. More accurate and reliable object detection and tracking system.	i. More expensive than lidars alone; ii. Can be computationally complex.

3. Systematic Literature Review

The main objective of this review is to determine the latest trends and approaches implemented for different ADAS methods in autonomous vehicles and discuss their achievements. This paper also attempts to evaluate the valuable basis of the methods, implementation, and applications to furnish a state-of-the-art understanding for new researchers in this computer vision and autonomous vehicles field.

The writing of this paper follows a planned, conducted, and observed process. The planning phase involved clarifying the research questions and review protocol, which comprised identifying the publications' sources, keywords to search for, and selection criteria. The conducting phase involved analyzing, extracting, and synthesizing the literature collection. This included identifying the key themes and findings from the literature and drawing conclusions that address the research questions and objectives. The observed stage contained the review results, addressing the summary of the key findings as well as any limitations or implications of the study.

3.1. Research Questions (RQs)

The main objective of this review is to determine the trend of the methods implemented for different ADAS methods in the field of autonomous vehicles, as well as the achievements of the latest techniques. Additionally, we aim to provide a valuable foundation for the methods, challenges, and opportunities, thus providing state-of-the-art knowledge to support new research in the field of computer vision and ADASs.

Two research questions (RQs) have been defined as follows:

1. What techniques have been implemented for different ADAS methods in an autonomous vehicle?
2. What dataset was applied for the network training, validation, and testing?

A focused approach has been adopted while scanning the literature. First, each article was reviewed to see if it answered the earlier questions. The information acquired was then presented comprehensively to achieve the vision of this article.

3.2. Review Protocol

Below, we have listed the literature search sources, search terms, and inclusion and exclusion selection criteria, as well as the technique of literature collection used for this systematic literature review (SLR).

3.2.1. Search Sources

IEEE Xplore and MDPI were chosen as the databases from which the data were extracted.

3.2.2. Search Terms

Different sets of search terms were used to investigate the various ADAS methods presented in this research. The OR, AND, and NOT operators were used to select and combine the most relevant and commonly used applicable phrases. The AND operator combined individual search strings into a search query. The databases included IEEE Xplore and MDPI. The search terms used for the respective different methods of ADASs are listed in the respective sections of this paper.

3.2.3. Inclusion Criteria

The study covered all primary publications published in English that discussed the different ADAS methods or any other task related to them discussed in this paper. There were no constraints on subject categories or time frames for a broad search spectrum. The selected articles were among the top most cited journal papers published across four years, from 2019 to 2022.

In addition, the below parameters were also considered while selecting the papers:

a. Relevance of the research papers to the topic of the review paper covering the most important aspects of the topic and providing a comprehensive overview of the current state of knowledge;
b. The quality of the research papers should be high. They should be well written, well argued, and well supported by implementation details and experimental results;
c. Coverage of the research papers should include a wide range of perspectives on the topic and not limited to a single viewpoint or approach;
d. The methodology presented in the research papers should be sound such that the research methods must be rigorous and provide clear evidence to support their conclusions;
e. The research papers should be well written and easy to understand in a clear and concise style so that the information is accessible and understandable to a wide audience;
f. The research papers should have had a significant impact on the field. They should have been cited by other researchers and used to inform new research.

3.2.4. Exclusion Criteria

Articles written in languages other than English were not considered. The exclusion criteria also included short papers, such as abstracts or expanded abstracts, earlier published versions of the detailed works, and survey/review papers.

4. Discussion—Methodology

4.1. Vehicle Detection

Vehicle detection, one of the key components and a critical task of ADASs, is the process of identifying and locating vehicles in the surrounding scenes using sensors such as cameras, radars, and lidar employing computer vision techniques. This information is used to provide drivers with warnings about potential hazards, such as cars that are too close or that are changing lanes and pedestrians or cyclists that might be in the vehicles' way. It is a crucial function for many ADAS features, such as ACC, LDWS, FCWS, and BSD, discussed in the later sections of the paper.

Vehicle detection is a challenging task, as vehicles vary in size, shape, and color, affecting their appearance in images and videos. They can be seen from a variety of different angles, which can also affect their appearance; furthermore, vehicle sizes can be too small or too big, they could be partially or fully occluded by other objects in the scene; there are different types of vehicles, each with a unique appearance, and the lighting conditions and possible background clutter also affect the appearance of vehicles. All of these factors make detection challenging.

Despite these challenges, the vehicle detection algorithm in ADASs has greatly evolved and is still evolving, and there have been significant advances in vehicle detection over the years. Early algorithms were based on relatively simple-to-implement image processing techniques, such as edge detection and color segmentation, but they were not very accurate. In the early 2000s, there was a shift towards using ML techniques that can learn from data, making them more accurate than simple image processing techniques. Some of the most common ML algorithms used for vehicle detection include support vector machines (SVMs), random forests, and DL NNs.

Deep learning NNs are the most effective machine learning algorithms for vehicle detection. Deep learning NNs can learn complex features from data, which makes them very accurate. Regardless, DL NNs are also more computationally expensive than other ML algorithms. In recent years, there has been a trend towards using sensor fusion for vehicle detection.

The vehicle detection algorithms in ADASs are still evolving. As sensor technology continues to improve, and as ML algorithms become more powerful, vehicle detection algorithms will become even more accurate and reliable.

Search Terms and Recent Trends in Vehicle Detection

'Vehicle detection', 'vehicle tracking', and 'vehicle detection and tracking' are three prominent search terms which were used to investigate the topic. The 'OR' operator was used to choose and combine the most relevant and regularly used applicable phrases; that is, the search phrases 'vehicle detection', 'vehicle tracking', and 'vehicle detection and tracking' were discovered. Figure 2 shows the complete search query for each of the databases. The databases include IEEE Xplore and MDPI.

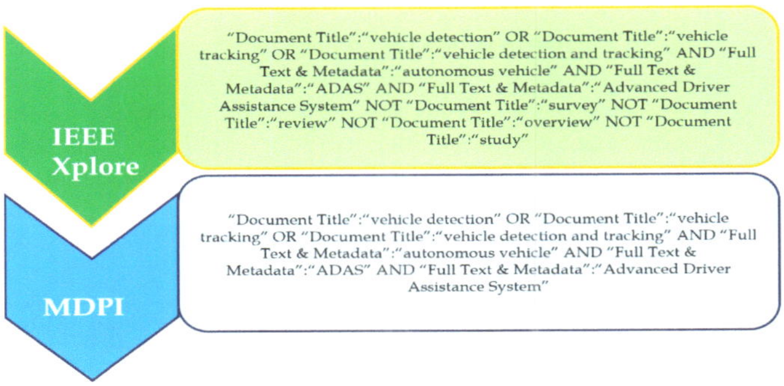

Figure 2. Search queries for each of the databases for vehicle detection. The databases include IEEE Xplore and MDPI.

Since the evolution of vehicle detection has been rapid, considering the detection, recognition, and tracking of other vehicles, pedestrians, and objects, plenty of different methods have been proposed in the past few years. Some of the recent prominent state-of-the-art vehicle detection methods are discussed in the following sections.

Ref. [33] presents a scale-insensitive CNN, SINet, which is designed for rapid and accurate vehicle detection. SINet employs two lightweight techniques: context-aware RoI pooling and multi-branch decision networks. These preserve small-scale object information and enhance classification accuracy. Ref. [34] introduces an integrated approach to monocular 3D vehicle detection and tracking. It utilizes a CNN for vehicle detection and employs a Kalman filter-based tracker for temporal continuity. The method incorporates multi-task learning, 3D proposal generation, and Kalman filter-based tracking. Combining radar and vision sensors, ref. [35] proposes a novel distant vehicle detection approach. Radar generates candidate bounding boxes for distant vehicles, which are classified using vision-based methods, ensuring accurate detection and localization. Ref. [36] focuses on multi-vehicle tracking, utilizing object detection and viewpoint estimation sensors. The CNN detects vehicles, while viewpoint estimation enhances tracking accuracy. Ref. [37] utilizes CNN with feature concatenation for urban vehicle detection, improving robustness through layer-wise feature combination. Ref. [38] presents a robust vehicle detection and counting method integrating CNN and optical flow, while [39] pioneers vehicle detection and classification via distributed fiber optic acoustic sensing. Ref. [40] introduces a vehicle tracking and speed estimation method using roadside lidar, incorporating a Kalman filter. Ref. [41] modifies Tiny-YOLOv3 for front vehicle detection with SPP-Net enhancement, excelling in challenging conditions. Ref. [42] proposes an Extended Kalman Filter (EKF) for vehicle tracking using radar and lidar data, while [43] enhances SSD for accurate front vehicle detection. Ref. [44] improves Faster RCNN for oriented vehicle detection in aerial images with feature amplification and oversampling. Ref. [45] employs reinforcement learning with partial vehicle detection for efficient intelligent traffic signal control. Ref. [46] presents a robust DL framework for vehicle detection in adverse weather conditions. Ref. [47] adopts GAN-based image style transfer for nighttime vehicle detection, while ref. [48] introduces MultEYE for real-time vehicle detection and tracking using UAV imagery. Ref. [49] analyzes traffic patterns during COVID-19 using Planet remote-sensing satellite images for vehicle detection. Ref. [50] proposes one-stage anchor-free 3D vehicle detection from lidar, ref. [51] fuses RGB-infrared images for accurate vehicle detection using uncertainty-aware learning. Ref. [52] optimizes YOLOv4 for improved vehicle detection and classification. Ref. [53] introduces a real-time foveal classifier-based system for nighttime vehicle detection. Ref. [54] combines YOLOv4 and SPP-Net for multi-scale vehicle detection in varying weather. Ref. [55] efficiently detects moving vehicles with a CNN-based method incorporating background subtraction. Ref. [56] refines YOLOv5 for vehicle detection in aerial infrared images, ensuring robustness against challenges like occlusion and low contrast.

Overall, the aforementioned papers represent a diverse set of approaches to vehicle detection and tracking. Each paper has its strengths and weaknesses, and it is important to consider the specific application when choosing a method. However, all of the papers represent significant advances in the field of vehicle detection and tracking. The list of reviewed papers on vehicle detection is summarized in Table 2.

Table 2. Chosen publications regarding vehicle detection, their source title, and their number of citations.

SI No.	Ref.	Year	Source Title	Citations
1	[33]	2019	IEEE Transactions on Intelligent Transportation Systems	165
2	[34]	2019	IEEE/CVF International Conference on Computer Vision (ICCV)	88
3	[35]	2019	IEEE International Conference on Robotics and Automation (ICRA)	79
4	[36]	2019	MDPI Intelligent Sensors	58
5	[37]	2019	MDPI Intelligent Sensors	42
6	[38]	2019	MDPI Remote Sensors	41
7	[39]	2020	IEEE Transactions on Vehicular Technology	47
8	[40]	2020	IEEE Journal of Selected Topics in Applied Earth Observations and Remote Sensing	44

Table 2. *Cont.*

SI No.	Ref.	Year	Source Title	Citations
9	[41]	2020	IEEE Access	38
10	[42]	2020	MDPI Sensors	56
11	[43]	2020	MDPI Sensors	27
12	[44]	2020	MDPI Remote Sensing	27
13	[45]	2021	IEEE Transactions on Intelligent Transportation Systems	52
14	[46]	2021	IEEE Transactions on Intelligent Transportation Systems	48
15	[47]	2021	IEEE Transactions on Intelligent Transportation Systems	47
16	[48]	2021	MDPI Remote Sensing	37
17	[49]	2021	MDPI Remote Sensing	30
18	[50]	2021	MDPI Sensors	11
19	[51]	2022	IEEE Transactions on Circuits and Systems for Video Technology	20
20	[52]	2022	IEEE Access	13
21	[53]	2022	IEEE Transactions on Intelligent Transportation Systems	12
22	[54]	2022	MDPI Electronics	21
23	[55]	2022	MDPI Sensors	10
24	[56]	2022	MDPI Electronics	6

4.2. Pedestrian Detection

Pedestrian detection is also a key component of ADASs that uses sensors to identify and track pedestrians in the surrounding environment and prevent collisions with pedestrians. The goal of pedestrian detection is to identify and track pedestrians in the surrounding environment, warn the driver of potential collisions with pedestrians, and take evasive action such as automatically applying brakes, if necessary.

Pedestrian detection systems typically use a combination of sensors, such as cameras, radar, and lidar. Cameras are often used to identify the shape and movement of pedestrians, while radar and lidar can be used to determine the distance and speed of pedestrians. Cameras can be susceptible to glare and shadows, whereas radar and lidars are less susceptible to these problems.

Pedestrian detection systems can be used to warn drivers of potential collisions in a variety of ways. Some systems simply alert the driver with a visual or audible warning. Others can take more active measures, such as automatically braking the vehicle or slightly steering it away from the pedestrian. However, pedestrian detection is more challenging, as pedestrians are often smaller and more difficult to distinguish from other objects in the environment. Thus, it is an important safety feature for ADASs, as it can help to prevent accidents involving pedestrians. According to the National Highway Traffic Safety Administration (NHTSA) [57], pedestrians are involved in about 17% of all traffic fatalities in the United States. Pedestrian detection systems can help to reduce this number by warning drivers of potential hazards and by automatically applying the brakes in emergencies.

Search Terms and Recent Trends in Pedestrian Detection

'Pedestrian detection', 'pedestrian tracking', and 'pedestrian detection and tracking' are three prominent search terms which were used to investigate this topic. The 'OR' operator was used to choose and combine the most relevant and regularly used applicable phrases; that is, the search phrases pedestrian detection, 'pedestrian tracking', and 'pedestrian detection and tracking' were discovered. Figure 3 shows the complete search query for each of the databases. The databases include IEEE Xplore and MDPI.

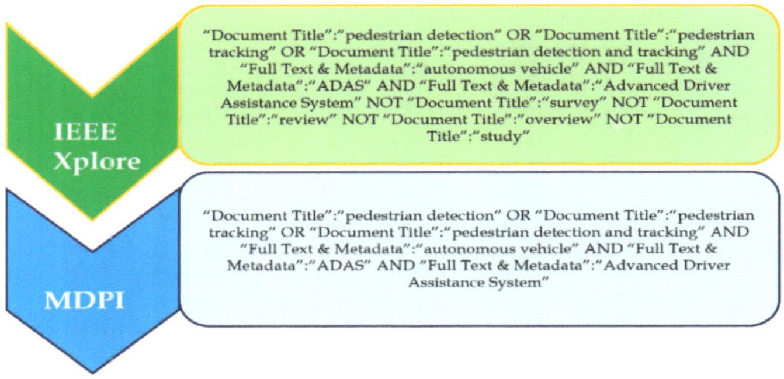

Figure 3. Search queries for each of the databases for pedestrian detection. The databases include IEEE Xplore and MDPI.

Ref. [58] introduces a novel approach to pedestrian detection, emphasizing high-level semantic features instead of traditional low-level features. This method employs context-aware RoI pooling and a multi-branch decision network to preserve small-scale object details and enhance classification accuracy. The CNN initially captures high-level semantic features from images, which are then used to train a classifier to distinguish pedestrians from other objects. Ref. [59] proposes an adaptive non-maximum suppression (NMS) technique tailored for refining pedestrian detection in crowded scenarios. Conventional NMS algorithms often eliminate valid detections along with duplicates in crowded scenes. The new 'Adaptive NMS' algorithm dynamically adjusts the NMS threshold based on crowd density, enabling the retention of more pedestrian candidates in congested areas. Ref. [60] introduces the 'Mask-Guided Attention Network' (MGAN) for detecting occluded pedestrians. Utilizing a CNN, MGAN extracts features from both pedestrians and backgrounds. Pedestrian features guide the network's focus towards occluded regions, improving the accuracy of detecting occluded pedestrians. Ref. [61] presents a real-time method to track pedestrians by utilizing camera and lidar sensors in a moving vehicle. Combining sensor features enables accurate pedestrian tracking. Features from the camera image, such as silhouette, clothing, and gait, are extracted. Additionally, features like height, width, and depth are obtained from the lidar point cloud. These details facilitate precise tracking of pedestrians' locations and poses over time. A Kalman filter enhances tracking performance through sensor data fusion, offering better insights into pedestrian behavior in dynamic environments. Ref. [62] proposes a computationally efficient single-template matching technique for accurate pedestrian detection in lidar point clouds. The method creates a pedestrian template from training data and uses it to identify pedestrians in new point clouds, even under partial occlusion. Ref. [63] focuses on tracking pedestrian flow and statistics using a monocular camera and a CNN–Kalman filter fusion. The CNN extracts features from the camera image, which is followed by a Kalman filter for trajectory estimation. This approach effectively tracks pedestrian flow and vital statistics, including count, speed, and direction.

Ref. [64] addresses hazy weather pedestrian detection with deep learning. DL models are trained on hazy weather datasets and use architectural modifications to handle challenging conditions. This approach achieves high pedestrian detection accuracy, even in hazy weather. Ref. [65] introduces the 'NMS by Representative Region' algorithm to refine pedestrian detection in crowded scenes. By employing representative regions, this method enhances crowded scene handling by comparing these regions and removing duplicate detections, resulting in reduced false positives. Ref. [66] proposes a graininess-aware deep feature learning approach, equipping DL models to handle grainy images. A DL model is trained using a graininess-aware loss function on a dataset containing grainy and non-grainy pedestrian images. This model effectively detects pedestrians in new images, even when they are grainy. Ref. [67] presents a DL framework for real-time vehicle and pedestrian detection on rural roads, optimized for embedded GPUs. Modified Faster R-CNN detects both vehicles and pedestrians simultaneously in rural road scenes. A new rural road image dataset is developed for training the model. Ref. [68] addresses infrared pedestrian detection at night using an attention-guided encoder–decoder CNN. Attention mechanisms focus on relevant regions in infrared images, enhancing detection accuracy in low-light conditions. Ref. [69] focuses on improved YOLOv3-based pedestrian detection in complex scenarios, incorporating modifications to handle various challenges like occlusions, lighting variations, and crowded environments.

Ref. [70] introduces Ratio-and-Scale-Aware YOLO (RASYOLO), handling pedestrians with varying sizes and occlusions through ratio-aware anchors and scale-aware feature fusion. Ref. [71] introduces Track Management and Occlusion Handling (TMOH), managing occlusions and multiple-pedestrian tracking through track suspension and resumption. Ref. [72] incorporates a Part-Aware Multi-Scale fully convolutional network (PAM-FCN) to enhance pedestrian detection accuracy by considering pedestrian body part information and addressing scale variation. Ref. [73] proposes Attention Fusion for One-Stage Multispectral Pedestrian Detection (AFOS-MSPD), combining attention fusion and a one-stage approach for multispectral pedestrian detection, improving efficiency and accuracy. Ref. [74] utilizes multispectral images for Multispectral Pedestrian Detection (MSPD), improving detection using a DNN designed for multispectral data. Ref. [75] presents Robust Pedestrian Detection Based on Multi-Spectral Image Fusion and Convolutional Neural Networks (RPOD-FCN), utilizing multi-spectral image fusion and a CNN-based model for accurate detection.

Ref. [76] introduces Uncertainty-Guided Cross-Modal Learning for Robust Multispectral Pedestrian Detection (UCM-RMPD), addressing multispectral detection challenges using uncertainty-guided cross-modal learning. Ref. [77] focuses on multimodal pedestrian detection for autonomous driving using a Spatio-Contextual Deep Network-Based Multimodal Pedestrian Detection (SCDN-PMD) approach. Ref. [78] proposes a Novel Approach to Model-Based Pedestrian Tracking Using Automotive Radar (NMPT radar), utilizing radar data for model-based pedestrian tracking. Ref. [79] adopts YOLOv4 Architecture for Low-Latency Multispectral Pedestrian Detection in Autonomous Driving (AYOLOv4), enhancing detection accuracy using multispectral images. Ref. [80] introduces modifications to [79] called AIR-YOLOv3, an improved network-pruned YOLOv3 for aerial infrared pedestrian detection, enhancing robustness and efficiency. Ref. [81] presents YOLOv5-AC, an attention mechanism-based lightweight YOLOv5 variant for efficient pedestrian detection on embedded devices. The list of reviewed papers on pedestrian detection is summarized in Table 3.

Table 3. Chosen publications regarding pedestrian detection, their source title, and their number of citations.

SI No.	Ref.	Year	Source Title	Citations
1	[58]	2019	IEEE/CVF Conference on Computer Vision and Pattern Recognition (CVPR)	186
2	[59]	2019	IEEE/CVF Conference on Computer Vision and Pattern Recognition (CVPR)	163
3	[60]	2019	2019 IEEE/CVF International Conference on Computer Vision (ICCV)	111
4	[61]	2019	MDPI Sensors	45
5	[62]	2019	MDPI Electronics	26
6	[63]	2019	MDPI Applied Sciences	15
7	[64]	2020	IEEE Transactions on Industrial Electronics	98
8	[65]	2020	2020 IEEE/CVF Conference on Computer Vision and Pattern Recognition (CVPR)	76
9	[66]	2020	IEEE Transactions on Image Processing	42
10	[67]	2020	MDPI Electronics	49
11	[68]	2020	MDPI Applied Science	28

Table 3. *Cont.*

SI No.	Ref.	Year	Source Title	Citations
12	[69]	2020	MDPI Sensors	14
13	[70]	2021	IEEE Transactions on Image Processing	54
14	[71]	2021	IEEE/CVF Conference on Computer Vision and Pattern Recognition (CVPR)	45
15	[72]	2021	IEEE Transactions on Intelligent Transportation Systems	27
16	[73]	2021	MDPI Sensors	21
17	[74]	2021	MDPI Sensors	19
18	[75]	2021	MDPI Electronics	15
19	[76]	2022	IEEE Transactions on Circuits and Systems for Video Technology	15
20	[77]	2022	IEEE Transactions on Intelligent Transportation Systems	12
21	[78]	2022	IEEE Transactions on Intelligent Transportation Systems	10
22	[79]	2022	MDPI Sensors	20
23	[80]	2022	MDPI Applied Sciences	11
24	[81]	2022	MDPI Sensors	11

4.3. Traffic Signs Detection

Traffic Signs Detection and Recognition (TSR) is another key component of ADASs that automatically detects and recognizes traffic signs on the road and provides information to the driver regarding speed limits, upcoming turns, and so on. TSR systems typically use cameras to capture images of traffic signs and then use computer vision algorithms to identify and classify the signs.

TSR systems can be a valuable safety feature, as they can help to prevent accidents caused by driver distraction or drowsiness. For example, TSR systems can alert drivers to speed limit changes, stop signs, and yield signs. They can also help drivers to stay in their lane and avoid crossing over into oncoming traffic. Although TSR can be challenging due to the variety of traffic signs, the different fonts and styles used, and the presence of noise and clutter, TSR systems are becoming increasingly common in new vehicles. The NHTSA has mandated that all new cars sold in the United States come equipped with TSR systems by 2023 [57].

Search Terms and Recent Trends in Traffic Signs Detection

'Traffic sign detection', 'traffic sign recognition, 'traffic sign classification', 'traffic sign detection and recognition', and 'traffic sign detection and recognition system' are some of the prominent search terms which were used to investigate this topic. The 'OR' operator was used to choose and combine the most relevant and regularly used applicable phrases; that is, the search phrases 'driver monitoring system' and 'driver monitoring and assistance system' were discovered. Figure 4 shows the complete search query for each of the databases. The databases include IEEE Xplore and MDPI.

Yuan et al. [82] introduce VSSA-NET, a novel architecture for traffic sign detection (TSD), which employs a vertical spatial sequence attention network to improve accuracy in complex scenes. VSSA-NET extracts features via CNN, followed by a vertical spatial sequence attention module to emphasize vertical locations crucial for TSD. The detection module outputs traffic sign bounding boxes. Li and Wang [83] present real-time traffic sign recognition using efficient CNNs, addressing diverse lighting and environmental conditions. MobileNet extracts features from input images, followed by SVM classification. Liu et al. [84] propose multi-scale region-based CNN (MR-CNN) for recognizing small traffic signs. MR-CNN extracts multi-scale features using CNN, generates proposals with RPN, and uses Fast R-CNN for classification and bounding box outputs. Tian et al. [85] introduce a multi-scale recurrent attention network for TSD. CNN extracts multi-scale features, the recurrent attention module prioritizes scale, and the detection module outputs bounding boxes for robust detection across scenarios. Cao et al. [86] present improved TSDR for intelligent vehicles. CNN performs feature extraction, RPN generates region proposals, and SVM classifies proposals, enhancing reliability in dynamic road environments. Shao et al. [87] improve Faster R-CNN TSD with a second RoI and HPRPN. CNN performs feature extraction, RPN generates region proposals, and the second RoI refines proposals, enhancing accuracy in complex scenarios.

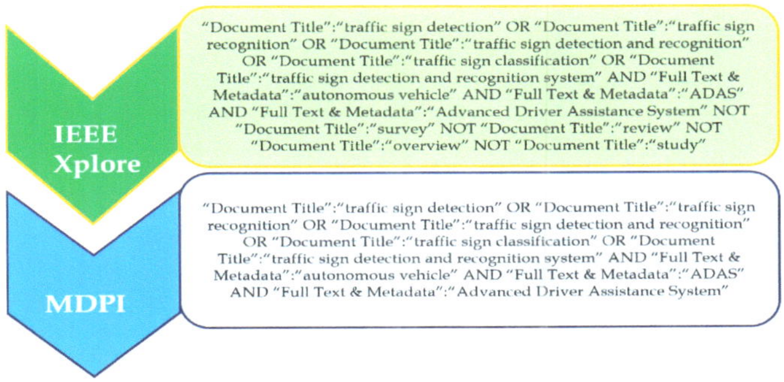

Figure 4. Search queries for each of the databases for traffic sign detection. The databases include IEEE Xplore and MDPI.

Zhang et al. [88] propose cascaded R-CNN with multiscale attention for TSD. RPN generates proposals, Fast R-CNN classifies, and multiscale attention improves detection performance, particularly when there is an imbalanced data distribution. Tabernik and Skočaj [89] explore the DL framework for large-scale TSDR. CNN performs feature extraction, RPN generates region proposals, and Fast R-CNN classifies, exploring DL's potential in handling diverse real-world scenarios. Kamal et al. [90] introduce automatic TSDR using SegU-Net and modified Tversky loss. SegU-Net segments traffic signs and modified loss function enhances detection and recognition, handling appearance variations. Tai et al. [91] propose a DL approach for TSR with spatial pyramid pooling and scale analysis. CNN performs feature extraction, while spatial pyramid pooling captures context and scales, enhancing recognition across scenarios. Dewi et al. [92] evaluate the spatial pyramid pooling technique on CNN for TSR system robustness. Assessing pooling sizes and strategies, they evaluate different CNN architectures for effective traffic sign recognition. Nartey et al. [93] propose robust semi-supervised TSR with self-training and weakly supervised learning. CNN performs feature extraction, self-training labels unlabeled data, and weakly supervised learning classifies labeled data, enhancing accuracy using limited labeled data.

Dewi et al. [94] leverage YOLOv4 with synthetic GAN-generated data for advanced TSR. YOLOv4 with synthetic data from BigGAN achieves top performance, enhancing detection on the GTSDB dataset. Wang et al. [95] improve YOLOv4-Tiny TSR with new features and classification modules. New data augmentation improves the performance on the GTSDB dataset, optimizing recognition while maintaining efficiency. Cao et al. [96] present improved sparse R-CNN for TSD with a new RPN and loss function. Enhancing detection accuracy using advanced techniques within the sparse R-CNN framework. Lopez-Montiel et al. [97] propose DL-based embedded system evaluation and synthetic data generation for TSD. Methods to assess DL system performance and efficiency for real-time TSD applications are developed. Zhou et al. [98] introduce a learning region-based attention network for TSR. The attention module emphasizes important image regions, potentially enhancing recognition accuracy. Koh et al. [99] evaluate senior adults' TSR recognition through EEG signals, utilizing EEG signals to gain unique insights into senior individuals' traffic sign perception.

Ahmed et al. [100] present a weather-adaptive DL framework for robust TSR. A cascaded detector with a weather classifier improves TSD performance in adverse conditions, enhancing road safety. Xie et al. [101] explore efficient federated learning in TSR with spike

NNs (SNNs). SNNs enable training on decentralized datasets, minimizing communication overhead and resources. Min et al. [102] propose semantic scene understanding and structural location for TSR, leveraging scene context and structural information for accurate traffic sign recognition. Gu and Si [103] introduce a lightweight real-time TSD integration framework based on YOLOv4. Novel data augmentation and YOLOv4 optimization are used for speed and accuracy, achieving real-time performance. Liu et al. [104] introduce the M-YOLO TSD algorithm for complex scenarios. M-YOLO detects and classifies traffic signs, addressing detection in intricate environments. Wang et al. [105] propose real-time multi-scale TSD for driverless cars. The multi-scale approach detects traffic signs of various sizes, enhancing performance in diverse scenarios. The list of reviewed papers on traffic signs detection is summarized in Table 4.

Table 4. Chosen publications, source title, and the number of citations for traffic signs detection.

SI No.	Ref.	Year	Source Title	Citations
1	[82]	2019	IEEE Transactions on Image Processing	118
2	[83]	2019	IEEE Transactions on Intelligent Transportation Systems	96
3	[84]	2019	IEEE Access	53
4	[85]	2019	IEEE Transactions on Intelligent Transportation Systems	50
5	[86]	2019	MDPI Sensors	66
6	[87]	2019	MDPI Sensors	44
7	[88]	2020	IEEE Access	151
8	[89]	2020	IEEE Transactions on Intelligent Transportation Systems	131
9	[90]	2020	IEEE Transactions on Intelligent Transportation Systems	52
10	[91]	2020	MDPI Applied Sciences	46
11	[92]	2020	MDPI Electronics	38
12	[93]	2020	MDPI Sensors	16
13	[94]	2021	IEEE Access	63
14	[95]	2021	IEEE Access	30
15	[96]	2021	IEEE Access	19
16	[97]	2021	IEEE Access	16
17	[98]	2021	MDPI Sensors	25
18	[99]	2020	MDPI Sensors	3
19	[100]	2022	IEEE Transactions on Intelligent Transportation Systems	15
20	[101]	2022	IEEE Transactions on Vehicular Technology	11
21	[102]	2022	IEEE Transactions on Intelligent Transportation Systems	11
22	[103]	2022	MDPI Entropy	13
23	[104]	2022	MDPI Symmetry	8
24	[105]	2022	MDPI Sensors	7

4.4. *Driver Monitoring System (DMS)*

A driver monitoring system (DMS), also called a driver monitoring and assistance system (DMAS), is a camera-based safety system used to assess the driver's alertness and attention. It monitors a driver's behavior by detecting and tracking the driver's face, eyes, and head position and warns or alerts them when they become distracted or drowsy for long enough to lose situational awareness or full attention to the task of driving. DMSs can also use other sensors, such as radar or infrared sensors, to gather additional information about the driver's state.

DMSs are becoming increasingly common in vehicles and are used to monitor the driver's alertness and attention. This information is then used to prevent accidents and save lives by warning the driver if they are starting to become drowsy or distracted. Some of the latest DMSs can even predict if drivers are eating and drinking while driving.

4.4.1. Driver Monitoring System Methods

There are a variety of methods used in DMSs. One common approach is to use a camera to monitor the driver's face, while the other approach is to use a sensor fusion approach, which combines data from multiple sensors, such as cameras, radar, and eye tracking sensors.

DMSs can use a variety of sensors to monitor the driver, including:

a. Facial recognition. This is the most common type of sensor used in DMSs. Facial recognition systems can track the driver's face and identify signs of distraction or drowsiness, such as eye closure, head tilt, and lack of facial expression.
b. A head pose sensor tracks the position of the driver's head and can identify signs of distraction or drowsiness, such as looking away from the road or nodding off.
c. An eye gaze sensor tracks the direction of the driver's eye gaze and can identify signs of distraction or drowsiness, such as looking at the phone or dashboard.
d. An eye blink rate sensor tracks the driver's eye blink rate and can identify signs of drowsiness, such as a decrease in the blink rate.
e. Speech recognition is used in DMSs to detect if the driver is talking on the phone or if they are not paying attention to the road.

The above sensors are used in DMSs to detect a variety of driver behaviors, such as (i) when a driver is distracted by looking away from the road, talking on the phone, or using a mobile device; (ii) when a driver is drowsy, which can be determined by tracking the driver's eye movements and eyelid closure; (iii) when a driver is inattentive, which can be determined by tracking the driver's head position and eye gaze.

When a DMS detects risky driver behavior, it can provide a variety of alerts to the driver, including alerts displayed on the dashboard or windshield, referred to as visual alerts; alerts played through the vehicle's speakers, which are called audio alerts; and hectic alerts, in which alerts are issued through vibrations of the steering wheel or the driver's seat. In some cases, the DMS may also take corrective action, such as applying the brakes or turning off the engine.

4.4.2. Search Terms and Recent Trends in Driver Monitoring System Methods

'Driver monitoring system' and 'driver monitoring and assistance system' are the two prominent search terms used to investigate this topic. The 'OR' operator was used to choose and combine the most relevant and regularly used applicable phrases. That is, the search phrases 'driver monitoring system' and 'driver monitoring and assistance system' were discovered. Figure 5 shows the complete search query for each of the databases. The databases include IEEE Xplore and MDPI.

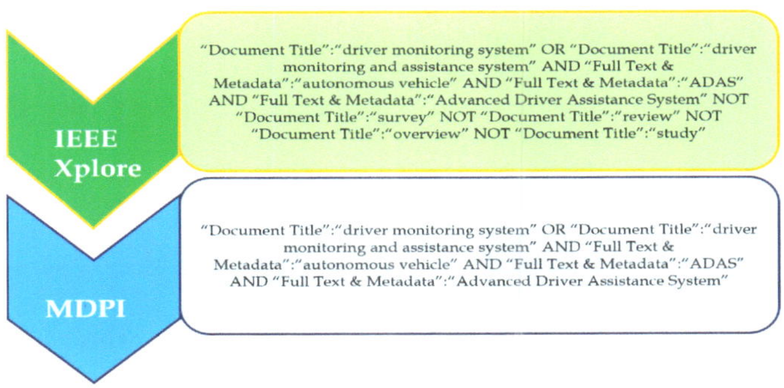

Figure 5. Search queries for each of the databases for the driver monitoring system. The databases include IEEE Xplore and MDPI.

The papers [106–114] discuss a variety of approaches to DMSs. These include some of the key methods like (i) the powerful technique employing DL, which is used to extract features from images and videos. These are used to identify driver behaviors such as eye closure, head pose, and facial expressions. (ii) A more general approach is using machine learning, which can be used to learn patterns from data. These are used to identify driver behaviors that are not easily captured using traditional methods, such as hand gestures and body language, and (iii) a technique that combines data from multiple sensors, referred to as sensor fusion, to improve the accuracy of DMSs. For instance, a DMS could combine data from a camera, an eye tracker, and a heart rate monitor to provide a more comprehensive assessment of the driver's state.

Y. Zhao et al. [106] propose a novel real-time DMSs based on deep CNN to monitor drivers' behavior and detect distractions. It uses video input from an in-car camera and employs CNNs to analyze the driver's facial expressions and head movements to assess their attentiveness. It can detect eye closure, head pose, and facial expressions with high accuracy. Ref. [107] works towards a DMS that uses machine learning to estimate driver situational awareness using eye-tracking data. It aims to predict driver attention and alertness to the road, enhancing road safety. Ref. [108] proposes a lightweight DMS based on Multi-Task Mobilenets architecture, which efficiently monitors drivers' behavior and attention using low computational resources. It can even run on a simple smartphone, making it suitable for real-time monitoring. Ref. [109] introduces an optimization algorithm for DMSs using DL. This algorithm improves the accuracy of the DMS by reducing the number of false positives and ensuring real-time performance.

Ref. [110] proposes a real-time DMS based on visual cues, leveraging facial expressions and eye movements to assess driver distraction and inattention. It is able to detect driver behaviors such as eye closure, head pose, and facial expressions using only a camera. Ref. [111] proposes an intelligent DMS that uses a combination of sensors and ML. It is capable of providing a comprehensive assessment of the driver's state, including their attention level, fatigue, and drowsiness, and provides timely alerts to improve safety. Ref. [112] proposes a hybrid DMS combining Internet of Things (IoT) and ML techniques for comprehensive driver monitoring. It collects data from multiple sensors and uses ML to identify driver behaviors. Ref. [113] focuses on a distracted DMS that uses AI to detect and prevent risky behaviors on the road. It detects distracted driving behaviors such as texting and talking on the phone while driving. Ref. [114] proposes a DMS based on a distracted driving decision algorithm which aims to assess and address potential distractions to ensure safe driving practices. It predicts whether the driver is distracted or not.

These papers provide a good overview of the current state of the art in DMS and contribute to the development of advanced DMS technologies, aiming to enhance driver safety, detect distractions, and improve situational awareness on the roads. They employ various techniques, including deep learning, IoT, and machine learning, to create efficient and effective driver monitoring solutions. However, before DMSs can be widely deployed, there are still some challenges that need to be addressed, such as:

a. Data collection: It is difficult to collect large datasets of driver behavior representative of the real world, as it is difficult to monitor drivers naturally without disrupting their driving experience.
b. Algorithm development: Since the driver behaviors can be subtle and vary from person to person, it is challenging to develop algorithms that can accurately identify driver behaviors in real time.
c. Cost: DMS demands the use of specialized sensors and software, making them expensive to implement and maintain.

Additionally, with the development and availability of new sensors, they could be used to improve the accuracy and performance of DMSs; for example, radar sensors could be used to track driver head movements and eye gaze. Besides, autonomous vehicles will not need DMSs in the same way that human-driven vehicles do. However, DMSs could still be used to monitor the state of the driver in autonomous vehicles and to provide feedback to the driver if necessary. Despite these challenges, there is a lot of potential for DMSs to improve road safety and the future of DMSs looks promising. As the technology continues to develop, DMSs could become an essential safety feature in vehicles, both human-driven and autonomous. The list of reviewed papers on driver monitoring system is summarized in Table 5.

Table 5. Chosen publications, source title, and the number of citations referring to the driver monitoring system.

SI No.	Ref.	Year	Source Title	Citations
1	[106]	2019	IEEE International Symposium on Robotic and Sensors Environments	2
2	[107]	2019	International Conference on Robot and Human Interactive Communication	1
3	[108]	2019	MDPI Sensors	28
4	[109]	2020	International Conference on Artificial Intelligence in Information and Communication	2
5	[110]	2020	6th International Conference on Interactive Digital Media	1
6	[111]	2021	2nd International Conference on Communication, Computing and Industry 4.0	1
7	[112]	2021	IEEE International Conference on Consumer Electronics and Computer Engineering	-
8	[113]	2022	Interdisciplinary Research in Technology and Management	-
9	[114]	2022	13th International Conference on Information and Communication Technology Convergence	-

4.5. Lane Departure Warning System

The Lane Departure Warning System (LDWS) is a type of ADAS that is designed to warn drivers when they are unintentionally drifting out of their lane. LDWSs typically use cameras, radar, lidar, or a combination of sensors to detect the lane markings on the road, and then they use this information to monitor the driver's position in the lane. If the driver starts to drift out of the lane, the LDWS will sound an audible alert or vibrate the steering wheel to warn the driver. These systems can be a valuable safety feature and are especially helpful for drivers, as they can help to prevent accidents caused by driver drowsiness or distraction and they can help to keep drivers alert and focused on the road.

LDWSs are becoming increasingly common in new vehicles. In fact, according to NHTSA, lane departure crashes account for about 5% of all fatal crashes in the United States and the NHTSA has mandated that all new vehicles sold in the United States be equipped with LDWSs by 2022 [115].

LDWSs can be a valuable safety feature, but they are not perfect. They can sometimes be fooled by objects that look like lane markings, such as shadows or road debris, and may not be accurate when the road markings are faded or obscured. Additionally, LDWS can only warn drivers; they cannot take corrective action on their own, which means they may not be effective for drivers who are drowsy or distracted.

Despite these limitations, LDWS can be a valuable tool for reducing the number of accidents, and are especially beneficial for long-distance driving, as they can help keep drivers alert and focused. They can: (i) help to prevent accidents by alerting drivers to unintentional lane departures, (ii) help drivers stay alert and focused on the road, (iii) be especially helpful for drivers who are drowsy or distracted, (iv) help to keep drivers in their lane, which can improve lane discipline and reduce the risk of sideswipe collisions, thus improving the driver safety and comfort. Therefore, LDWSs are becoming increasingly common in new vehicles, as they greatly reduce drivers' stress and fatigue.

Overall, LDWSs are a valuable safety feature that can help to prevent accidents, though they are not guaranteed to do so. It is important to remember that these systems are not a substitute for safe driving practices. Drivers should always be alert and focused on the road, aware of their surroundings and use safe driving practices at all times, even when they are using an LDWS.

Search Terms and Recent Trends in LDWS

'Lane departure warning', 'lane deflection warning', 'lane detection', and 'lane detection and tracking' are four prominent search terms used to investigate the topic. The 'OR' operator was used to choose and combine the most relevant and regularly used applicable phrases. The search phrases 'lane departure warning', 'lane deflection warning', 'lane detection', and 'lane detection and tracking' were discovered. Figure 6 shows the complete search query for each of the databases. The databases include IEEE Xplore and MDPI.

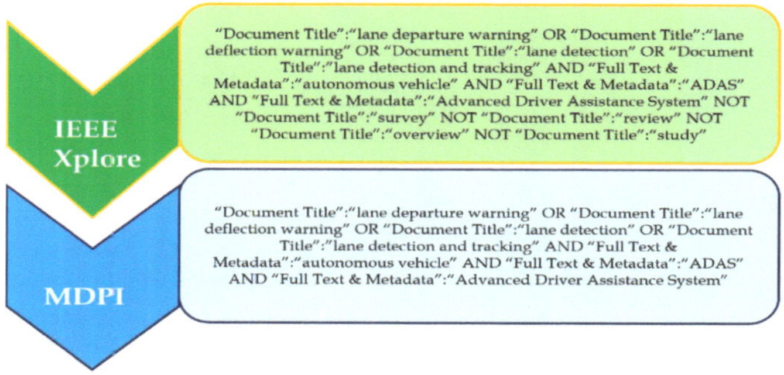

Figure 6. Search queries for each of the databases for the lane departure warning system. The databases include IEEE Xplore and MDPI.

Lane detection is a critical task in computer vision and autonomous driving systems. These review papers explore various lane detection techniques proposed in recent research papers. The reviewed papers cover diverse approaches, including lightweight CNNs, sequential prediction networks, 3D lane detection, and algorithms for intelligent vehicles in complex environments. The existing lane detection algorithms are not robust to challenging road conditions, such as shadows, rain, and snow, along with occlusion and illumination, and scenarios where lane markings are not visible and are limited in their ability to detect multiple lanes and to accurately estimate the 3D position of the lanes.

This research review paper examines recent advancements in lane detection techniques, focusing on the integration of DNNs and sensor fusion methodologies. The review encompasses papers published between 2019 and 2022, exploring innovative approaches to improve the robustness, accuracy, and performance of lane detection systems in various challenging scenarios.

The reviewed papers present various innovative approaches for lane detection in the context of autonomous driving systems. Lee et al. [116] introduce a self-attention distillation method to improve the efficiency of lightweight lane detection CNNs without compromising accuracy. FastDraw [117] addresses the long tail of lane detection using a sequential prediction network to consider contextual information for better predictions. 3D-LaneNet [118] incorporates depth information from stereo cameras for end-to-end 3D multiple lane detection. Wang et al. [119] propose a data enhancement technique called Light Conditions Style Transfer for lane detection in low-light conditions, improving model robustness. Other methods explore techniques such as ridge detectors [120], LSTM networks [121], and multitask attention networks [122] to enhance lane detection accuracy in various challenging scenarios. Additionally, some papers integrate multiple sensor data [123–126] or use specific sensors like radar [127] and light photometry systems [128] to achieve more robust and accurate lane detection for autonomous vehicles. These research contributions provide valuable insights into the development of advanced lane detection systems for safer and more reliable autonomous driving applications.

In their recent research, Lee et al. [116] proposed a novel approach for learning lightweight lane detection CNNs by applying self-attention distillation. FastDraw [117] addressed the long tail of lane detection by using a sequential prediction network to better predict lane markings in challenging conditions. Garnett et al. [118] presented 3D-LaneNet, an end-to-end method incorporating depth information from stereo cameras for 3D multiple lane detection. Additionally, Cao et al. [123] tailored a lane detection algorithm for intelligent vehicles in complex road conditions, enhancing real-world driving reliability. Kuo et al. [129] optimized image sensor processing techniques for lane detection in vehicle lane-keeping systems. Lu et al. [120] improved lane detection accuracy using a ridge detector and regional G-RANSAC. Zou et al. [130] achieved robust lane detection from continuous driving scenes using deep neural networks. Liu et al. [119] introduced Light Conditions Style Transfer for lane detection in low-light conditions. Wang et al. [124] used a map to enhance ego-lane detection in missing feature scenarios. Khan et al. [127] utilized impulse radio ultra-wideband radar and metal lane reflectors for robust lane detection in adverse weather conditions. Yang et al. [121] employed long short-term memory (LSTM) networks for lane position detection. Gao et al. [131] minimized false alarms in lane departure warnings using an Extreme Learning Residual Network and ϵ-greedy LSTM. Moreover, ref. [132] proposed a real-time attention-guided DNN-based lane detection framework and CondLaneNet [133] used conditional convolution for top-to-down lane detection. Dewangan and Sahu [134] analyzed driving behavior using vision-sensor-based lane detection. Haris and Glowacz [135] utilized object feature distillation for lane line detection. Lu et al. [136] combined semantic segmentation and optical flow estimation for fast and robust lane detection. Suder et al. [128] designed low-complexity lane detection methods for light photometry systems. Ko et al. [137] combined key points estimation and point instance segmentation for lane detection. Zheng et al. [138] introduced CLRNet for lane detection, while Wang et al. [122] proposed a multitask attention network (MAN). Khan et al. [139] developed LLDNet, a lightweight lane detection approach for autonomous cars. Chen and Xiang [125] incorporated pre-aligned spatial–temporal attention for lane mark detection. Nie et al. [126] integrated a camera with dual light sensors to improve lane-detection performance in autonomous vehicles. These studies collectively present diverse and effective methodologies, contributing to the advancement of lane-detection systems in autonomous driving and intelligent vehicle applications. The list of reviewed papers on lane-departure warning system is summarized in Table 6.

Table 6. Chosen publications, source title, and the number of citations related to a lane-departure warning system.

SI No.	Ref.	Year	Source Title	Cited by
1	[116]	2019	IEEE/CVF International Conference on Computer Vision	253
2	[117]	2019	IEEE/CVF Conference on Computer Vision and Pattern Recognition	78
3	[118]	2019	IEEE/CVF International Conference on Computer Vision	57
4	[123]	2019	MDPI Sensors	34
5	[129]	2019	MDPI Sensors	16
6	[120]	2019	MDPI Sensors	12
7	[130]	2020	IEEE Transactions on Vehicular Technology	165
8	[119]	2020	IEEE Intelligent Vehicles Symposium (IV)	32
9	[124]	2020	IEEE Access	9
10	[127]	2020	MDPI Sensors	14
11	[121]	2020	MDPI Sensors	9
12	[131]	2020	MDPI Sensors	6
13	[132]	2021	IEEE/CVF Conference on Computer Vision and Pattern Recognition	60
14	[133]	2021	IEEE/CVF International Conference on Computer Vision	44
15	[134]	2021	IEEE Sensors Journal	40

Table 6. *Cont.*

SI No.	Ref.	Year	Source Title	Cited by
16	[135]	2021	MDPI Electronics	17
17	[136]	2021	MDPI Sensors	14
18	[128]	2021	MDPI Electronics	12
19	[137]	2022	IEEE Transactions on Intelligent Transportation Systems	54
20	[138]	2022	IEEE/CVF Conference on Computer Vision and Pattern Recognition	17
21	[122]	2022	IEEE Transactions on Neural Networks and Learning Systems	15
22	[139]	2022	MDPI Sensors	4
23	[125]	2022	MDPI Sensors	2
24	[126]	2022	MDPI Electronics	-

4.6. Forward-Collision Warning System

A Forward-Collision Warning System (FCWS) is a type of ADAS that warns drivers of potential collisions with other vehicles or objects in front of them. FCWSs typically use radar, cameras, or lidar to track the distance and speed of vehicles in front of the vehicle, and they alert the driver if the vehicle is getting too close to the vehicle in front. When the system detects that a collision is imminent, it alerts the driver with a visual or audible warning.

FCWSs can be an invaluable safety feature, as they can help prevent accidents caused by driver distraction or drowsiness. According to the NHTSA, rear-end collisions account for about 25% of all fatal crashes in the United States [140].

FCWSs are becoming increasingly common in new vehicles. The NHTSA has mandated that all new cars sold in the United States come equipped with FCWS systems by 2022.

FCWSs: (i) help prevent accidents caused by driver distraction or drowsiness, (ii) help drivers to brake sooner, which can reduce the severity of rear-end crashes and accidents, (iii) help improve the driver awareness of the surrounding traffic, (iv) help to reduce driver stress and fatigue.

Although FCWSs offer many advantages, they have limitations such as: (i) being less effective in certain conditions, such as heavy rain or snow, (ii) being prone to false alarms, which can lead to driver desensitization, (iii) are not a substitute for safe driving practices, such as paying attention to the road and using turn signals.

Overall, FCWSs can be a valuable safety feature, but they are not guaranteed to prevent accidents. Drivers should still be aware of their surroundings and use safe driving practices at all times.

Search Terms and Recent Trends in FCWS

'Forward collision warning', 'forward collision', 'pre-crash', 'collision mitigating', and 'forward crash' are the prominent search terms used to investigate this topic. The 'OR' operator was used to choose and combine the most relevant and regularly used applicable phrases. That is, the search phrases 'forward collision warning', 'forward collision', 'pre-crash', 'collision mitigating', and 'forward crash' were discovered. Figure 7 shows the complete search query for each of the databases. The databases include IEEE Xplore and MDPI.

The papers listed discuss the development of FCWSs for autonomous vehicles in recent years. Ref. [141] suggests an autonomous vehicle collision avoidance system that employs predictive occupancy maps to estimate other vehicles' future positions, enabling collision-free motion planning. Ref. [142] introduces a forward collision prediction system using online visual tracking to anticipate potential collisions based on other vehicles' positions. Ref. [143] proposes an FCWS that combines driving intention recognition and V2V communication to predict and warn about potential collisions with front vehicles. Ref. [144]

presents an FCWS for autonomous vehicles that deploys a CNN to detect and track nearby vehicles. Ref. [145] introduces a real-time FCW technique involving detection and depth estimation networks to identify nearby vehicles and estimate distances. Ref. [146] proposes a vision-based FCWS merging camera and radar data for real-time multi-vehicle detection, addressing challenging conditions like occlusions and lighting variations. Tang et al. [147] introduce a monocular range estimation system using a single camera for precise FCWS, especially in difficult scenarios. Lim et al. [148] suggest a smartphone-based FCWS for motorcyclists utilizing phone sensors to predict collision risks. Farhat et al. [149] present a cooperative FCWS using DL to predict collision likelihood in real time by considering data from both vehicles' sensors. Hong and Park [150] offer a lightweight FCWS for low-power embedded systems, combining cameras and radar for real-time multi-vehicle detection. Albarella et al. [151] and Lin et al. [152] propose V2X communication-based FCWS, with [151] for electric vehicles and [152] targeting curve scenarios. Yu and Ai [153] suggest a hybrid DL approach employing CNN and recurrent NN for robust FCWS predictions. Olou et al. [154] introduce an efficient CNN model for accurate forward collision prediction, even in challenging conditions. Pak [155] presents a hybrid filtering method that improves radar-based FCWS by fusing data from multiple sensors, enhancing reliability.

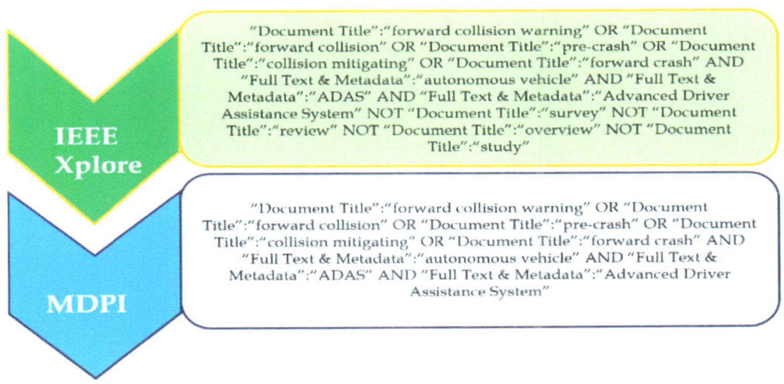

Figure 7. Search queries for each of the databases for the lane-departure warning system. The databases include IEEE Xplore and MDPI.

This compilation of research papers demonstrates the extensive efforts in the field of forward-collision warning and avoidance systems, which are crucial for enhancing vehicular safety. Lee and Kum [141] propose a 'Collision Avoidance/Mitigation System' incorporating predictive occupancy maps for autonomous vehicles. Manghat and El-Sharkawy [142] present 'Forward Collision Prediction with Online Visual Tracking', utilizing online visual tracking for collision prediction. Yang, Wan, and Qu [143] introduce 'A Forward Collision Warning System Using Driving Intention Recognition', integrating driving intention recognition and V2V communication. Kumar, Shaw, Maitra, and Karmakar [144] offer 'FCW: A Forward Collision Warning System Using Convolutional Neural Network', deploying CNN for warning generation. Wang and Lin [145] present 'A Real-Time Forward Collision Warning Technique', integrating detection and depth estimation networks for real-time warnings. Lin, Dai, Wu, and Chen [146] introduce a 'Driver Assistance System with Forward Collision and Overtaking Detection'. Tang and Li [147] propose 'End-to-End Monocular Range Estimation' for collision warning. Lim et al. [148] created a 'Forward Collision Warning System for Motorcyclists' using smartphone sensors. Farhat, Rhaiem, Faiedh, and Souani [149] present a 'Cooperative Forward Collision Avoidance System Based on Deep Learning'. Hong and Park [150] propose a 'Lightweight Collaboration of Detecting and Tracking Algorithm' for embedded systems. Albarella et al. [151] present a 'Forward-Collision Warning System for Electric Vehicles', validated both virtually and in real environments. Liu et al. [152] focus on 'Forward Collision on a Curve based on V2X' with a target selection method. Yu and Ai [153] present 'Vehicle Forward Collision

Warning based upon Low-Frequency Video Data' using hybrid deep learning. Olou, Ezin, Dembele, and Cambier [154] propose 'FCPNet: A Novel Model to Predict Forward Collision' based on CNN. Pak [155] contributes 'Hybrid Interacting Multiple Model Filtering' to improve radar-based warning reliability. Together, these papers collectively advance the understanding and development of forward collision warning and avoidance systems. The list of reviewed papers on forward-collision warning system is summarized in Table 7.

Table 7. Chosen publications, source title, and the number of citations related to forward-collision warning systems.

SI No.	Ref.	Year	Source Title	Cited by
1	[141]	2019	IEEE Access	48
2	[142]	2019	IEEE International Conference on Vehicular Electronics and Safety (ICVES)	2
3	[143]	2020	IEEE Access	31
4	[144]	2020	IEEE International Conference on Electrical and Electronics Engineering (ICE3)	2
5	[145]	2020	IEEE International Conference on Systems, Man, and Cybernetics (SMC)	-
6	[146]	2020	MDPI Sensors	26
7	[147]	2020	MDPI Sensors	4
8	[148]	2021	IEEE Journal of Intelligent and Connected Vehicles	1
9	[149]	2021	IEEE International Conference on Developments in eSystems Engineering (DeSE)	-
10	[150]	2021	IEEE Twelfth International Conference on Ubiquitous and Future Networks (ICUFN)	-
11	[151]	2021	MDPI Energies	-
12	[152]	2022	7th International Conference on Intelligent Informatics and Biomedical Science (ICIIBMS)	1
13	[153]	2022	IEEE 25th International Conference on Intelligent Transportation Systems (ITSC)	-
14	[154]	2022	22nd International Conference on Control, Automation and Systems (ICCAS)	-
15	[155]	2022	MDPI Sensors	3

4.7. Blind Spot Detection

Blind spot detection (BSD) is a type of ADAS that helps to prevent accidents by alerting drivers to vehicles, pedestrians, or objects that are in their blind spots. Blind spots are the areas around a vehicle that cannot be seen by the driver when looking in the rear-view or side mirrors. These areas can be especially dangerous when changing lanes, merging onto a highway, or while parking, and it is necessary to present accidents caused by lane changes into the blind spot of other vehicles.

When a vehicle is detected in the blind spot, the system alerts the driver with a visual or audible warning. Some systems will also illuminate a light in the side mirror to indicate that there is a vehicle in the blind spot, while some systems also provide a graphic representation of the vehicle in the blind spot on the dashboard.

BSD systems can be a valuable safety feature and are becoming increasingly common in new vehicles, as they can help to prevent accidents caused by driver inattention or driving changing lanes into other vehicles. They help to reduce the severity of accidents that do occur, thereby reducing drivers' stress and fatigue and helping drivers to stay alert and more aware of their surroundings. According to the NHTSA, blind spot crashes account for about 2% of all fatal crashes in the United States [57], and the NHTSA has mandated that all new cars sold in the United States come equipped with BSD systems by 2022.

Although BSD has many advantages, it has certain limitations such as: (i) it is less effective in certain conditions, such as heavy rain or snow, (ii) it is prone to false alarms, which can lead to driver desensitization, (iii) it is not a substitute for safe driving practices, such as using turn signals and checking blind spots before changing lanes.

Overall, BSD systems can be a valuable safety feature, but they are not a guarantee against accidents. Drivers should still be aware of their surroundings and use safe driving practices at all times.

Search Terms and Recent Trends in Blind Spot Detection

'Blind spot', 'blind spot detection', and 'blind spot warning', are the three prominent search terms used to investigate this topic. The 'OR' operator was used to choose and combine the most relevant and regularly used applicable phrases. That is, the search phrases 'blind spot', 'blind spot detection', and 'blind spot warning', were discovered. Figure 8 shows the complete search query for each of the databases. The databases include IEEE Xplore and MDPI.

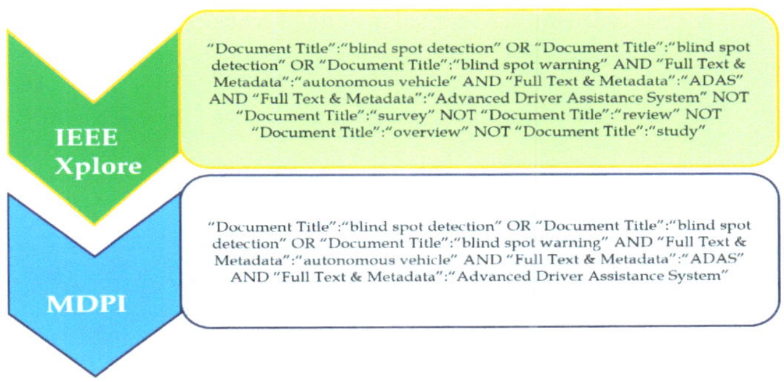

Figure 8. Search queries for each of the databases for blind spot detection. The databases include IEEE Xplore and MDPI.

The papers mentioned discuss the development of blind-spot detection systems (BSDSs) for vehicles. BSDSs are designed to alert drivers to vehicles that are in their blind spots, where they cannot be seen in their mirrors.

The Gale Bagi et al. [156] paper discusses a BSDS combining radar and cameras for accurate vehicle detection in blind spots. Radar detects vehicles and cameras identify them. Details about sensors and system architecture are necessary for a comprehensive understanding.

Ref. [157] introduces a probabilistic BSDS estimating blind spot risks using vehicle speed, direction, and driver's blind spot angle. It offers nuanced insights into collision potential, enhancing safe driving.

Zhao et al. [158] propose a promising BSDS using a lightweight NN and cameras for real-time detection. This approach improves detection capabilities with practical design. Chang et al. [159] present an AI-based BSDS warning for motorcyclists using various sensors, proactively detecting blind spot vehicles and enhancing rider safety. Naik et al. [160] propose lidar-based early BSDS, creating a 3D map to detect blind-spot vehicles in advance.

The authors of [161] describe a real-time two-wheeler BSDS using computer vision and ultrasonic sensors, confirming blind spot vehicles. Shete et al. [162] suggest a forklift-specific BSDS using ultrasonic sensors to detect blind spot vehicles and warn drivers. Schlegel et al. [163] propose an optimization-based planner for robots, considering blind spots and other vehicles to ensure safe navigation. Kundid et al. [164] introduce an ADAS algorithm creating a wider view to enhance driver awareness, mitigating blind spot issues.

Sui et al. [165] propose an A-pillar blind spot display algorithm using cameras to show blind spot information on the A-pillar and side mirrors. Wang et al. [166] present a vision-based BSDS using depth cameras to identify blind spot vehicles in a 3D map. Zhou et al. [167] focus on high-speed pedestrians in blind spots, using cameras and radar to detect pedestrians and pre-detection to avoid collisions. Ref. [168] introduces a multi-sensor BSDS for micro e-mobility vehicles, using cameras, radar, ultrasonic sensors, and gesture recognition for better blind-spot awareness. Ref. [169] suggests a multi-deep CNN-based BSDS for commercial vehicles using cameras, effectively addressing blind-spot challenges.

Overall, these papers present a variety of promising methods for developing BSDS. The systems proposed in these papers can detect vehicles in a variety of conditions, and they can be used in a variety of vehicles. The collection of research papers explores a broad spectrum of approaches to address blind spots in various domains, including robotics, automotive applications, and micro e-mobility. The focus ranges from sensor technologies such as cameras, lidar, and ultrasonic sensors to methodologies including AI, probabilistic estimation, and computer vision, introducing innovative algorithms, technologies, and architectures to enhance blind-spot detection, awareness, and collision prevention. The studies emphasize real-time detection, early warning, and proactive risk prediction, all contributing to enhance vehicular safety. The common thread among these studies is their commitment to improving safety by addressing the visibility limitations posed by blind spots. The list of reviewed papers on driver monitoring system is summarized in Table 8.

Table 8. Chosen publications, source title, and the number of citations related to driver monitoring systems.

SI No.	Ref.	Year	Source Title	Number of Citations
1	[156]	2019	2019 International Conference on Control, Automation and Information Sciences (ICCAIS)	3
2	[157]	2019	IEEE Intelligent Transportation Systems Conference (ITSC)	1
3	[158]	2019	MDPI Electronics	16
4	[159]	2020	International Symposium on Computer, Consumer, and Control (IS3C)	1
5	[160]	2020	International Conference on Smart Electronics and Communication (ICOSEC)	-
6	[161]	2021	5th International Conference on Electronics, Communication and Aerospace Technology (ICECA)	1
7	[162]	2021	IEEE International Conference on Technology, Research, and Innovation for Betterment of Society (TRIBES)	-
8	[163]	2021	European Conference on Mobile Robots (ECMR)	-
9	[164]	2021	Zooming Innovation in Consumer Technologies Conference (ZINC)	-
10	[165]	2022	IEEE 5th International Conference on Computer and Communication Engineering Technology (CCET)	-
11	[166]	2022	IEEE Intl Conf on Dependable, Autonomic and Secure Computing, Intl Conf on Pervasive Intelligence and Computing, Intl Conf on Cloud and Big Data Computing, Intl Conf on Cyber Science and Technology Congress (DASC/PiCom/CBDCom/CyberSciTech)	-
12	[167]	2022	IEEE 25th International Conference on Intelligent Transportation Systems (ITSC)	-
13	[168]	2022	MDPI Sensors	2
14	[169]	2022	MDPI Sensors	1

4.8. Emergency Braking System

The Emergency Braking System (EBS), also referred to as automatic emergency braking (AEB), is an ADAS that detects and tracks other vehicles in the vicinity, calculates the risk of a collision, and automatically applies the brakes in the event of an imminent collision to prevent or mitigate a collision. EBS helps to prevent accidents caused by the driver's inattention, drowsiness, or reaction time. EBSs can be a valuable safety feature, typically using radar, camera, or laser sensors to detect vehicles or objects in front of the car. According to the NHTSA [140], rear-end crashes account for about 25% of all fatal crashes in the United States.

EBSs are becoming increasingly common in new vehicles. In fact, the NHTSA has mandated that all new cars sold in the United States come equipped with EBSs by 2022. EBSs have numerous benefits, as they help to (i) prevent accidents caused by driver distraction or drowsiness, (ii) reduce the severity of accidents that do occur, and (iii) keep drivers alert and focused on the road.

With these benefits comes certain limitations, as these systems are (i) less effective in certain conditions, such as heavy rain or snow, (ii) prone to false alarms, which can lead to driver desensitization, and (iii) not a substitute for safe driving practices, such as paying attention to the road and using turn signals.

Overall, EBSs can be a valuable safety feature, but they are not guaranteed to prevent accidents. Drivers should still be aware of their surroundings and use safe driving practices at all times.

Search Terms and Recent Trends in Emergency Braking Systems

'Emergency braking system', 'autonomous emergency braking', 'EBS', and 'AEB', are the prominent search terms used to investigate this topic. The 'OR' operator was used to choose and combine the most relevant and regularly used applicable phrases. That is, the search phrases 'emergency braking system', 'autonomous emergency braking', 'EBS', and 'AEB', were discovered. Figure 9 shows the complete search query for each of the databases. The databases include IEEE Xplore and MDPI.

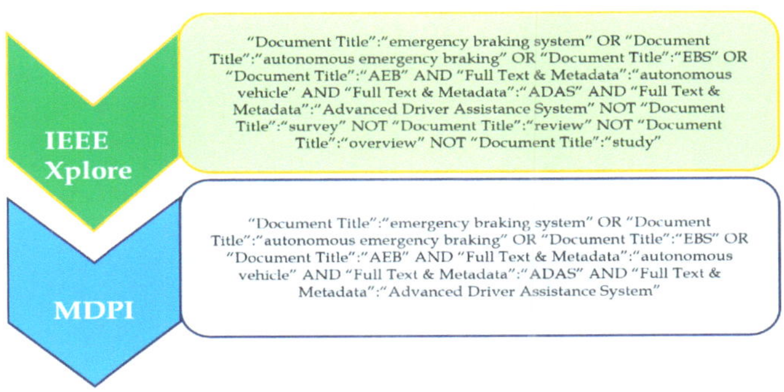

Figure 9. Search queries for each of the databases for the emergency braking system. The databases include IEEE Xplore and MDPI.

Flores et al. [170] propose a cooperative car-following and emergency braking system using radar, lidar, and cameras to detect and predict vehicle and pedestrian movements. It automatically applies the brakes to prevent collisions while also facilitating vehicle-to-vehicle communication. Shin et al. [171] introduce an adaptive AEB strategy utilizing radar and cameras to detect and calculate braking forces for front and rear vehicle collision avoidance. It considers speed, distance, and vehicle dynamics for effective collision prevention.

Yang et al. [172] have developed an AEB-P system with radar and cameras, using advanced control to determine braking forces for pedestrian collision avoidance, accounting for pedestrian speed, distance, and vehicle dynamics. Gao et al. [173] present a hardware-in-the-loop simulation platform for AEB system testing across various scenarios, ensuring reliability and effectiveness. Guo et al. [174] introduce a variable time headway AEB algorithm using predictive modeling, combining radar and cameras. It adapts time headway for braking by considering speed, distance, and vehicle dynamics.

Leyrer et al. [175] propose a simulation-based robust AEBS design using optimization techniques to enhance system performance and reliability. Yu et al. [176] introduce an AEBC system utilizing radar and cameras, applying control algorithms to prevent collisions at intersections considering vehicle and pedestrian speed, distance, and dynamics. Izquierdo et al. [177] explore using MEMS microphone arrays for AEBS, improving pedestrian detection through audio cues in a variety of environments.

Jin et al. [178] present an adaptive AEBC strategy for driverless vehicles in campus environments, utilizing radar and cameras to prevent collisions by considering vehicle and pedestrian characteristics and dynamics. Mannam and Rajalakshmi [179] assess AEBS scenarios for autonomous vehicles using radar and cameras, determining collision interventions based on vehicle and pedestrian detection, speed, and distance. Guo et al. [180] study AEBS control for commercial vehicles, considering driving conditions alongside radar and camera-based detection and control algorithms to avoid collisions based on vehicle and pedestrian dynamics.

These papers all represent significant advances in the field of AEB systems. They propose new methods for detecting and tracking vehicles, pedestrians, and environmental features. They also propose new control algorithms for determining the optimal braking force to apply to avoid a collision. These advances have the potential to make AEB systems more effective and reliable and to help prevent traffic accidents.

All the systems discussed were evaluated in a variety of traffic scenarios, and they were shown to be able to significantly reduce the number of accidents. The reviewed papers collectively explore a diverse range of topics within the realm of autonomous emergency braking (AEB) systems for enhanced road safety.

These topics include cooperative car-following, pedestrian avoidance, collision avoidance with rear vehicles, longitudinal active collision avoidance, hardware-in-the-loop simulation, variable time headway control, environmental feature recognition, simulation-based robust design, inevitable collision state-based control, innovative sensor utilization (MEMS microphone array), adaptive strategies for specific scenarios, determination of AEB-relevant scenarios, and specialized AEB algorithms for commercial vehicles. These contributions highlight the multi-faceted nature of AEB research, highlighting advancements in simulation, sensing, control strategies, and contextual optimization and emphasizing safety, prediction, algorithm optimization, and system validation. As autonomous vehicles continue to evolve, these papers will collectively contribute to enhancing the effectiveness and reliability of AEB systems, thereby advancing road safety in modern transportation and ultimately promoting safer and more reliable autonomous driving experiences. The list of reviewed papers on emergency braking system is summarized in Table 9.

4.9. Adaptive Cruise Control

Adaptive cruise control (ACC) is a driver assistance system that automatically adjusts a vehicle's speed when there are slow-moving vehicles ahead to maintain a safe following distance. When the road ahead is clear, ACC automatically accelerates to the driver's pre-set speed.

ACC is a Level 1 ADAS feature, which means that it requires some driver input. The driver still needs to be alert and ready to take over if necessary. However, ACC can help to reduce driver fatigue and stress, and it can also help to prevent accidents.

ACC systems typically use a radar sensor to detect the speed and distance of vehicles ahead. The sensor is mounted in the front of the vehicle, and it can typically detect vehicles up to several hundred feet away. The sensor sends this information to a control unit, which then calculates the appropriate speed for the vehicle to maintain a safe following distance.

Table 9. Chosen publications, source title, and the number of citations related to the emergency braking system.

SI No.	Ref.	Year	Source Title	Cited by
1	[170]	2019	IEEE Transactions on Intelligent Transportation Systems	31
2	[171]	2019	IEEE Intelligent Transportation Systems Conference (ITSC)	5
3	[172]	2019	MDPI Sensors	43
4	[173]	2019	IEEE 23rd International Conference on Intelligent Transportation Systems (ITSC)	4
5	[174]	2019	Chinese Automation Congress (CAC)	4
6	[175]	2019	IEEE Intelligent Vehicles Symposium (IV)	-
7	[176]	2020	American Control Conference (ACC)	2
8	[177]	2020	MDPI Sensors	2
9	[178]	2020	International Conference on Advanced Mechatronic Systems (ICAMechS)	-
10	[179]	2020	IEEE Global Conference on Computing, Power, and Communication Technologies (GlobConPT)	-
11	[180]	2020	MDPI Machines	4

ACC systems can be either speed-only or full-range systems. Speed-only systems only adjust the vehicle's speed, while full-range systems can also brake the vehicle to maintain a safe following distance. Full-range systems are more advanced, and they are typically more expensive. ACC systems can be set to a specific speed, or they can be set to follow the speed of the vehicle ahead. ACC systems can also be set to a maximum following distance, and the system will not allow the vehicle to get closer than the set distance to the vehicle ahead.

ACC systems are becoming increasingly common in vehicles, as they offer several safety and convenience benefits such as reducing traffic congestion and improving fuel efficiency. ACC systems can also help to prevent accidents by reducing the risk of rear-end collisions. They are especially beneficial for long-distance driving, as they can help to reduce driver fatigue. The benefits of ACC systems are as follows:

a. Reduced driver fatigue: ACC can help to reduce driver fatigue by taking over the task of maintaining a safe following distance. This can be especially beneficial for long-distance driving.
b. Increased safety: ACC can help prevent accidents by automatically adjusting the vehicle's speed to maintain a safe following distance.
c. Improved convenience: ACC can make driving more convenient by allowing the driver to set a cruising speed and then relax.
d. Improved fuel efficiency: ACC systems can help to improve fuel efficiency by allowing drivers to maintain a constant speed, which can reduce unnecessary acceleration and braking.

Despite these benefits, ACC systems face numerous challenges, as they are (i) expensive, especially in high-end vehicles, (ii) complex to install and calibrate, which can increase the cost of ownership, and (iii) unreliable in poor weather conditions, such as rain or snow.

Overall, ACC systems are a valuable safety feature that can help to prevent accidents and make driving more convenient. However, they are not without their challenges, such as cost and complexity. As ACC systems become more affordable and reliable, they are likely to become more widespread in vehicles.

Search Terms and Recent Trends in Adaptive Cruise Control

'Adaptive cruise control', 'ACC', 'autonomous cruise control', and 'intelligent cruise control' are the prominent search terms used to investigate this topic. The 'OR' operator was used to choose and combine the most relevant and regularly used applicable phrases. That is, the search phrases 'adaptive cruise control', 'ACC', 'autonomous cruise control', and 'intelligent cruise control' were discovered. Figure 10 shows the complete search query for each of the databases. The databases include IEEE Xplore and MDPI.

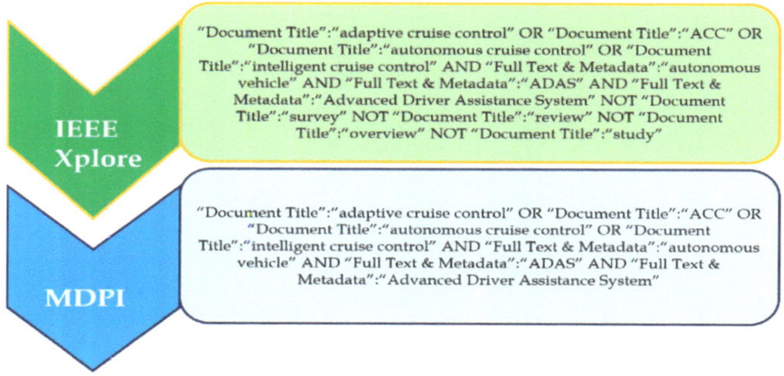

Figure 10. Search queries for each of the databases for the adaptive cruise control system. The databases include IEEE Xplore and MDPI.

G. Li and D. Görges [181] propose an innovative approach combining ecological ACC and energy management for HEVs using heuristic dynamic programming. The algorithm optimizes speed profiles, considering traffic conditions, state of charge, and driver preferences for fuel efficiency and comfort. S. Cheng et al. [182] discuss a multiple-objective ACC with dynamic velocity obstacle (DYC) prediction, optimizing speed, acceleration, safety, comfort, and fuel efficiency by forecasting surrounding vehicle trajectories. J. Lunze [183] introduces an ACC strategy ensuring collision avoidance through predictive control using a combination of predictive control and MPC to optimize vehicle speed profiles. Woo, H. et al. [184] enhance ACC safety and efficiency through operation characteristic estimation and trajectory prediction. Their work adjusts speed and acceleration considering vehicles' dynamics and surroundings.

Zhang, S. and Zhuan, X. [185] developed an ACC for BEVs that accounts for weight changes. Weight adjustments based on battery discharge and passenger load are used to ensure safe and comfortable driving. C. Zhai et al. [186] present an ecological CAC strategy for HDVs with time delays using distributed algorithms for platoon coordination, achieving fuel efficiency and ecological benefits. Li and Görges [187] designed an ecological ACC for step-gear transmissions using reinforcement learning. It optimizes fuel efficiency while maintaining safety through learned intelligent control strategies. Jia, Jibrin, and Görges [188] propose an energy-optimal ACC for EVs using linear and nonlinear MPC techniques, minimizing energy consumption based on dynamic driving and traffic conditions. Nie and Farzaneh [189] focus on eco-driving ACC with an MPC algorithm for reduced fuel consumption and emissions while ensuring safety and comfort. Guo, Ge, Sun, and Qiao [190] introduce an MPC-based ACC with relaxed constraints to enhance fuel efficiency while considering speed limits and safety distances for driving comfort.

Liu, Wang, Hua, and Wang [191] analyze CACC safety with communication delays using MPC and fuzzy logic to ensure stable and effective CACC operation under real-world communication conditions. Lin et al. [192] compare DRL and MPC for ACC, suggesting a hybrid approach for improved fuel efficiency, comfort, and stability. Gunter et al. [193] investigate the string stability of commercial ACC systems, highlighting potential collision risks in platooning situations and recommending improvements. Sawant et al. [194] present a robust CACC control algorithm using MPC and fuzzy logic to ensure safe operation even with limited data on preceding vehicle acceleration. Yang, Wang, and Yan [195] optimize ACC through a combination of MPC and ADRC, enhancing fuel efficiency and robustness to disturbances. Anselma [196] proposes a powertrain-oriented ACC considering fuel efficiency and passenger comfort using MPC and powertrain modeling.

Chen [197] designed an ACC tailored to cut-in scenarios using MPC for fuel efficiency optimization during lane changes. Hu and Wang [198] introduce a trust-based ACC with individualization using a CBF approach, allowing vehicles to have personalized safety requirements. Yan et al. [199] hybridized DDPG and CACC for optimized traffic flow, leveraging learning-based and cooperative techniques. Zhang et al. [200] created a human-lead-platooning CACC to integrate human-driven vehicles into platoons. The author of [201] presents a resilient CACC using ML to enhance robustness and adaptability to uncertainties and disruptions. Kamal et al. [202] propose an ACC with look-ahead anticipation for freeway driving, adjusting control inputs based on predicted traffic conditions. Li et al. [203] leverage variable compass operator pigeon-inspired optimization (VCPO-PIO) for ACC control input optimization. Petri et al. [204] address ACC for EVs with FOC, considering unique characteristics and energy management needs. The list of reviewed papers on adaptive cruise control is summarized in Table 10.

Table 10. Chosen publications, source title, and the number of citations related to adaptive cruise control.

SI No.	Ref.	Year	Source Title	Number of Citations
1	[181]	2019	IEEE Transactions on Intelligent Transportation Systems	57
2	[182]	2019	IEEE Transactions on Vehicular Technology	54
3	[183]	2019	IEEE Transactions on Intelligent Transportation Systems	39
4	[184]	2019	MDPI Applied Sciences	9
5	[185]	2019	MDPI Symmetry	9
6	[186]	2020	IEEE Access	39
7	[187]	2020	IEEE Transactions on Intelligent Transportation Systems	29
8	[188]	2020	IEEE Transactions on Vehicular Technology	25
9	[189]	2020	MDPI Applied Sciences	29
10	[190]	2020	MDPI Applied Sciences	12
11	[191]	2020	MDPI Sustainability	11
12	[192]	2021	IEEE Transactions on Intelligent Vehicles	69
13	[193]	2021	IEEE Transactions on Intelligent Transportation Systems	68
14	[194]	2021	IEEE Transactions on Intelligent Transportation Systems	31
15	[195]	2021	MDPI Actuators	16
16	[196]	2021	MDPI Energies	13
17	[197]	2021	MDPI Applied Sciences	12
18	[198]	2022	IEEE Transactions on Intelligent Transportation Systems	12
19	[199]	2022	IEEE Transactions on Automation Science and Engineering	10
20	[200]	2022	IEEE Transactions on Intelligent Transportation Systems	8
21	[201]	2022	IEEE Transactions on Intelligent Transportation Systems	8
22	[202]	2022	MDPI Applied Sciences	5
23	[203]	2022	MDPI Electronics	1
24	[204]	2022	MDPI Applied Sciences	1

4.10. Around-View Monitoring (AVM)

Around-View Monitoring (AVM) is an ADAS that uses multiple cameras to provide a 360-degree view of the vehicle's surroundings. This helps drivers to see more of what is around them, which can improve safety and make it easier to park. It is especially helpful in tight spaces or when backing up.

AVM systems typically use four cameras, one mounted on each side of the vehicle and one in the rear. The cameras are connected to a central computer, which stitches the images together to create a panoramic view of the vehicle's surroundings. This view is displayed on a screen in the vehicle's cabin, giving the driver a bird's-eye view of what is around them and preventing blind spots. Thus, AVM systems are a valuable safety feature and can be used for a variety of purposes, including parking, backing up, maneuvering in tight spaces, monitoring blind spots, and overall enhancing safety by giving drivers a better view of their surroundings and preventing accidents, especially in low-visibility conditions. The challenges of AVM in ADAS are their high cost and complexity of installation.

The ADAS features with which AVM are often combined include blind-spot detection, lane departure warning system, a forward collision warning system, and parking assistance systems. Overall, these features can work together to provide drivers with a more comprehensive view of their surroundings, help them avoid accidents, and make it easier to park.

Search Terms and Recent Trends in Around-View Monitoring

'Around view monitoring', 'AVM', and 'surround view monitoring' are the prominent search terms used to investigate this topic. The 'OR' operator was used to choose and combine the most relevant and regularly used applicable phrases. That is the search phrases 'around view monitoring', 'AVM', and 'surround view monitoring' were discovered. Figure 11 shows the complete search query for each of the databases. The databases include IEEE Xplore and MDPI.

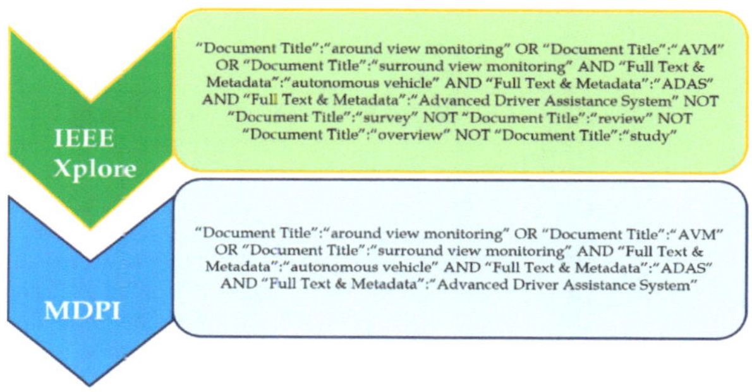

Figure 11. Search queries for each of the databases for around view monitoring. The databases include IEEE Xplore and MDPI.

Ref. [205] introduces a novel method by integrating semantic segmentation with AVM for lane-level localization. Utilizing visual data and semantic information, a DL model segments lanes and localizes the vehicle, enhancing navigation precision and safety. Refs. [206,207] integrate motion estimation into an AVM for ADAS. The author of [206] employs a Kalman filter to estimate motion, improving AVM image accuracy by up to 20%. The author of [207] focuses on homogeneous surfaces, achieving 90% accuracy with image registration and optical flow. Ref. [208] discusses AVM/lidar sensor fusion for parking-based SLAM. The fusion creates a map for SLAM and parking detection, with an improved loop closure accuracy of 95%.

Ref. [209] proposes AVM-based parking space detection using image processing and machine learning, providing an effective solution. Ref. [210] presents automatic AVM camera calibration using image processing and machine learning, streamlining the process without a physical calibration rig. Ref. [211] enhances AVM image quality via synthetic image learning for deblurring, addressing blurriness and distortion. Ref. [212] introduces AVM calibration using unaligned square boards, simplifying the process and increasing accuracy without a physical rig. Ref. [213] proposes an AVM-based automatic parking system using parking line detection, offering an accurate and efficient solution. Ref. [214] suggests a DL-based approach to detect parking and collision risk areas in autonomous parking scenarios, improving accuracy and collision assessment.

The papers discussed above provide a good overview of the current state-of-the-art approaches using AVM systems for lane-level localization, motion estimation, parking space detection, and collision risk area detection and improving the performance of AVM systems. The methods proposed in these papers have the potential to significantly improve the safety and efficiency of AVM systems, which in turn improves driving and parking efficiencies, and they are likely to become increasingly common in the future.

These amalgamations of these research papers collectively introduce innovative approaches ranging from semantic segmentation for lane-level localization to motion estimation techniques for enhancing monitoring accuracy, and collectively focus on crucial aspects such as automatic calibration, image-quality enhancement, parking-line detection, and collision-risk assessment. Additionally, by employing advanced techniques like supervised deblurring and DL, the integration of sensor fusion, such as AVM and lidar, significantly improves AVM systems' reliability, accuracy, and safety, offering promising outcomes for applications like autonomous parking. The synthesis of these diverse techniques showcases the recent advancements and growing potential of AVM in improving vehicle navigation, parking, and overall safety, thus revolutionizing vehicle navigation, parking, and overall driving experiences. The list of reviewed papers on around view monitoring is summarized in Table 11.

Table 11. Chosen publications, source title, and the number of citations related to around-view monitoring.

SI No.	Ref.	Year	Source Title	Cited by
1	[205]	2019	IEEE Sensors Journal	18
2	[206]	2019	7th International Conference on Mechatronics Engineering (ICOM)	-
3	[207]	2019	7th International Conference on Mechatronics Engineering (ICOM)	-
4	[208]	2019	MDPI Sensors	10
5	[209]	2019	MDPI Applied Sciences	9
6	[210]	2020	IEEE Access	3
7	[211]	2021	17th International Conference on Machine Vision and Applications (MVA)	1
8	[212]	2021	MDPI Sensors	2
9	[213]	2021	MDPI Applied Sciences	1
10	[214]	2022	MDPI Sensors	1

5. Discussion Datasets

The input data are the most important factor for the ADAS functionalities discussed in this paper. The preparation of the dataset is essential for the DL approaches, particularly in the training phase. The quality of the dataset preparation in the network model determines how well the autonomous car can manage its behavior and make decisions.

A review of journal articles, conference papers, and book chapters found that many studies used self-collected data or collected data online. Some researchers compiled their own dataset for training and then compared it to a publicly available benchmark dataset. Others only used self-collected data for training and validation. Still, others relied only on publicly available datasets for training and validation.

The choice of dataset preparation method depends on the specific research and the availability of resources. Self-collected data can be more representative of the specific environment in which the autonomous car will be operating, but it can be more time-consuming and expensive to collect. Publicly available datasets are more convenient to use, but they may not be as representative of the specific environment. Table 12 lists various public datasets used for different state-of-the-art methods discussed in Sections 4.1–4.10.

Table 12. Datasets employed by the references chosen in this review paper.

SI No.	Name.	Categories	No. of Objects	Papers Used
1	KITTI Vision Benchmark Suite [215,216]	Vehicles, pedestrians, cyclists, and road objects	Over 70,000 images & 30,000 Lidar scans	[33–35,37,41,43,46,50, 58,59,65,77,78,118, 121,124,126,129,133, 141,145,151,154,157, 170,208]
2	Argoverse [217]	Vehicles, pedestrians, cyclists, traffic lights, road objects, and more	Over 1M	[34]
3	nuScenes [218]	Vehicles, pedestrians, cyclists, traffic signs, lights, road markings, and more	Over 1.4M	[35,142,146,147,150, 153,163,165]
4	GRAM [38]	Vehicles, pedestrians, cyclists	Around 1M	[38]
5	GRAM-RTM [36]	Vehicles, pedestrians, cyclists, traffic signs, lights, road markings, and more	-	[36]
6	UA-DETRAC [36,219,220]	Car, bus, van, and others	8550	[37]
7	CDNet [221]	Cars, pedestrians, animals, buildings, trees, traffic signs, background scenes, and more	93,702	[38]
8	VEDAI [222]	Car, bus, truck, motorcycle, bicycle, pedestrian, traffic light, signs, buildings, vegetation, background	33,360	[44]
9	DAWN [223]	Person, car, bus, truck, motorcycle, bicycle, pedestrian, traffic light, signs, trailer, pole, buildings, vegetation, sky, ground, and unknown	275,350	[46,54]
10	MS-COCO [224]	Car, person, bicycle, motorcycle, bus, truck, train, stop sign, fire hydrant, traffic light	Over 2M	[46,55,105]
11	OSM [225]	No fixed categories	-	[49]
12	DroneVehicle [226]	Car, truck, bus, van, freight car	24,358	[51]
13	Highway Dataset [227]	Vehicles, pedestrians, bicycles, traffic signs, construction, and other objects	42,000	[33,55]
14	Space Cup Competition [228]			[228]
15	CityPersons pedestrian detection benchmark [229]	Pedestrians	3475	[60,70]
16	PETS2009 [230]	People, bicycles, motorcycles, cars, vans, trucks, and other vehicles	4005	[71]
17	CalTech Lanes Dataset [231]	People, bicycles, motorcycles, cars, vans, airplanes, faces, Frisbee, trucks, and more	30,607	[72,131]
18	Multispectral pedestrian detection [232]	Pedestrians	86,152	[73–76,79]
19	Aerial Infrared Pedestrian Detection Benchmark [80]	Pedestrians	Over 100K	[80]
20	GTSRB [233]	Traffic signs	51,839	[82–89,93,98]
21	BTSC [234]	Traffic signs	3740	[93]
22	LISA [235]	Traffic signs	6160	[97,169]
23	ITSRB & ITSDB [98]	Traffic signs	500	[98]
24	Cure-TSD [236]	Traffic signs	1080	[100]
25	Tsinghua-Tencent 100K [237]	Traffic signs	100,000	[102]

Table 12. *Cont.*

SI No.	Name.	Categories	No. of Objects	Papers Used
26	CCTSDB [238]	Traffic signs	7717	[104]
27	HRRSD [239]	Traffic signs	58,290	[104]
28	CuLane [240]	Lane marking, traffic signs, dazzle lights, and more	10,2448	[116,117,119,122,124, 128,132,134,135,137]
29	TUSimple [241]	Vehicles, lane markings, traffic signs, pedestrians, cyclists, and more	12,224	[116,119,122–126,130, 132,133,137,138]
30	BDD100K [242]	Pedestrians, riders, cars, trucks, buses, traffic signs, and more	1,407,782	[116]
31	Udacity Machine Learning Nanodegree Project Dataset [243]	Vehicles, lane markings, traffic signs, pedestrians, cyclists, and more	242,999	[139,144]
32	LLAMAS Dataset [244]	Car, bus, truck, motorcycle, bicycle, pedestrian, traffic lights and signs, yield light, and more	1300	[122]
33	Cracks and Potholes in Road Images Dataset [245]	Cracks and potholes	3235	[139]
34	Waymo Open Dataset [246]	Vehicles, pedestrians, cyclists, and signs	5,447,059	[148]
35	ETH Pedestrian Dataset [247]	Pedestrians, cyclists, cars, and van	61,764	[170]

Besides employing publicly available, free-to-use open-source datasets, the most recent state-of-the-art work uses a self-collected dataset and proposes datasets suitable for their proposed works and makes their proposed dataset available for other researchers. For instance, ref. [40] manually constructed a dataset containing 316 vehicle clusters and 224 non-vehicle clusters, ref. [47] used datasets generated from the transformed results that demonstrate significant improvement, and ref. [62] initially generated a template of a pedestrian from a training dataset. The template was then used to match pedestrians in the lidar point cloud. The authors of the paper evaluated their method based on a dataset of lidar point clouds. Additionally, ref. [63] was evaluated using their dataset and [67] was evaluated using a dataset of images captured in hazy weather, ref. [66] was trained and tested on a dataset of images captured in different weather conditions, ref. [67] was trained on a dataset of images from rural roads, ref. [68] was trained on infrared images captured during nighttime, and ref. [69] was trained on a dataset of images collected from different scenarios, including urban roads, highways, and intersections. If a public dataset is unavailable and the target is specific to a country, as was the case for [91], in which a public dataset for Taiwan was not available, the author evaluated their method based on a locally built dataset [248]. On the other hand, many publications do not mention exactly which dataset was used, instead highlighting that 'the proposed method was evaluated on a publicly available dataset' [94–96].

In addition to the state-of-the-art methods discussed in the above sections, some of the other notable publications are:

The paper [249] provides a comprehensive overview of the advancements and techniques in object detection facilitated by DL methodologies. The authors survey the state-of-the-art approaches up to the time of publication in 2019, and discuss various DL architectures and algorithms used for object detection, including two-stage detectors, one-stage detectors, anchor-based and anchor-free methods, RetinaNet, and FPNs, along with methodologies handling small objects, occlusions, and cluttered backgrounds. Additionally, they present some promising research directions for future work, such as multi-task learning, attention mechanisms, weakly supervised learning, and domain adaptation. Additionally, their paper explores the architectural evolution of DL models for object detection, discussing the transition from traditional methods to the emergence of region-based and anchor-based detectors, as well as the introduction of feature pyramid networks. The review also covers commonly used datasets for object detection, highlighting their significance in benchmarking algorithms, and discusses the evaluation metrics used to assess the performance of object detection models.

The paper [250] serves as a thorough survey of driving monitoring and assistance systems (DMAS), covering a wide range of technologies and methodologies such as driver monitoring systems (DMS), advanced driver assistance systems (ADAS), autonomous emergency braking (AEB), lane-departure warning systems (LDWS), adaptive cruise control (ACC), and blind spot monitoring (BSM). It explores various aspects of systems designed to monitor driver behavior and provide assistance, contributing to the understanding of advancements in the field of intelligent transportation systems. The comprehensive nature of the survey suggests an in-depth examination of existing technologies, challenges, and potential future directions for driving monitoring and assistance systems.

The paper [251] proposes a novel approach to 3D object detection utilizing monocular images. The key focus is on the use of a Proposal Generation Network tailored for 3D object detection, which integrates depth information derived from monocular images to generate proposals efficiently, contributing to improve the overall accuracy and efficiency of 3D object detection. The paper addresses the challenge of 3D object detection using only monocular images, which is a significant contribution, as many real-world applications rely on single-camera setups.

The paper [252] presents an innovative one-stage approach to monocular 3D object detection, streamlining the detection pipeline and potentially improving real-time performance compared to traditional two-stage approaches, emphasizing the use of discrete depth and orientation representations that suggest a departure from continuous representations, potentially leading to more interpretable and efficient models of the detection process.

The paper [253] explores the integration of AI techniques for object detection and distance measurement in which the algorithms are employed to identify and locate objects in images or videos. Once the objects have been detected, the model estimates their distance from the camera using various techniques, such as depth estimation networks, monocular depth estimation, and stereo depth estimation. This AI-based approach to object detection and distance measurement has the potential to revolutionize various fields. It offers high accuracy, real-time performance, and low cost, making it a promising solution for a wide range of applications.

6. Conclusions and Future Trends

Various ADASs discussed in the previous section have the potential to revolutionize the way we drive. By improving road safety, reducing driver workload, and providing a more comfortable and enjoyable driving experience, ADASs can make our roads safer and our journeys more enjoyable.

These DL algorithms are still under development, but they have the potential to revolutionize the way ADASs are designed and implemented. As these algorithms become more powerful and efficient, they will become more widely used in ADASs. Some of the advantages of using deep learning for object detection, recognition, and tracking in ADAS are as follows:

a. Accuracy: Deep learning algorithms have been shown to be more accurate than traditional algorithms, especially in challenging conditions.
b. Speed: Deep learning algorithms can be very fast, which is important for real-time applications.
c. Scalability: Deep learning algorithms can be scaled to handle large datasets and complex tasks.
d. Robustness: Deep learning algorithms are relatively robust to noise and other disturbances.

These advantages come with some of the challenges of using DL for object detection, recognition, and tracking in ADAS:

a. Data requirements: Deep learning algorithms require large datasets of labeled data to train. This can be a challenge to obtain, especially for rare or unusual objects.
b. Computational requirements: Deep learning algorithms can be computationally expensive, which can limit their use in real-time applications.
c. Interpretability: Deep learning algorithms are often difficult to interpret, which can make it difficult to understand why they make certain decisions.

Researchers are working on developing newer algorithms and improvising the existing algorithms and techniques to address these challenges. As a result, ADASs are becoming increasingly capable of detecting and tracking objects in a variety of challenging conditions.

ADASs are still under development, but they have the potential to revolutionize the way we drive. By making our roads safer and more efficient, ADASs can help to create a better future for transportation.

ADASs are not without their drawbacks. They can be expensive, and they can sometimes malfunction. Additionally, drivers may become too reliant on ADASs and become less attentive to their driving.

Overall, ADASs offer numerous potential benefits for safety and convenience. However, it is important to be aware of the drawbacks and to use these systems responsibly.

The ongoing continuous advancements and researches are focusing on overcoming the existing drawbacks and the same can be foreseen as the future trends of ADAS.

a. Multi-sensor fusion: ADASs are increasingly using multiple sensors, such as cameras, radar, and lidar, to improve the accuracy and reliability of object detection. Multi-sensor fusion can help to overcome the limitations of individual sensors, such as occlusion and poor weather conditions.
b. Deep learning: DL is rapidly becoming the dominant approach for object detection, recognition, and tracking in ADAS. Deep learning algorithms are very effective at learning the features that are important for identifying different objects.
c. Real-time performance: ADASs must be able to detect, recognize, and track objects in real time. This is essential for safety-critical applications, as delays in detection or tracking can lead to accidents.
d. Robustness to challenging conditions: ADASs must be able to operate in a variety of challenging conditions, such as different lighting conditions, weather conditions, and road conditions. Researchers are working on developing new algorithms and techniques to improve the robustness of ADASs to challenging conditions.
e. Integration with other ADAS features: ADASs are seeing increased integration with other ADAS features, such as collision avoidance, lane departure warning, and adaptive cruise control. This integration can help to improve the overall safety of vehicles.

These are just some of the future trends in object detection, recognition, and tracking for ADAS. As research in this area continues, ADASs are becoming increasingly capable of detecting and tracking objects in a variety of challenging conditions. This will help to make vehicles safer and more reliable.

Some additional trends that are worth mentioning could be:

a. The use of synthetic data: Synthetic data are being used increasingly often to train object detection, recognition, and tracking algorithms. Synthetic data are generated by computer simulations, and they can be used to create training datasets that are more diverse and challenging than the real-world datasets. This might enhance the efficiency of the neural networks, as they can be trained with a combination of real-world datasets supplemented with the synthetic datasets.
b. The use of edge computing: Edge computing is a distributed computing paradigm that brings computation and storage closer to the edge of the network. Edge computing can be used to improve the performance and efficiency of ADASs by performing object detection, recognition, and local tracking on the vehicle, implying that the greater the storage on the ADAS implement vehicles, the better the performance of the ADASs.

c. The use of 5G: 5G is the next generation of cellular network technology. 5G will offer much higher bandwidth and lower latency than 4G, which will make it possible to stream high-definition video from cameras to cloud-based servers for object detection, recognition, and tracking. Thus, a better cellular network will aid in the continuous training of the NNs and greatly improve the performance with newer data from real environments.

These are just some of the future trends that are likely to shape the development of object detection, recognition, and tracking for ADAS in the years to come.

Author Contributions: Conceptualization, V.M.S. and J.-I.G.; methodology, V.M.S. and J.-I.G.; validation, V.M.S. and J.-I.G.; formal analysis, V.M.S.; investigation, V.M.S.; resources, V.M.S. and J.-I.G.; data curation, V.M.S.; writing—original draft preparation, V.M.S.; writing—review and editing, V.M.S.; visualization, V.M.S.; supervision, J.-I.G.; project administration, J.-I.G.; funding acquisition, J.-I.G. All authors have read and agreed to the published version of the manuscript.

Funding: This work is supported by the National Science and Technology Council (NSTC), Taiwan R.O.C. projects with grants 112-2218-E-A49-027-, 112-2218-E-002-042-, 111-2622-8-A49-023-, 111-2221-E-A49-126-MY3, 111-2634-F-A49-013-, and 110-2221-E-A49-145-MY3, and by the Satellite Communications and AIoT Research Center/The Co-operation Platform of the Industry-Academia Innovation School, National Yang Ming Chiao Tung University (NYCU), Taiwan R.OC. projects with grants 111UC2N006 and 112UC2N006.

Institutional Review Board Statement: Not applicable.

Informed Consent Statement: Not applicable.

Data Availability Statement: No data used in the article but only the state-of-the-art publications as listed in the 'References' section.

Acknowledgments: We extend of sincere thanks to the National Yang Ming Chiao Tung University (NYCU), Taiwan R.OC., National Science and Technology Council (NSTC), Taiwan R.O.C., and the Satellite Communications and AIoT Research Center/The Co-operation Platform of the Industry-Academia Innovation School, National Yang Ming Chiao Tung University (NYCU), Taiwan R.OC. for their valuable support. We extend our heartfelt thanks to all the members and staff of the Intelligent Vision System Laboratory (iVSL), National Yang Ming Chiao Tung University, Taiwan R.O.C.

Conflicts of Interest: Author Jiun-In Guo was employed by the company eNeural Technologies Inc. All the authors declare that the research was conducted in the absence of any commercial or financial relationships that could be construed as a potential conflict of interest.

References

1. Dewesoft. What Is ADAS? Dewesoft Blog. 8 March 2022. Available online: https://dewesoft.com/blog/what-is-adas (accessed on 12 March 2022).
2. FEV Consulting. Forbes Honors FEV Consulting as One of the World's Best Management Consulting Firms. FEV Media Center. 20 July 2022. Available online: https://www.fev.com/en/media-center/press/press-releases/news-article/article/forbes-honors-fev-consulting-as-one-of-the-worlds-best-management-consulting-firms-2022.html (accessed on 17 March 2022).
3. Insurance Institute for Highway Safety. Effectiveness of advanced driver assistance systems in preventing fatal crashes. *Traffic Inj. Prev.* **2019**, *20*, 849–858.
4. Traffic Safety Facts: 2021 Data. Available online: https://crashstats.nhtsa.dot.gov/Api/Public/ViewPublication/813001 (accessed on 1 October 2022).
5. Palat, B.; Delhomme, P.; Saint Pierre, G. Numerosity heuristic in route choice based on the presence of traffic lights. *Transp. Res. Part F Traffic Psychol. Behav.* **2014**, *22*, 104–112. [CrossRef]
6. Papadimitriou, E.; Lassarre, S.; Yannis, G. Introducing human factors in pedestrian crossing behaviour models. *Transp. Res. Part F Traffic Psychol. Behav.* **2016**, *36*, 69–82. [CrossRef]
7. King, E.; Bourdeau, E.; Zheng, X.; Pilla, F A combined assessment of air and noise pollution on the High Line, New York City. *Transp. Res. Part D Transp. Environ.* **2016**, *42*, 91–103. [CrossRef]
8. Woodburn, A. An analysis of rail freight operational efficiency and mode share in the British port-hinterland container market. *Transp. Res. Part D Transp. Environ.* **2017**, *51*, 190–202. [CrossRef]
9. Haybatollahi, M.; Czepkiewicz, M.; Laatikainen, T.; Kyttä, M. Neighbourhood preferences, active travel behaviour, and built environment: An exploratory study. *Transp. Res. Part F Traffic Psychol. Behav.* **2015**, *29*, 57–69. [CrossRef]

10. Honda Worldwide. Honda Motor Co. Advanced Brake Introduced for Motorcycles by Honda ahead of Others. Available online: https://web.archive.org/web/20160310200739/http://world.honda.com/motorcycle-technology/brake/p2.html (accessed on 30 November 2022).
11. American Honda. Combined Braking System (CBS). 9 December 2013. Available online: https://web.archive.org/web/20180710010624/http://powersports.honda.com/experience/articles/090111c08139be28.aspx (accessed on 16 September 2022).
12. Blancher, A.; Zuby, D. Interview: Into the Future with ADAS and Vehicle Autonomy. Visualize, Verisk. 8 March 2023. Available online: https://www.verisk.com/insurance/visualize/interview-into-the-future-with-adas-and-vehicle-autonomy/ (accessed on 16 September 2022).
13. Yeong, D.J.; Velasco-Hernandez, G.; Barry, J.; Walsh, J. Sensor and Sensor Fusion Technology in Autonomous Vehicles: A Review. *Sensors* **2021**, *21*, 2140. [CrossRef] [PubMed]
14. Continental, A.G. ADAS Challenges and Solutions. 2022. Available online: https://conf.laas.fr/WORCS13/Slides/WORCS-13_2013-SergeBoverie.pdf (accessed on 8 March 2023).
15. Blanco, S. Advanced Driver-Assistance Systems. What the Heck Are They Anyway? Forbes. 26 May 2022. Available online: https://www.forbes.com/wheels/advice/advanced-driver-assistance-systems-what-are-they/ (accessed on 20 May 2023).
16. Cortes, C.; Vapnik, V. Support-vector networks. *Mach. Learn.* **1995**, *20*, 273–297. [CrossRef]
17. Breiman, L. Random Forests. *Mach. Learn.* **2001**, *45*, 5–32. [CrossRef]
18. Swain, M.J.; Ballard, D.H. Color indexing. *Int. J. Comput. Vis.* **1991**, *7*, 11–32. [CrossRef]
19. Sobel, I.; Feldman, G. *A 3 × 3 Isotropic Gradient Operator for Edge Detection*; Presented at the Stanford Artificial Project; Stanford University: Stanford, CA, USA, 1968.
20. Belongie, S.; Malik, J.; Puzicha, J. Shape matching and object recognition using shape contexts. *IEEE Trans. Pattern Anal. Mach. Intell.* **2002**, *24*, 509–522. [CrossRef]
21. Kalman, R.E. A new approach to linear filtering and prediction problems. *J. Basic Eng.* **1960**, *82*, 35–45. [CrossRef]
22. Simon, D. *Optimal State Estimation: Kalman, H Infinity, and Nonlinear Approaches*; John Wiley & Sons: Hoboken, NJ, USA, 2006.
23. Wu, J.K.; Wong, Y.F. Bayesian Approach for Data Fusion in Sensor Networks. In Proceedings of the 2006 9th International Conference on Information Fusion, Florence, Italy, 10–13 July 2006; pp. 1–5. [CrossRef]
24. Sun, Y.-Q.; Tian, J.-W.; Liu, J. Target Recognition using Bayesian Data Fusion Method. In Proceedings of the 2006 International Conference on Machine Learning and Cybernetics, Dalian, China, 13–16 August 2006; pp. 3288–3292. [CrossRef]
25. Le Hegarat-Mascle, S.L.; Bloch, I.; Vidal-Madjar, D. Application of Dempster-Shafer evidence theory to unsupervised classification in multisource remote sensing. *IEEE Trans. Geosci. Remote Sens.* **1997**, *35*, 1018–1031. [CrossRef]
26. Chen, C.; Jafari, R.; Kehtarnavaz, N. Improving Human Action Recognition Using Fusion of Depth Camera and Inertial Sensors. *IEEE Trans. Hum. Mach. Syst.* **2015**, *45*, 51–61. [CrossRef]
27. Ding, B.; Wen, G.; Huang, X.; Ma, C.; Yang, X. Target Recognition in Synthetic Aperture Radar Images via Matching of Attributed Scattering Centers. *IEEE J. Sel. Top. Appl. Earth Obs. Remote Sens.* **2017**, *10*, 3334–3347. [CrossRef]
28. Gu, J.; Lind, A.; Chhetri, T.R.; Bellone, M.; Sell, R. End-to-End Multimodal Sensor Dataset Collection Framework for Autonomous Vehicles. *Sensors* **2023**, *23*, 6783. [CrossRef] [PubMed]
29. RGBSI. What Is Sensor Fusion for Autonomous Driving Systems?—Part 1. RGBSI Blog. 15 February 2023. Available online: https://blog.rgbsi.com/sensor-fusion-autonomous-driving-systems-part-1 (accessed on 30 April 2023).
30. Sasken. Sensor Fusion Paving the Way for Autonomous Vehicles. *Sasken Blog*. 22 February 2023. Available online: https://blog.sasken.com/sensor-fusion-paving-the-way-for-autonomous-vehicles (accessed on 18 May 2023).
31. Haider, A.; Pigniczki, M.; Köhler, M.H.; Fink, M.; Schardt, M.; Cichy, Y.; Zeh, T.; Haas, L.; Poguntke, T.; Jakobi, M.; et al. Development of High-Fidelity Automotive LiDAR Sensor Model with Standardized Interfaces. *Sensors* **2022**, *22*, 7556. [CrossRef]
32. Waymo. The Waymo Driver Handbook: Teaching an Autonomous Vehicle How to Perceive and Understand the World around It. Waymo Blog. 11 October 2021. Available online: https://waymo.com/blog/2021/10/the-waymo-driver-handbook-perception.html (accessed on 18 May 2023).
33. Hu, X.; Xu, X.; Xiao, Y.; Chen, H.; He, S.; Qin, J.; Heng, P.-A. SINet: A Scale-Insensitive Convolutional Neural Network for Fast Vehicle Detection. *IEEE Trans. Intell. Transp. Syst.* **2019**, *20*, 1010–1019. [CrossRef]
34. Hu, X.; Xu, X.; Xiao, Y.; Chen, H.; He, S.; Qin, J.; Heng, P.-A. Joint Monocular 3D Vehicle Detection and Tracking. In Proceedings of the 2019 IEEE/CVF International Conference on Computer Vision (ICCV), Seoul, Republic of Korea, 27 October–2 November 2019; pp. 5389–5398. [CrossRef]
35. Chadwick, S.; Maddern, W.; Newman, P. Distant Vehicle Detection Using Radar and Vision. In Proceedings of the 2019 International Conference on Robotics and Automation (ICRA), Montreal, QC, Canada, 20–24 May 2019; pp. 8311–8317. [CrossRef]
36. López-Sastre, R.J.; Herranz-Perdiguero, C.; Guerrero-Gómez-Olmedo, R.; Oñoro-Rubio, D.; Maldonado-Bascón, S. Boosting Multi-Vehicle Tracking with a Joint Object Detection and Viewpoint Estimation Sensor. *Sensors* **2019**, *19*, 4062. [CrossRef]
37. Zhang, F.; Li, C.; Yang, F. Vehicle Detection in Urban Traffic Surveillance Images Based on Convolutional Neural Networks with Feature Concatenation. *Sensors* **2019**, *19*, 594. [CrossRef]
38. Gomaa, A.; Abdelwahab, M.M.; Abo-Zahhad, M.; Minematsu, T.; Taniguchi, R.-I. Robust Vehicle Detection and Counting Algorithm Employing a Convolution Neural Network and Optical Flow. *Sensors* **2019**, *19*, 4588. [CrossRef] [PubMed]
39. Liu, H.; Ma, J.; Xu, T.; Yan, W.; Ma, L.; Zhang, X. Vehicle Detection and Classification Using Distributed Fiber Optic Acoustic Sensing. *IEEE Trans. Veh. Technol.* **2020**, *69*, 1363–1374. [CrossRef]

40. Zhang, J.; Xiao, W.; Coifman, B.; Mills, J.P. Vehicle Tracking and Speed Estimation From Roadside Lidar. *IEEE J. Sel. Top. Appl. Earth Obs. Remote Sens.* **2020**, *13*, 5597–5608. [CrossRef]
41. Wang, X.; Wang, S.; Cao, J.; Wang, Y. Data-Driven Based Tiny-YOLOv3 Method for Front Vehicle Detection Inducing SPP-Net. *IEEE Access* **2020**, *8*, 110227–110236. [CrossRef]
42. Kim, T.; Park, T.-H. Extended Kalman Filter (EKF) Design for Vehicle Position Tracking Using Reliability Function of Radar and Lidar. *Sensors* **2020**, *20*, 4126. [CrossRef] [PubMed]
43. Cao, J.; Song, C.; Song, S.; Peng, S.; Wang, D.; Shao, Y.; Xiao, F. Front Vehicle Detection Algorithm for Smart Car Based on Improved SSD Model. *Sensors* **2020**, *20*, 4646. [CrossRef] [PubMed]
44. Mo, N.; Yan, L. Improved Faster RCNN Based on Feature Amplification and Oversampling Data Augmentation for Oriented Vehicle Detection in Aerial Images. *Remote Sens.* **2020**, *12*, 2558. [CrossRef]
45. Zhang, R.; Ishikawa, A.; Wang, W.; Striner, B.; Tonguz, O.K. Using Reinforcement Learning with Partial Vehicle Detection for Intelligent Traffic Signal Control. *IEEE Trans. Intell. Transp. Syst.* **2021**, *22*, 404–415. [CrossRef]
46. Hassaballah, M.; Kenk, M.A.; Muhammad, K.; Minaee, S. Vehicle Detection and Tracking in Adverse Weather Using a Deep Learning Framework. *IEEE Trans. Intell. Transp. Syst.* **2021**, *22*, 4230–4242. [CrossRef]
47. Lin, C.-T.; Huang, S.-W.; Wu, Y.-Y.; Lai, S.-H. GAN-Based Day-to-Night Image Style Transfer for Nighttime Vehicle Detection. *IEEE Trans. Intell. Transp. Syst.* **2021**, *22*, 951–963. [CrossRef]
48. Balamuralidhar, N.; Tilon, S.; Nex, F. MultEYE: Monitoring System for Real-Time Vehicle Detection, Tracking and Speed Estimation from UAV Imagery on Edge-Computing Platforms. *Remote Sens.* **2021**, *13*, 573. [CrossRef]
49. Chen, Y.; Qin, R.; Zhang, G.; Albanwan, H. Spatial-Temporal Analysis of Traffic Patterns during the COVID-19 Epidemic by Vehicle Detection Using Planet Remote-Sensing Satellite Images. *Remote Sens.* **2021**, *13*, 208. [CrossRef]
50. Li, H.; Zhao, S.; Zhao, W.; Zhang, L.; Shen, J. One-Stage Anchor-Free 3D Vehicle Detection from LiDAR Sensors. *Sensors* **2021**, *21*, 2651. [CrossRef] [PubMed]
51. Sun, Y.; Cao, B.; Zhu, P.; Hu, Q. Drone-Based RGB-Infrared Cross-Modality Vehicle Detection Via Uncertainty-Aware Learning. *IEEE Trans. Circuits Syst. Video Technol.* **2022**, *32*, 6700–6713. [CrossRef]
52. Zhao, J.; Hao, S.; Dai, C.; Zhang, H.; Zhao, L.; Ji, Z.; Ganchev, I. Improved Vision-Based Vehicle Detection and Classification by Optimized YOLOv4. *IEEE Access* **2022**, *10*, 8590–8603. [CrossRef]
53. Bell, A.; Mantecon, T.; Diaz, C.; Del-Blanco, C.R.; Jaureguizar, F.; Garcia, N. A Novel System for Nighttime Vehicle Detection Based on Foveal Classifiers with Real-Time Performance. *IEEE Trans. Intell. Transp. Syst.* **2022**, *23*, 5421–5433. [CrossRef]
54. Humayun, M.; Ashfaq, F.; Jhanjhi, N.Z.; Alsadun, M.K. Traffic Management: Multi-Scale Vehicle Detection in Varying Weather Conditions Using YOLOv4 and Spatial Pyramid Pooling Network. *Electronics* **2022**, *11*, 2748. [CrossRef]
55. Charouh, Z.; Ezzouhri, A.; Ghogho, M.; Guennoun, Z. A Resource-Efficient CNN-Based Method for Moving Vehicle Detection. *Sensors* **2022**, *22*, 1193. [CrossRef]
56. Fan, Y.; Qiu, Q.; Hou, S.; Li, Y.; Xie, J.; Qin, M.; Chu, F. Application of Improved YOLOv5 in Aerial Photographing Infrared Vehicle Detection. *Electronics* **2022**, *11*, 2344. [CrossRef]
57. National Highway Traffic Safety Administration. Traffic Safety Facts 2021 Data: Pedestrians. [Fact Sheet]; 27 June 2023. Available online: https://crashstats.nhtsa.dot.gov/Api/Public/ViewPublication/813450 (accessed on 2 May 2023).
58. Liu, W.; Liao, S.; Ren, W.; Hu, W.; Yu, Y. High-Level Semantic Feature Detection: A New Perspective for Pedestrian Detection. In Proceedings of the 2019 IEEE/CVF Conference on Computer Vision and Pattern Recognition (CVPR), Long Beach, CA, USA, 16–20 June 2019; pp. 5182–5191. [CrossRef]
59. Liu, S.; Huang, D.; Wang, Y. Adaptive NMS: Refining Pedestrian Detection in a Crowd. In Proceedings of the 2019 IEEE/CVF Conference on Computer Vision and Pattern Recognition (CVPR), Long Beach, CA, USA, 16–20 June 2019; pp. 6452–6461. [CrossRef]
60. Pang, Y.; Xie, J.; Khan, M.H.; Anwer, R.M.; Khan, F.S.; Shao, L. Mask-Guided Attention Network for Occluded Pedestrian Detection. In Proceedings of the 2019 IEEE/CVF International Conference on Computer Vision (ICCV), Seoul, Republic of Korea, 27 October–2 November 2019; pp. 4966–4974. [CrossRef]
61. Dimitrievski, M.; Veelaert, P.; Philips, W. Behavioral Pedestrian Tracking Using a Camera and LiDAR Sensors on a Moving Vehicle. *Sensors* **2019**, *19*, 391. [CrossRef]
62. Liu, K.; Wang, W.; Wang, J. Pedestrian Detection with Lidar Point Clouds Based on Single Template Matching. *Electronics* **2019**, *8*, 780. [CrossRef]
63. He, M.; Luo, H.; Hui, B.; Chang, Z. Pedestrian Flow Tracking and Statistics of Monocular Camera Based on Convolutional Neural Network and Kalman Filter. *Appl. Sci.* **2019**, *9*, 1624. [CrossRef]
64. Li, G.; Yang, Y.; Qu, X. Deep Learning Approaches on Pedestrian Detection in Hazy Weather. *IEEE Trans. Ind. Electron.* **2020**, *67*, 8889–8899. [CrossRef]
65. Huang, X.; Ge, Z.; Jie, Z.; Yoshie, O. NMS by Representative Region: Towards Crowded Pedestrian Detection by Proposal Pairing. In Proceedings of the 2020 IEEE/CVF Conference on Computer Vision and Pattern Recognition (CVPR), Seattle, WA, USA, 14–19 June 2020; pp. 10747–10756. [CrossRef]
66. Lin, C.; Lu, J.; Wang, G.; Zhou, J. Graininess-Aware Deep Feature Learning for Robust Pedestrian Detection. *IEEE Trans. Image Process.* **2020**, *29*, 3820–3834. [CrossRef]

67. Barba-Guaman, L.; Eugenio Naranjo, J.; Ortiz, A. Deep Learning Framework for Vehicle and Pedestrian Detection in Rural Roads on an Embedded GPU. *Electronics* **2020**, *9*, 589. [CrossRef]
68. Chen, Y.; Shin, H. Pedestrian Detection at Night in Infrared Images Using an Attention-Guided Encoder-Decoder Convolutional Neural Network. *Appl. Sci.* **2020**, *10*, 809. [CrossRef]
69. Cao, J.; Song, C.; Peng, S.; Song, S.; Zhang, X.; Shao, Y.; Xiao, F. Pedestrian Detection Algorithm for Intelligent Vehicles in Complex Scenarios. *Sensors* **2020**, *20*, 3646. [CrossRef]
70. Hsu, W.-Y.; Lin, W.-Y. Ratio-and-Scale-Aware YOLO for Pedestrian Detection. *IEEE Trans. Image Process.* **2021**, *30*, 934–947. [CrossRef]
71. Stadler, D.; Beyerer, J. Improving Multiple Pedestrian Tracking by Track Management and Occlusion Handling. In Proceedings of the 2021 IEEE/CVF Conference on Computer Vision and Pattern Recognition (CVPR), Nashville, TN, USA, 20–25 June 2021; pp. 10953–10962. [CrossRef]
72. Yang, P.; Zhang, G.; Wang, L.; Xu, L.; Deng, Q.; Yang, M.-H. A Part-Aware Multi-Scale Fully Convolutional Network for Pedestrian Detection. *IEEE Trans. Intell. Transp. Syst.* **2021**, *22*, 1125–1137. [CrossRef]
73. Cao, Z.; Yang, H.; Zhao, J.; Guo, S.; Li, L. Attention Fusion for One-Stage Multispectral Pedestrian Detection. *Sensors* **2021**, *21*, 4184. [CrossRef]
74. Nataprawira, J.; Gu, Y.; Goncharenko, I.; Kamijo, S. Pedestrian Detection Using Multispectral Images and a Deep Neural Network. *Sensors* **2021**, *21*, 2536. [CrossRef] [PubMed]
75. Chen, X.; Liu, L.; Tan, X. Robust Pedestrian Detection Based on Multi-Spectral Image Fusion and Convolutional Neural Networks. *Electronics* **2022**, *11*, 1. [CrossRef]
76. Kim, J.U.; Park, S.; Ro, Y.M. Uncertainty-Guided Cross-Modal Learning for Robust Multispectral Pedestrian Detection. *IEEE Trans. Circuits Syst. Video Technol.* **2022**, *32*, 1510–1523. [CrossRef]
77. Dasgupta, K.; Das, A.; Das, S.; Bhattacharya, U.; Yogamani, S. Spatio-Contextual Deep Network-Based Multimodal Pedestrian Detection for Autonomous Driving. *IEEE Trans. Intell. Transp. Syst.* **2022**, *23*, 15940–15950. [CrossRef]
78. Held, P.; Steinhauser, D.; Koch, A.; Brandmeier, T.; Schwarz, U.T. A Novel Approach for Model-Based Pedestrian Tracking Using Automotive Radar. *IEEE Trans. Intell. Transp. Syst.* **2022**, *23*, 7082–7095. [CrossRef]
79. Roszyk, K.; Nowicki, M.R.; Skrzypczyński, P. Adopting the YOLOv4 Architecture for Low-Latency Multispectral Pedestrian Detection in Autonomous Driving. *Sensors* **2022**, *22*, 1082. [CrossRef]
80. Shao, Y.; Zhang, X.; Chu, H.; Zhang, X.; Zhang, D.; Rao, Y. AIR-YOLOv3: Aerial Infrared Pedestrian Detection via an Improved YOLOv3 with Network Pruning. *Appl. Sci.* **2022**, *12*, 3627. [CrossRef]
81. Lv, H.; Yan, H.; Liu, K.; Zhou, Z.; Jing, J. YOLOv5-AC: Attention Mechanism-Based Lightweight YOLOv5 for Track Pedestrian Detection. *Sensors* **2022**, *22*, 5903. [CrossRef]
82. Yuan, Y.; Xiong, Z.; Wang, Q. VSSA-NET: Vertical Spatial Sequence Attention Network for Traffic Sign Detection. *IEEE Trans. Image Process.* **2019**, *28*, 3423–3434. [CrossRef]
83. Li, J.; Wang, Z. Real-Time Traffic Sign Recognition Based on Efficient CNNs in the Wild. *IEEE Trans. Intell. Transp. Syst.* **2019**, *20*, 975–984. [CrossRef]
84. Liu, Z.; Du, J.; Tian, F.; Wen, J. MR-CNN: A Multi-Scale Region-Based Convolutional Neural Network for Small Traffic Sign Recognition. *IEEE Access* **2019**, *7*, 57120–57128. [CrossRef]
85. Tian, Y.; Gelernter, J.; Wang, X.; Li, J.; Yu, Y. Traffic Sign Detection Using a Multi-Scale Recurrent Attention Network. *IEEE Trans. Intell. Transp. Syst.* **2019**, *20*, 4466–4475. [CrossRef]
86. Cao, J.; Song, C.; Peng, S.; Xiao, F.; Song, S. Improved Traffic Sign Detection and Recognition Algorithm for Intelligent Vehicles. *Sensors* **2019**, *19*, 4021. [CrossRef] [PubMed]
87. Shao, F.; Wang, X.; Meng, F.; Zhu, J.; Wang, D.; Dai, J. Improved Faster R-CNN Traffic Sign Detection Based on a Second Region of Interest and Highly Possible Regions Proposal Network. *Sensors* **2019**, *19*, 2288. [CrossRef] [PubMed]
88. Zhang, J.; Xie, Z.; Sun, J.; Zou, X.; Wang, J. A Cascaded R-CNN with Multiscale Attention and Imbalanced Samples for Traffic Sign Detection. *IEEE Access* **2020**, *8*, 29742–29754. [CrossRef]
89. Tabernik, D.; Skočaj, D. Deep Learning for Large-Scale Traffic-Sign Detection and Recognition. *IEEE Trans. Intell. Transp. Syst.* **2020**, *21*, 1427–1440. [CrossRef]
90. Kamal, U.; Tonmoy, T.I.; Das, S.; Hasan, M.K. Automatic Traffic Sign Detection and Recognition Using SegU-Net and a Modified Tversky Loss Function with L1-Constraint. *IEEE Trans. Intell. Transp. Syst.* **2020**, *21*, 1467–1479. [CrossRef]
91. Tai, S.-K.; Dewi, C.; Chen, R.-C.; Liu, Y.-T.; Jiang, X.; Yu, H. Deep Learning for Traffic Sign Recognition Based on Spatial Pyramid Pooling with Scale Analysis. *Appl. Sci.* **2020**, *10*, 6997. [CrossRef]
92. Dewi, C.; Chen, R.-C.; Tai, S.-K. Evaluation of Robust Spatial Pyramid Pooling Based on Convolutional Neural Network for Traffic Sign Recognition System. *Electronics* **2020**, *9*, 889. [CrossRef]
93. Nartey, O.T.; Yang, G.; Asare, S.K.; Wu, J.; Frempong, L.N. Robust Semi-Supervised Traffic Sign Recognition via Self-Training and Weakly-Supervised Learning. *Sensors* **2020**, *20*, 2684. [CrossRef]
94. Dewi, C.; Chen, R.-C.; Liu, Y.-T.; Jiang, X.; Hartomo, K.D. Yolo V4 for Advanced Traffic Sign Recognition with Synthetic Training Data Generated by Various GAN. *IEEE Access* **2021**, *9*, 97228–97242. [CrossRef]
95. Wang, L.; Zhou, K.; Chu, A.; Wang, G.; Wang, L. An Improved Light-Weight Traffic Sign Recognition Algorithm Based on YOLOv4-Tiny. *IEEE Access* **2021**, *9*, 124963–124971. [CrossRef]

96. Cao, J.; Zhang, J.; Jin, X. A Traffic-Sign Detection Algorithm Based on Improved Sparse R-cnn. *IEEE Access* **2021**, *9*, 22774–122788. [CrossRef]
97. Lopez-Montiel, M.; Orozco-Rosas, U.; Sánchez-Adame, M.; Picos, K.; Ross, O.H.M. Evaluation Method of Deep Learning-Based Embedded Systems for Traffic Sign Detection. *IEEE Access* **2021**, *9*, 101217–101238. [CrossRef]
98. Zhou, K.; Zhan, Y.; Fu, D. Learning Region-Based Attention Network for Traffic Sign Recognition. *Sensors* **2021**, *21*, 686. [CrossRef]
99. Koh, D.-W.; Kwon, J.-K.; Lee, S.-G. Traffic Sign Recognition Evaluation for Senior Adults Using EEG Signals. *Sensors* **2021**, *21*, 4607. [CrossRef] [PubMed]
100. Ahmed, S.; Kamal, U.; Hasan, M.K. DFR-TSD: A Deep Learning Based Framework for Robust Traffic Sign Detection Under Challenging Weather Conditions. *IEEE Trans. Intell. Transp. Syst.* **2022**, *23*, 5150–5162. [CrossRef]
101. Xie, K.; Zhang, Z.; Li, B.; Kang, J.; Niyato, D.; Xie, S.; Wu, Y. Efficient Federated Learning with Spike Neural Networks for Traffic Sign Recognition. *IEEE Trans. Veh. Technol.* **2022**, *71*, 9980–9999. [CrossRef]
102. Min, W.; Liu, R.; He, D.; Han, Q.; Wei, Q.; Wang, Q. Traffic Sign Recognition Based on Semantic Scene Understanding and Structural Traffic Sign Location. *IEEE Trans. Intell. Transp. Syst.* **2022**, *23*, 15794–15807. [CrossRef]
103. Gu, Y.; Si, B. A Novel Lightweight Real-Time Traffic Sign Detection Integration Framework Based on YOLOv4. *Entropy* **2022**, *24*, 487. [CrossRef]
104. Liu, Y.; Shi, G.; Li, Y.; Zhao, Z. M-YOLO: Traffic Sign Detection Algorithm Applicable to Complex Scenarios. *Symmetry* **2022**, *14*, 952. [CrossRef]
105. Wang, X.; Guo, J.; Yi, J.; Song, Y.; Xu, J.; Yan, W.; Fu, X. Real-Time and Efficient Multi-Scale Traffic Sign Detection Method for Driverless Cars. *Sensors* **2022**, *22*, 6930. [CrossRef] [PubMed]
106. Zhao, Y.; Mammeri, A.; Boukerche, A. A Novel Real-time Driver Monitoring System Based on Deep Convolutional Neural Network. In Proceedings of the 2019 IEEE International Symposium on Robotic and Sensors Environments (ROSE), Ottawa, ON, Canada, 17–18 June 2019; pp. 1–7. [CrossRef]
107. Hijaz, A.; Louie, W.-Y.G.; Mansour, I. Towards a Driver Monitoring System for Estimating Driver Situational Awareness. In Proceedings of the 2019 28th IEEE International Conference on Robot and Human Interactive Communication (RO-MAN), New Delhi, India, 14–18 October 2019; pp. 1–6. [CrossRef]
108. Kim, W.; Jung, W.-S.; Choi, H.K. Lightweight Driver Monitoring System Based on Multi-Task Mobilenets. *Sensors* **2019**, *19*, 3200. [CrossRef] [PubMed]
109. Yoo, M.W.; Han, D.S. Optimization Algorithm for Driver Monitoring System using Deep Learning Approach. In Proceedings of the 2020 International Conference on Artificial Intelligence in Information and Communication (ICAIIC), Fukuoka, Japan, 19–21 February 2020; pp. 043–046. [CrossRef]
110. Pondit, A.; Dey, A.; Das, A. Real-time Driver Monitoring System Based on Visual Cues. In Proceedings of the 2020 6th International Conference on Interactive Digital Media (ICIDM), Bandung, Indonesia, 14–15 December 2020; pp. 1–6. [CrossRef]
111. Supraja, P.; Revati, P.; Ram, K.S.; Jyotsna, C. An Intelligent Driver Monitoring System. In Proceedings of the 2021 2nd International Conference on Communication, Computing and Industry 4.0 (C2I4), Bangalore, India, 16–17 December 2021; pp. 1–5. [CrossRef]
112. Zhu, L.; Xiao, Y.; Li, X. Hybrid driver monitoring system based on Internet of Things and machine learning. In Proceedings of the 2021 IEEE International Conference on Consumer Electronics and Computer Engineering (ICCECE), Guangzhou, China, 15–17 January 2021; pp. 635–638. [CrossRef]
113. Darapaneni, N.; Parikh, B.; Paduri, A.R.; Kumar, S.; Beedkar, T.; Narayanan, A.; Tripathi, N.; Khoche, T. Distracted Driver Monitoring System Using AI. In Proceedings of the 2022 Interdisciplinary Research in Technology and Management (IRTM), Kolkata, India, 24–26 February 2022; pp. 1–8. [CrossRef]
114. Jeon, S.; Lee, S.; Lee, E.; Shin, J. Driver Monitoring System based on Distracted Driving Decision Algorithm. In Proceedings of the 2022 13th International Conference on Information and Communication Technology Convergence (ICTC), Jeju Island, Republic of Korea, 19–21 October 2022; pp. 2280–2283. [CrossRef]
115. National Highway Traffic Safety Administration. NHTSA Orders Crash Reporting for Vehicles Equipped with Advanced Driver Assistance Systems. 31 May 2023. Available online: https://www.nhtsa.gov/press-releases/nhtsa-orders-crash-reporting-vehicles-equipped-advanced-driver-assistance-systems (accessed on 24 June 2023).
116. Hou, Y.; Ma, Z.; Liu, C.; Loy, C.C. Learning Lightweight Lane Detection CNNs by Self Attention Distillation. In Proceedings of the 2019 IEEE/CVF International Conference on Computer Vision (ICCV), Seoul, Republic of Korea, 27 October–2 November 2019; pp. 1013–1021. [CrossRef]
117. Philion, J. FastDraw: Addressing the Long Tail of Lane Detection by Adapting a Sequential Prediction Network. In Proceedings of the 2019 IEEE/CVF Conference on Computer Vision and Pattern Recognition (CVPR), Long Beach, CA, USA, 16–20 June 2019; pp. 11574–11583. [CrossRef]
118. Garnett, N.; Cohen, R.; Pe, T.; Lahav, R.; Levi, D. 3D-LaneNet: End-to-End 3D Multiple Lane Detection. In Proceedings of the 2019 IEEE/CVF International Conference on Computer Vision (ICCV), Seoul, Republic of Korea, 27 October–2 November 2019; pp. 2921–2930. [CrossRef]
119. Liu, T.; Chen, Z.; Yang, Y.; Wu, Z.; Li, H. Lane Detection in Low-light Conditions Using an Efficient Data Enhancement: Light Conditions Style Transfer. In Proceedings of the 2020 IEEE Intelligent Vehicles Symposium (IV), Las Vegas, NV, USA, 19 October–13 November 2020; pp. 1394–1399. [CrossRef]

120. Lu, Z.; Xu, Y.; Shan, X.; Liu, L.; Wang, X.; Shen, J. A Lane Detection Method Based on a Ridge Detector and Regional G-RANSAC. *Sensors* **2019**, *19*, 4028. [CrossRef] [PubMed]
121. Yang, W.; Zhang, X.; Lei, Q.; Shen, D.; Xiao, P.; Huang, Y. Lane Position Detection Based on Long Short-Term Memory (LSTM). *Sensors* **2020**, *20*, 3115. [CrossRef] [PubMed]
122. Wang, Q.; Han, T.; Qin, Z.; Gao, J.; Li, X. Multitask Attention Network for Lane Detection and Fitting. *IEEE Trans. Neural Netw. Learn. Syst.* **2022**, *33*, 1066–1078. [CrossRef] [PubMed]
123. Cao, J.; Song, C.; Song, S.; Xiao, F.; Peng, S. Lane Detection Algorithm for Intelligent Vehicles in Complex Road Conditions and Dynamic Environments. *Sensors* **2019**, *19*, 3166. [CrossRef]
124. Wang, X.; Qian, Y.; Wang, C.; Yang, M. Map-Enhanced Ego-Lane Detection in the Missing Feature Scenarios. *IEEE Access* **2020**, *8*, 107958–107968. [CrossRef]
125. Chen, Y.; Xiang, Z. Lane Mark Detection with Pre-Aligned Spatial-Temporal Attention. *Sensors* **2022**, *22*, 794. [CrossRef]
126. Lee, Y.; Park, M.-k.; Park, M. Improving Lane Detection Performance for Autonomous Vehicle Integrating Camera with Dual Light Sensors. *Electronics* **2022**, *11*, 1474. [CrossRef]
127. Kim, D.-H. Lane Detection Method with Impulse Radio Ultra-Wideband Radar and Metal Lane Reflectors. *Sensors* **2020**, *20*, 324. [CrossRef] [PubMed]
128. Suder, J.; Podbucki, K.; Marciniak, T.; Dąbrowski, A. Low Complexity Lane Detection Methods for Light Photometry System. *Electronics* **2021**, *10*, 1665. [CrossRef]
129. Kuo, C.Y.; Lu, Y.R.; Yang, S.M. On the Image Sensor Processing for Lane Detection and Control in Vehicle Lane Keeping Systems. *Sensors* **2019**, *19*, 1665. [CrossRef] [PubMed]
130. Zou, Q.; Jiang, H.; Dai, Q.; Yue, Y.; Chen, L.; Wang, Q. Robust Lane Detection From Continuous Driving Scenes Using Deep Neural Networks. *IEEE Trans. Veh. Technol.* **2020**, *69*, 41–54. [CrossRef]
131. Gao, Q.; Yin, H.; Zhang, W. Lane Departure Warning Mechanism of Limited False Alarm Rate Using Extreme Learning Residual Network and ϵ-Greedy LSTM. *Sensors* **2020**, *20*, 644. [CrossRef]
132. Tabelini, L.; Berriel, R.; Paixão, T.M.; Badue, C.; De Souza, A.F.; Oliveira-Santos, T. Keep your Eyes on the Lane: Real-time Attention-guided Lane Detection. In Proceedings of the 2021 IEEE/CVF Conference on Computer Vision and Pattern Recognition (CVPR), Nashville, TN, USA, 20–25 June 2021; pp. 294–302. [CrossRef]
133. Liu, L.; Chen, X.; Zhu, S.; Tan, P. CondLaneNet: A Top-to-down Lane Detection Framework Based on Conditional Convolution. In Proceedings of the 2021 IEEE/CVF International Conference on Computer Vision (ICCV), Montreal, QC, Canada, 10–17 October 2021; pp. 3753–3762. [CrossRef]
134. Dewangan, D.K.; Sahu, S.P. Driving Behavior Analysis of Intelligent Vehicle System for Lane Detection Using Vision-Sensor. *IEEE Sens. J.* **2021**, *21*, 6367–6375. [CrossRef]
135. Haris, M.; Glowacz, A. Lane Line Detection Based on Object Feature Distillation. *Electronics* **2021**, *10*, 1102. [CrossRef]
136. Lu, S.; Luo, Z.; Gao, F.; Liu, M.; Chang, K.; Piao, C. A Fast and Robust Lane Detection Method Based on Semantic Segmentation and Optical Flow Estimation. *Sensors* **2021**, *21*, 400. [CrossRef]
137. Ko, Y.; Lee, Y.; Azam, S.; Munir, F.; Jeon, M.; Pedrycz, W. Key Points Estimation and Point Instance Segmentation Approach for Lane Detection. *IEEE Trans. Intell. Transp. Syst.* **2022**, *23*, 8949–8958. [CrossRef]
138. Zheng, T.; Huang, Y.; Liu, Y.; Tang, W.; Yang, Z.; Cai, D.; He, X. CLRNet: Cross-Layer Refinement Network for Lane Detection. In Proceedings of the 2022 IEEE/CVF Conference on Computer Vision and Pattern Recognition (CVPR), New Orleans, LA, USA, 18–24 June 2022; pp. 888–897. [CrossRef]
139. Khan, M.A.-M.; Haque, M.F.; Hasan, K.R.; Alajmani, S.H.; Baz, M.; Masud, M.; Nahid, A.-A. LLDNet: A Lightweight Lane Detection Approach for Autonomous Cars Using Deep Learning. *Sensors* **2022**, *22*, 5595. [CrossRef]
140. National Highway Traffic Safety Administration. Traffic Safety Facts 2020 Data: Crashes. 20 September 2021. Available online: https://crashstats.nhtsa.dot.gov/Api/Public/ViewPublication/812801 (accessed on 19 July 2023).
141. Lee, K.; Kum, D. Collision Avoidance/Mitigation System: Motion Planning of Autonomous Vehicle via Predictive Occupancy Map. *IEEE Access* **2019**, *7*, 52846–52857. [CrossRef]
142. Manghat, S.K.; El-Sharkawy, M. Forward Collision Prediction with Online Visual Tracking. In Proceedings of the 2019 IEEE International Conference on Vehicular Electronics and Safety (ICVES), Cairo, Egypt, 4–6 September 2019; pp. 1–5. [CrossRef]
143. Yang, W.; Wan, B.; Qu, X. A Forward Collision Warning System Using Driving Intention Recognition of the Front Vehicle and V2V Communication. *IEEE Access* **2020**, *8*, 11268–11278. [CrossRef]
144. Kumar, S.; Shaw, V.; Maitra, J.; Karmakar, R. FCW: A Forward Collision Warning System Using Convolutional Neural Network. In Proceedings of the 2020 International Conference on Electrical and Electronics Engineering (ICE3), Gorakhpur, India, 14–15 February 2020; pp. 1–5. [CrossRef]
145. Wang, H.-M.; Lin, H.-Y. A Real-Time Forward Collision Warning Technique Incorporating Detection and Depth Estimation Networks. In Proceedings of the 2020 IEEE International Conference on Systems, Man, and Cybernetics (SMC), Toronto, ON, Canada, 11–14 October 2020; pp. 1966–1971. [CrossRef]
146. Lin, H.-Y.; Dai, J.-M.; Wu, L.-T.; Chen, L.-Q. A Vision-Based Driver Assistance System with Forward Collision and Overtaking Detection. *Sensors* **2020**, *20*, 5139. [CrossRef] [PubMed]
147. Tang, J.; Li, J. End-to-End Monocular Range Estimation for Forward Collision Warning. *Sensors* **2020**, *20*, 5941. [CrossRef] [PubMed]

148. Lim, Q.; Lim, Y.; Muhammad, H.; Tan, D.W.M.; Tan, U.-X. Forward collision warning system for motorcyclist using smartphone sensors based on time-to-collision and trajectory prediction. *J. Intell. Connect. Veh.* **2021**, *4*, 93–103. [CrossRef]
149. Farhat, W.; Rhaiem, O.B.; Faiedh, H.; Souani, C. Cooperative Forward Collision Avoidance System Based on Deep Learning. In Proceedings of the 2021 14th International Conference on Developments in eSystems Engineering (DeSE), Sharjah, United Arab Emirates, 7–10 December 2021; pp. 515–519. [CrossRef]
150. Hong, S.; Park, D. Lightweight Collaboration of Detecting and Tracking Algorithm in Low-Power Embedded Systems for Forward Collision Warning. In Proceedings of the 2021 Twelfth International Conference on Ubiquitous and Future Networks (ICUFN), Jeju Island, Republic of Korea, 17–20 August 2021; pp. 159–162. [CrossRef]
151. Albarella, N.; Masuccio, F.; Novella, L.; Tufo, M.; Fiengo, G. A Forward-Collision Warning System for Electric Vehicles: Experimental Validation in Virtual and Real Environment. *Energies* **2021**, *14*, 4872. [CrossRef]
152. Liu, Y.; Wang, X.; Zhang, Y.; Wang, Y. An effective target selection method for forward collision on a curve based on V2X. In Proceedings of the 2022 7th International Conference on Intelligent Informatics and Biomedical Science (ICIIBMS), Nara, Japan, 24–26 November 2022; pp. 110–114. [CrossRef]
153. Yu, R.; Ai, H. Vehicle Forward Collision Warning based upon Low-Frequency Video Data: A hybrid Deep Learning Modeling Approach. In Proceedings of the 2022 IEEE 25th International Conference on Intelligent Transportation Systems (ITSC), Macau, China, 8–12 October 2022; pp. 59–64. [CrossRef]
154. Olou, H.B.; Ezin, E.C.; Dembele, J.M.; Cambier, C. FCPNet: A novel model to predict forward collision based upon CNN. In Proceedings of the 2022 22nd International Conference on Control, Automation, and Systems (ICCAS), Jeju, Republic of Korea, 27 November–1 December 2022; pp. 1327–1332. [CrossRef]
155. Pak, J.M. Hybrid Interacting Multiple Model Filtering for Improving the Reliability of Radar-Based Forward Collision Warning Systems. *Sensors* **2022**, *22*, 875. [CrossRef]
156. Bagi, S.S.G.; Garakani, H.G.; Moshiri, B.; Khoshnevisan, M. Sensing Structure for Blind Spot Detection System in Vehicles. In Proceedings of the 2019 International Conference on Control, Automation and Information Sciences (ICCAIS), Chengdu, China, 24–27 October 2019; pp. 1–6. [CrossRef]
157. Sugiura, T.; Watanabe, T. Probable Multi-hypothesis Blind Spot Estimation for Driving Risk Prediction. In Proceedings of the 2019 IEEE Intelligent Transportation Systems Conference (ITSC), Auckland, New Zealand, 27–30 October 2019; pp. 4295–4302. [CrossRef]
158. Zhao, Y.; Bai, L.; Lyu, Y.; Huang, X. Camera-Based Blind Spot Detection with a General Purpose Lightweight Neural Network. *Electronics* **2019**, *8*, 233. [CrossRef]
159. Chang, I.-C.; Chen, W.-R.; Kuo, X.-M.; Song, Y.-J.; Liao, P.-H.; Kuo, C. An Artificial Intelligence-based Proactive Blind Spot Warning System for Motorcycles. In Proceedings of the 2020 International Symposium on Computer, Consumer and Control (IS3C), Taichung City, Taiwan, 13–16 November 2020; pp. 404–407. [CrossRef]
160. Naik, A.; Naveen, G.V.V.S.; Satardhan, J.; Chavan, A. LiEBiD—A LIDAR based Early Blind Spot Detection and Warning System for Traditional Steering Mechanism. In Proceedings of the 2020 International Conference on Smart Electronics and Communication (ICOSEC), Trichy, India, 10–12 September 2020; pp. 604–609. [CrossRef]
161. Singh, N.; Ji, G. Computer vision assisted, real-time blind spot detection based collision warning system for two-wheelers. In Proceedings of the 2021 5th International Conference on Electronics, Communication and Aerospace Technology (ICECA), Coimbatore, India, 2–4 December 2021; pp. 1179–1184. [CrossRef]
162. Shete, R.G.; Kakade, S.K.; Dhanvijay, M. A Blind-spot Assistance for Forklift using Ultrasonic Sensor. In Proceedings of the 2021 IEEE International Conference on Technology, Research, and Innovation for Betterment of Society (TRIBES), Raipur, India, 17–19 December 2021; pp. 1–4. [CrossRef]
163. Schlegel, K.; Weissig, P.; Protzel, P. A blind-spot-aware optimization-based planner for safe robot navigation. In Proceedings of the 2021 European Conference on Mobile Robots (ECMR), Bonn, Germany, 31 August–3 September 2021; pp. 1–8. [CrossRef]
164. Kundid, J.; Vranješ, M.; Lukač, Ž.; Popović, M. ADAS algorithm for creating a wider view of the environment with a blind spot display for the driver. In Proceedings of the 2021 Zooming Innovation in Consumer Technologies Conference (ZINC), Novi Sad, Serbia, 26–27 May 2021; pp. 219–224. [CrossRef]
165. Sui, S.; Li, T.; Chen, S. A-pillar Blind Spot Display Algorithm Based on Line of Sight. In Proceedings of the 2022 IEEE 5th International Conference on Computer and Communication Engineering Technology (CCET), Beijing, China, 19–21 August 2022; pp. 100–105. [CrossRef]
166. Wang, Z.; Jin, Q.; Wu, B. Design of a Vision Blind Spot Detection System Based on Depth Camera. In Proceedings of the 2022 IEEE Intl Conf on Dependable, Autonomic and Secure Computing, Intl Conf on Pervasive Intelligence and Computing, Intl Conf on Cloud and Big Data Computing, Intl Conf on Cyber Science and Technology Congress (DASC/PiCom/CBDCom/CyberSciTech), Falerna, Italy, 12–15 September 2022; pp. 1–5. [CrossRef]
167. Zhou, J.; Hirano, M.; Yamakawa, Y. High-Speed Recognition of Pedestrians out of Blind Spot with Pre-detection of Potentially Dangerous Regions. In Proceedings of the 2022 IEEE 25th International Conference on Intelligent Transportation Systems (ITSC), Macau, China, 8–12 October 2022; pp. 945–950. [CrossRef]
168. Seo, H.; Kim, H.; Lee, K.; Lee, K. Multi-Sensor-Based Blind-Spot Reduction Technology and a Data-Logging Method Using a Gesture Recognition Algorithm Based on Micro E-Mobility in an IoT Environment. *Sensors* **2022**, *22*, 1081. [CrossRef]

169. Muzammel, M.; Yusoff, M.Z.; Saad, M.N.M.; Sheikh, F.; Awais, M.A. Blind-Spot Collision Detection System for Commercial Vehicles Using Multi Deep CNN Architecture. *Sensors* **2022**, *22*, 6088. [CrossRef]
170. Flores, C.; Merdrignac, P.; de Charette, R.; Navas, F.; Milanés, V.; Nashashibi, F. A Cooperative Car-Following/Emergency Braking System with Prediction-Based Pedestrian Avoidance Capabilities. *IEEE Trans. Intell. Transp. Syst.* **2019**, *20*, 1837–1846. [CrossRef]
171. Shin, S.-G.; Ahn, D.-R.; Baek, Y.-S.; Lee, H.-K. Adaptive AEB Control Strategy for Collision Avoidance Including Rear Vehicles. In Proceedings of the 2019 IEEE Intelligent Transportation Systems Conference (ITSC), Auckland, New Zealand, 27–30 October 2019; pp. 2872–2878. [CrossRef]
172. Yang, W.; Zhang, X.; Lei, Q.; Cheng, X. Research on Longitudinal Active Collision Avoidance of Autonomous Emergency Braking Pedestrian System (AEB-P). *Sensors* **2019**, *19*, 4671. [CrossRef] [PubMed]
173. Gao, Y.; Xu, Z.; Zhao, X.; Wang, G.; Yuan, Q. Hardware-in-the-Loop Simulation Platform for Autonomous Vehicle AEB Prototyping and Validation. In Proceedings of the 2020 IEEE 23rd International Conference on Intelligent Transportation Systems (ITSC), Rhodes, Greece, 20–23 September 2020; pp. 1–6. [CrossRef]
174. Guo, L.; Ge, P.; Sun, D. Variable Time Headway Autonomous Emergency Braking Control Algorithm Based on Model Predictive Control. In Proceedings of the 2020 Chinese Automation Congress (CAC), Shanghai, China, 6–8 November 2020; pp. 1794–1798. [CrossRef]
175. Leyrer, M.L.; Stöckle, C.; Herrmann, S.; Dirndorfer, T.; Utschick, W. An Efficient Approach to Simulation-Based Robust Function and Sensor Design Applied to an Automatic Emergency Braking System. In Proceedings of the 2020 IEEE Intelligent Vehicles Symposium (IV), Las Vegas, NV, USA, 19 October–13 November 2020; pp. 617–622. [CrossRef]
176. Yu, L.; Wang, R.; Lu, Z. Autonomous Emergency Braking Control Based on Inevitable Collision State for Multiple Collision Scenarios at Intersection. In Proceedings of the 2021 American Control Conference (ACC), New Orleans, LA, USA, 25–28 May 2021; pp. 148–153. [CrossRef]
177. Izquierdo, A.; Val, L.D.; Villacorta, J.J. Feasibility of Using a MEMS Microphone Array for Pedestrian Detection in an Autonomous Emergency Braking System. *Sensors* **2021**, *21*, 4162. [CrossRef] [PubMed]
178. Jin, X.; Zhang, J.; Wu, Y.; Gao, J. Adaptive AEB control strategy for driverless vehicles in campus scenario. In Proceedings of the 2022 International Conference on Advanced Mechatronic Systems (ICAMechS), Toyama, Japan, 17–20 December 2022; pp. 47–52. [CrossRef]
179. Mannam, N.P.B.; Rajalakshmi, P. Determination of ADAS AEB Car to Car and Car to Pedestrian Scenarios for Autonomous Vehicles. In Proceedings of the 2022 IEEE Global Conference on Computing, Power and Communication Technologies (GlobConPT), New Delhi, India, 23–25 September 2022; pp. 1–7. [CrossRef]
180. Guo, J.; Wang, Y.; Yin, X.; Liu, P.; Hou, Z.; Zhao, D. Study on the Control Algorithm of Automatic Emergency Braking System (AEBS) for Commercial Vehicle Based on Identification of Driving Condition. *Machines* **2022**, *10*, 895. [CrossRef]
181. Li, G.; Görges, D. Ecological Adaptive Cruise Control and Energy Management Strategy for Hybrid Electric Vehicles Based on Heuristic Dynamic Programming. *IEEE Trans. Intell. Transp. Syst.* **2019**, *20*, 3526–3535. [CrossRef]
182. Cheng, S.; Li, L.; Mei, M.-M.; Nie, Y.-L.; Zhao, L. Multiple-Objective Adaptive Cruise Control System Integrated with DYC. *IEEE Trans. Veh. Technol.* **2019**, *68*, 4550–4559. [CrossRef]
183. Lunze, J. Adaptive Cruise Control with Guaranteed Collision Avoidance. *IEEE Trans. Intell. Transp. Syst.* **2019**, *20*, 1897–1907. [CrossRef]
184. Woo, H.; Madokoro, H.; Sato, K.; Tamura, Y.; Yamashita, A.; Asama, H. Advanced Adaptive Cruise Control Based on Operation Characteristic Estimation and Trajectory Prediction. *Appl. Sci.* **2019**, *9*, 4875. [CrossRef]
185. Zhang, S.; Zhuan, X. Study on Adaptive Cruise Control Strategy for Battery Electric Vehicle Considering Weight Adjustment. *Symmetry* **2019**, *11*, 1516. [CrossRef]
186. Zhai, C.; Chen, X.; Yan, C.; Liu, Y.; Li, H. Ecological Cooperative Adaptive Cruise Control for a Heterogeneous Platoon of Heavy-Duty Vehicles with Time Delays. *IEEE Access* **2020**, *8*, 146208–146219. [CrossRef]
187. Li, G.; Görges, D. Ecological Adaptive Cruise Control for Vehicles with Step-Gear Transmission Based on Reinforcement Learning. *IEEE Trans. Intell. Transp. Syst.* **2020**, *21*, 4895–4905. [CrossRef]
188. Jia, Y.; Jibrin, R.; Görges, D. Energy-Optimal Adaptive Cruise Control for Electric Vehicles Based on Linear and Nonlinear Model Predictive Control. *IEEE Trans. Veh. Technol.* **2020**, *69*, 14173–14187. [CrossRef]
189. Nie, Z.; Farzaneh, H. Adaptive Cruise Control for Eco-Driving Based on Model Predictive Control Algorithm. *Appl. Sci.* **2020**, *10*, 5271. [CrossRef]
190. Guo, L.; Ge, P.; Sun, D.; Qiao, Y. Adaptive Cruise Control Based on Model Predictive Control with Constraints Softening. *Appl. Sci.* **2020**, *10*, 1635. [CrossRef]
191. Liu, Y.; Wang, W.; Hua, X.; Wang, S. Safety Analysis of a Modified Cooperative Adaptive Cruise Control Algorithm Accounting for Communication Delay. *Sustainability* **2020**, *12*, 7568. [CrossRef]
192. Lin, Y.; McPhee, J.; Azad, N.L. Comparison of Deep Reinforcement Learning and Model Predictive Control for Adaptive Cruise Control. *IEEE Trans. Intell. Veh.* **2021**, *6*, 221–231. [CrossRef]
193. Gunter, G.; Gloudemans, D.; Stern, R.E.; McQuade, S.; Bhadani, R.; Bunting, M.; Monache, M.L.D.; Lysecky, R.; Seibold, B.; Sprinkle, J.; et al. Are Commercially Implemented Adaptive Cruise Control Systems String Stable? *IEEE Trans. Intell. Transp. Syst.* **2021**, *22*, 6992–7003. [CrossRef]

194. Sawant, J.; Chaskar, U.; Ginoya, D. Robust Control of Cooperative Adaptive Cruise Control in the Absence of Information About Preceding Vehicle Acceleration. *IEEE Trans. Intell. Transp. Syst.* **2021**, *22*, 5589–5598. [CrossRef]
195. Yang, Z.; Wang, Z.; Yan, M. An Optimization Design of Adaptive Cruise Control System Based on MPC and ADRC. *Actuators* **2021**, *10*, 110. [CrossRef]
196. Anselma, P.G. Optimization-Driven Powertrain-Oriented Adaptive Cruise Control to Improve Energy Saving and Passenger Comfort. *Energies* **2021**, *14*, 2897. [CrossRef]
197. Chen, C.; Guo, J.; Guo, C.; Chen, C.; Zhang, Y.; Wang, J. Adaptive Cruise Control for Cut-In Scenarios Based on Model Predictive Control Algorithm. *Appl. Sci.* **2021**, *11*, 5293. [CrossRef]
198. Hu, C.; Wang, J. Trust-Based and Individualizable Adaptive Cruise Control Using Control Barrier Function Approach with Prescribed Performance. *IEEE Trans. Intell. Transp. Syst.* **2022**, *23*, 6974–6984. [CrossRef]
199. Yan, R.; Jiang, R.; Jia, B.; Huang, J.; Yang, D. Hybrid Car-Following Strategy Based on Deep Deterministic Policy Gradient and Cooperative Adaptive Cruise Control. *IEEE Trans. Autom. Sci. Eng.* **2022**, *19*, 2816–2824. [CrossRef]
200. Zhang, Y.; Wu, Z.; Zhang, Y.; Shang, Z.; Wang, P.; Zou, Q.; Zhang, X.; Hu, J. Human-Lead-Platooning Cooperative Adaptive Cruise Control. *IEEE Trans. Intell. Transp. Syst.* **2022**, *23*, 18253–18272. [CrossRef]
201. Boddupalli, S.; Rao, A.S.; Ray, S. Resilient Cooperative Adaptive Cruise Control for Autonomous Vehicles Using Machine Learning. *IEEE Trans. Intell. Transp. Syst.* **2022**, *23*, 15655–15672. [CrossRef]
202. Kamal, M.A.S.; Hashikura, K.; Hayakawa, T.; Yamada, K.; Imura, J.-i. Adaptive Cruise Control with Look-Ahead Anticipation for Driving on Freeways. *Appl. Sci.* **2022**, *12*, 929. [CrossRef]
203. Li, Z.; Deng, Y.; Sun, S. Adaptive Cruise Predictive Control Based on Variable Compass Operator Pigeon-Inspired Optimization. *Electronics* **2022**, *11*, 1377. [CrossRef]
204. Petri, A.-M.; Petreuș, D.M. Adaptive Cruise Control in Electric Vehicles with Field-Oriented Control. *Appl. Sci.* **2022**, *12*, 7094. [CrossRef]
205. Deng, L.; Yang, M.; Hu, B.; Li, T.; Li, H.; Wang, C. Semantic Segmentation-Based Lane-Level Localization Using Around View Monitoring System. *IEEE Sens. J.* **2019**, *19*, 10077–10086. [CrossRef]
206. Rasdi, M.H.F.B.; Hashim, N.N.W.B.N.; Hanizam, S. Around View Monitoring System with Motion Estimation in ADAS Application. In Proceedings of the 2019 7th International Conference on Mechatronics Engineering (ICOM), Putrajaya, Malaysia, 30–31 October 2019; pp. 1–5. [CrossRef]
207. Hanizam, S.; Hashim, N.N.W.N.; Abidin, Z.Z.; Zaki, H.F.M.; Rahman, H.A.; Mahamud, N.H. Motion Estimation on Homogenous Surface for Around View Monitoring System. In Proceedings of the 2019 7th International Conference on Mechatronics Engineering (ICOM), Putrajaya, Malaysia, 30–31 October 2019; pp. 1–6. [CrossRef]
208. Im, G.; Kim, M.; Park, J. Parking Line Based SLAM Approach Using AVM/LiDAR Sensor Fusion for Rapid and Accurate Loop Closing and Parking Space Detection. *Sensors* **2019**, *19*, 4811. [CrossRef]
209. Hsu, C.-M.; Chen, J.-Y. Around View Monitoring-Based Vacant Parking Space Detection and Analysis. *Appl. Sci.* **2019**, *9*, 3403. [CrossRef]
210. Lee, Y.H.; Kim, W.-Y. An Automatic Calibration Method for AVM Cameras. *IEEE Access* **2020**, *8*, 192073–192086. [CrossRef]
211. Akita, K.; Hayama, M.; Kyutoku, H.; Ukita, N. AVM Image Quality Enhancement by Synthetic Image Learning for Supervised Deblurring. In Proceedings of the 2021 17th International Conference on Machine Vision and Applications (MVA), Aichi, Japan, 25–27 July 2021; pp. 1–5. [CrossRef]
212. Lee, J.H.; Lee, D.-W. A Novel AVM Calibration Method Using Unaligned Square Calibration Boards. *Sensors* **2021**, *21*, 2265. [CrossRef] [PubMed]
213. Lee, Y.; Park, M. Around-View-Monitoring-Based Automatic Parking System Using Parking Line Detection. *Appl. Sci.* **2021**, *11*, 11905. [CrossRef]
214. Lee, S.; Lee, D.; Kee, S.-C. Deep-Learning-Based Parking Area and Collision Risk Area Detection Using AVM in Autonomous Parking Situation. *Sensors* **2022**, *22*, 1986. [CrossRef]
215. Geiger, A.; Lenz, P.; Urtasun, R. Are we ready for autonomous driving? The KITTI vision benchmark suite. In Proceedings of the 2012 IEEE Conference on Computer Vision and Pattern Recognition, Providence, RI, USA, 16–21 June 2012; pp. 3354–3361. [CrossRef]
216. Geiger, A.; Lenz, P.; Stiller, C.; Urtasun, R. Vision meets robotics: The KITTI dataset. *Int. J. Robot. Res.* **2013**, *32*, 1231–1237. [CrossRef]
217. Chang, M.-F.; Ramanan, D.; Hays, J.; Lambert, J.; Sangkloy, P.; Singh, J.; Bak, S.; Hartnett, A.; Wang, D.; Carr, P.; et al. Argoverse: 3D Tracking and Forecasting with Rich Maps. In Proceedings of the 2019 IEEE/CVF Conference on Computer Vision and Pattern Recognition (CVPR), Long Beach, CA, USA, 15–20 June 2019; pp. 8740–8749. [CrossRef]
218. Caesar, H.; Bankiti, V.; Lang, A.H.; Vora, S.; Liong, V.E.; Xu, Q.; Krishnan, A.; Pan, Y.; Baldan, G.; Beijbom, O. nuScenes: A Multimodal Dataset for Autonomous Driving. In Proceedings of the 2020 IEEE/CVF Conference on Computer Vision and Pattern Recognition (CVPR), Seattle, WA, USA, 13–19 June 2020; pp. 11618–11628. [CrossRef]
219. Lyu, S.; Chang, M.-C.; Du, D.; Wen, L.; Qi, H.; Li, Y.; Wei, Y.; Ke, L.; Hu, T.; Del Coco, M.; et al. UA-DETRAC 2017: Report of AVSS2017 & IWT4S Challenge on Advanced Traffic Monitoring. In Proceedings of the 2017 14th IEEE International Conference on Advanced Video and Signal Based Surveillance (AVSS), Lecce, Italy, 29 August–1 September 2017; pp. 1–7. [CrossRef]

220. Wen, L.; Du, D.; Cai, Z.; Lei, Z.; Chang, M.C.; Qi, H.; Lim, J.; Yang, M.H.; Lyu, S. UA-DETRAC: A New Benchmark and Protocol for Multi-Object Detection and Tracking. *Comput. Vis. Image Underst.* **2020**, *193*, 102907. [CrossRef]
221. Goyette, N.; Jodoin, P.-M.; Porikli, F.; Konrad, J.; Ishwar, P. Changedetection.net: A new change detection benchmark dataset. In Proceedings of the 2012 IEEE Computer Society Conference on Computer Vision and Pattern Recognition Workshops, Providence, RI, USA, 16–21 June 2012; pp. 1–8. [CrossRef]
222. Razakarivony, S.; Jurie, F. Vehicle detection in aerial imagery: A small target detection benchmark. *J. Vis. Commun. Image Represent.* **2015**, *26*, 2289–2302. [CrossRef]
223. Kenk, M.A.; Hassaballah, M. DAWN: Vehicle detection in adverse weather nature. *arXiv* **2020**, arXiv:2008.05402.
224. Lin, T.Y.; Maire, M.; Belongie, S.; Bourdev, L.; Girshick, R.; Hays, J.; Perona, P.; Zitnick, C.L.; Dollár, P. Microsoft COCO: Common Objects in Context. In *Computer Vision—ECCV 2014 Lecture Notes in Computer Science*; Fleet, D., Pajdla, T., Schiele, B., Tuytelaars, T., Eds.; Springer: Cham, Switzerland, 2014; Volume 8693. [CrossRef]
225. OpenStreetMap contributors. *OpenStreetMap Database [PostgreSQL Via API]*; OpenStreetMap Foundation: Cambridge, UK, 2023.
226. Li, J.; Sun, W. Drone-based RGB-Infrared Cross-Modality Vehicle Detection via Uncertainty-Aware Learning. *arXiv* **2020**, arXiv:2003.02437.
227. Song, H.; Liang, H.; Li, H.; Dai, Z.; Yun, X. Vision-based vehicle detection and counting system using deep learning in highway scenes. *Eur. Transp. Res. Rev.* **2019**, *11*, 51. [CrossRef]
228. The Third "Aerospace Cup" National Innovation and Creativity Competition Preliminary Round, Proposition 2, Track 2, Optical Target Recognition, Preliminary Data Set. Available online: https://www.atrdata.cn/#/customer/match/2cdfe76d-de6c-48f1-abf9-6e8b7ace1ab8/bd3aac0b-4742-438d-abca-b9a84ca76cb3?questionType=model (accessed on 15 March 2023).
229. Zhang, S.; Benenson, R.; Schiele, B. CityPersons: A Diverse Dataset for Pedestrian Detection. In Proceedings of the 2017 IEEE Conference on Computer Vision and Pattern Recognition (CVPR), Honolulu, HI, USA, 21–26 July 2017; pp. 4457–4465. [CrossRef]
230. Ferryman, J.; Shahrokni, A. PETS2009: Dataset and challenge. In Proceedings of the 2009 Twelfth IEEE International Workshop on Performance Evaluation of Tracking and Surveillance, Snowbird, UT, USA, 7–12 December 2009; pp. 1–6. [CrossRef]
231. Dollar, P.; Wojek, C.; Schiele, B.; Perona, P. Pedestrian Detection: An Evaluation of the State of the Art. *IEEE Trans. Pattern Anal. Mach. Intell.* **2012**, *34*, 743–761. [CrossRef] [PubMed]
232. Hwang, S.; Park, J.; Kim, N.; Choi, Y.; Kweon, I.S. Multispectral pedestrian detection: Benchmark dataset and baseline. In Proceedings of the 2015 IEEE Conference on Computer Vision and Pattern Recognition (CVPR), Boston, MA, USA, 7–12 June 2015; pp. 1037–1045. [CrossRef]
233. Houben, S.; Stallkamp, J.; Salmen, J.; Schlipsing, M.; Igel, C. Detection of traffic signs in real-world images: The German traffic sign detection benchmark. In Proceedings of the 2013 International Joint Conference on Neural Networks (IJCNN), Dallas, TX, USA, 4–9 August 2013; pp. 1–8.
234. Mathias, M.; Timofte, R.; Benenson, R.; Van Gool, L. Traffic sign recognition—How far are we from the solution? In Proceedings of the 2013 International Joint Conference on Neural Networks (IJCNN), Dallas, TX, USA, 4–9 August 2013; pp. 1–8.
235. Sivaraman, S.; Trivedi, M.M. A general active-learning framework for on-road vehicle recognition and tracking. *IEEE Trans. Intell. Transp. Syst.* **2010**, *11*, 267–276. [CrossRef]
236. Temel, D.; Kwon, G.; Prabhushankar, M.; AlRegib, G. CURE-TSD: Challenging unreal and real environments for traffic sign recognition. In Proceedings of the NeurIPS Workshop on Machine Learning for Intelligent Transportation Systems, Long Beach, CA, USA, 4–9 December 2017; pp. 1–6.
237. Zhu, Z.; Liang, D.; Zhang, S.; Huang, X.; Li, B.; Hu, S. Traffic-Sign Detection and Classification in the Wild. In Proceedings of the 2016 IEEE Conference on Computer Vision and Pattern Recognition (CVPR), Las Vegas, NV, USA, 27–30 June 2016; pp. 2110–2118. [CrossRef]
238. Zhang, J.; Zou, X.; Kuang, L.D.; Wang, J.; Sherratt, R.S.; Yu, X. CCTSDB 2021: A more comprehensive traffic sign detection benchmark. *Hum.-Centric Comput. Inf. Sci.* **2022**, *12*, 23. [CrossRef]
239. Bai, C.; Wu, K.; Wang, D.; Yan, M. A Small Object Detection Research Based on Dynamic Convolution Neural Network. Available online: https://assets.researchsquare.com/files/rs-1116930/v1_covered.pdf?c=1639594752 (accessed on 14 August 2023).
240. Pan, X.; Shi, J.; Luo, P.; Wang, X.; Tang, X. Spatial as deep: Spatial CNN for traffic scene understanding. In Proceedings of the AAAI Conference on Artificial Intelligenc, New Orleans, LA, USA, 2–7 February 2018. [CrossRef]
241. Tusimple Benchmark. Available online: https://github.com/%0ATuSimple/tusimple-benchmark (accessed on 1 January 2021).
242. Yu, F.; Chen, H.; Wang, X.; Xian, W.; Chen, Y.; Liu, F.; Madhavan, V.; Darrell, T. BDD100K: A Diverse Driving Dataset for Heterogeneous Multitask Learning. In Proceedings of the 2020 IEEE/CVF Conference on Computer Vision and Pattern Recognition (CVPR), Seattle, WA, USA, 13–19 June 2020; pp. 2633–2642. [CrossRef]
243. Mvirgo. Mvirgo/MLND-Capstone: Lane Detection with Deep Learning—My Capstone Project for Udacity's ML Nanodegree. *GitHub*. Available online: https://github.com/mvirgo/MLND-Capstone (accessed on 12 July 2022).
244. Bosch Automated Driving, Unsupervised Llamas Lane Marker Dataset. 2020. Available online: https://unsupervised-llamas.com/llamas/ (accessed on 2 April 2023).
245. Passos, B.T.; Cassaniga, M.; Fernandes, A.M.R.; Medeiros, K.B.; Comunello, E. Cracks and Potholes in Road Images. Mendeley Data, V4. 2020. Available online: https://data.mendeley.com/datasets/t576ydh9v8/4 (accessed on 13 August 2023).
246. Waymo LLC. Waymo Open Dataset. Available online: https://waymo.com/open (accessed on 29 July 2023).

247. Ess, A.; Leibe, B.; Van Gool, L. Depth and Appearance for Mobile Scene Analysis. In Proceedings of the 2007 IEEE 11th International Conference on Computer Vision, Rio de Janeiro, Brazil, 14–20 October 2007; -8, pp. 1–8. [CrossRef]
248. Yen-Zhang, H. Building Traffic Signs Opens the Dataset in Taiwan and Verifies It by Convolutional Neural Network. Ph.D. Thesis, National Taichung University of Science and Technology, Taichung, Taiwan, 2018.
249. Zhao, Z.-Q.; Zheng, P.; Xu, S.-T.; Wu, X. Object Detection with Deep Learning: A Review. *IEEE Trans. Neural Netw. Learn. Syst.* **2019**, *30*, 3212–3232. [CrossRef]
250. Khan, M.Q.; Lee, S. A Comprehensive Survey of Driving Monitoring and Assistance Systems. *Sensors* **2019**, *19*, 2574. [CrossRef]
251. Haq, Q.M.U.; Haq, M.A.; Ruan, S.-J.; Liang, P.-J.; Gao, D.-Q. 3D Object Detection Based on Proposal Generation Network Utilizing Monocular Images. *IEEE Consum. Electron. Mag.* **2022**, *11*, 47–53. [CrossRef]
252. Haq, M.A.; Ruan, S.-J.; Shao, M.-E.; Haq, Q.M.U.; Liang, P.-J.; Gao, D.-Q. One Stage Monocular 3D Object Detection Utilizing Discrete Depth and Orientation Representation. *IEEE Trans. Intell. Transp. Syst.* **2022**, *23*, 21630–21640. [CrossRef]
253. Faisal, M.M.; Mohammed, M.S.; Abduljabar, A.M.; Abdulhussain, S.H.; Mahmmod, B.M.; Khan, W.; Hussain, A. Object Detection and Distance Measurement Using AI. In Proceedings of the 2021 14th International Conference on Developments in eSystems Engineering (DeSE), Sharjah, United Arab Emirates, 7–10 December 2021; pp. 559–565. [CrossRef]

MDPI AG
Grosspeteranlage 5
4052 Basel
Switzerland
Tel.: +41 61 683 77 34

Sensors Editorial Office
E-mail: sensors@mdpi.com
www.mdpi.com/journal/sensors

www.ingramcontent.com/pod-product-compliance
Lightning Source LLC
Chambersburg PA
CBHW041419210926
51850CB00013B/3314
* 9 7 8 3 7 2 5 8 3 2 6 1 3 *